Emotion and Reason

Emotion and Reason
The cognitive science of decision making

Alain Berthoz
Director
Laboratory of Perception and Action
UMR/CNRS
Collège de France
Paris, France

Translated by
Giselle Weiss

OXFORD
UNIVERSITY PRESS

Great Clarendon Street, Oxford OX2 6DP

Oxford University Press is a department of the University of Oxford.
It furthers the University's objective of excellence in research, scholarship,
and education by publishing worldwide in

Oxford New York

Auckland Cape Town Dar es Salaam Hong Kong Karachi
Kuala Lumpur Madrid Melbourne Mexico City Nairobi
New Delhi Shanghai Taipei Toronto

With offices in

Argentina Austria Brazil Chile Czech Republic France Greece
Guatemala Hungary Italy Japan Poland Portugal Singapore
South Korea Switzerland Thailand Turkey Ukraine Vietnam

Oxford is a registered trade mark of Oxford University Press
in the UK and in certain other countries

Published in the United States
by Oxford University Press Inc., New York

© Odile Jacob, January 2003

Ouvrage publié avec le concours du Ministère français chargé de la culture—Centre national du livre

Work published with the collaboration of the French Ministry for Culture—National Book Centre

This book is supported by the French Ministry for Foreign Affairs, as part of the Burgess Programme headed for the French Embassy in London by the Institut Français du Royaume-Uni

ii institut français

The moral rights of the author have been asserted
Database right Oxford University Press (maker)

First published (French) 2003
First published (English) 2006

All rights reserved. No part of this publication may be reproduced,
stored in a retrieval system, or transmitted, in any form or by any means,
without the prior permission in writing of Oxford University Press,
or as expressly permitted by law, or under terms agreed with the appropriate
reprographics rights organization. Enquiries concerning reproduction
outside the scope of the above should be sent to the Rights Department,
Oxford University Press, at the address above

You must not circulate this book in any other binding or cover
and you must impose the same condition on any acquirer

British Library Cataloguing in Publication Data

Data available

Library of Congress Cataloging in Publication Data

Data available

Typeset by Newgen Imaging Systems (P) Ltd., Chennai, India
Printed in Great Britain
on acid-free paper by
Biddles Ltd., King's Lynn

ISBN 0–19–856626–3 (Hbk.: alk.paper)   978–0–19–856626–7 (Hbk.)
ISBN 0–19–856627–1 (Pbk.: alk.paper)   978–0–19–856627–4 (Pbk.)

10 9 8 7 6 5 4 3 2 1

Contents

Introduction *ix*

Part 1 **Is decision making rational or irrational?**

1 The brain—gambler and logician *3*
2 Decision making and emotion *23*
3 The pathology of decision making *51*

Part 2 **Decision making with my second self**

4 Fight or flight *73*
5 Walking and balance *95*
6 Deliberating with one's body: me and my second self *109*

Part 3 **Perception, preference, and decision making**

7 To perceive visually is to decide: the physiology of doubt *137*
8 Decision making and shape recognition: ambiguity and rivalry *155*
9 Sensory conflict: perception of movement *173*
10 Fountains *185*

Part 4 **Magical thinking**

11 The physiology of preference *199*
12 'I think, therefore I suppress' *233*
13 The brain as emulator and generator of strategies: the vagabond thought *251*

Epilogue *279*

Picture credits *285*

Index *289*

Acknowledgements

Many thanks to Odile Jacob for having freed me once again from the burden of scientific articles and allowed me the privilege of leaving my brain free to consider ideas from different fields in the world of knowledge. Her advice was decisive in giving coherence to the text and in clarifying both its organization and content. Her enthusiasm and professionalism have supported my efforts and added to my pleasure in writing, and her great patience with my delays allowed me to reconcile the editing of this work with the demands of scientific life.

This book would not exist without the contribution of Gérard Jorland, who guided me through the production process. The editing was the subject of an intellectual debate between us that helped me to sort out my thoughts and to find a style that, while suitably rigorous, maintained the lightness required to be understood by everyone, or almost everyone. His knowledge in history and philosophy of science, and his deep interest in how the brain works and what constitutes creativity make his function as editor integral to meeting the challenges posed by a book like this one. Most of all, he gave me confidence, and I hope I have not disappointed him.

The editing of this book benefited also from the very generous contributions of colleagues and collaborators whose names follow and whom I would like to thank either for having corrected chapters or for having helped in researching sources and illustrations. I would like first of all to mention Solange Fanjat de Saint-Font. Her premature death following a long and painful illness saddened all of us. She contributed to this work her passion for knowledge and for just the right word. I dedicate this book to her in honour of the enthusiasm with which she led her entire career at the Collège de France. She was proud of working there; her competence and her joy supported us immeasurably.

Many thanks also to France Maloumian who did the digital imaging for the illustrations with exceptional competence and creativity. Michel Francheteau helped to organize the courses and seminars at the Collège de France, of which this book is partly the result, and worked efficiently in assembling the manuscript. Sandrine Mouret, documentarist in the Institute of Biology at the Collège de France, managed to find even the most exotic references. Delphine Coudert worked with Valérie Blondel at various stages of editing. Françoise Sotelo introduced me to Etruscan script.

I thank my colleagues, who reread and critiqued chapters. Professors Roger Guesnerie and Massimo Piatelli-Palmarini critiqued the chapters on economics and theories of decision making. Ian Hacking's book clarified the origin of the concept of probability for me. Paolo Crenna corrected the sections on posture and locomotion.

Yves Trotter and Jean Bullier corrected the chapters on visual perception. Sylvie Berthoz and Léon Tremblay commented helpfully on the chapters about emotion and the physiology of preference. Oliver Houdé corrected and commented on the chapters on inhibition and logical thinking. Theodor Landis corrected the chapters on pathology and magical thinking. Maya Berthoz reread several chapters and helped me to select words judiciously and to prune superfluous ones. She was willing to let the editing of the book consume large chunks of our rare and precious leisure time for two years. Guy Bertrand led me to Poussin's painting of the *Judgement of Solomon*. I am nevertheless solely responsible for any errors or imperfections in this book.

My warmest thanks also to Odile Jacob's entire team: Dominique Renoux, who chose many images with impeccable taste, Emilie Barian, who insisted on rereading the manuscript and contributed to organizing it, and Claudine Roth-Islert, who knows how to get results with a smile.

Thanks also to Giselle Weiss, who did the translation. Her remarkable professionalism, her ability to understand the work's many facets, and her enthusiasm were essential in trying to convey my thoughts. The exchanges we had gave me a marvelous opportunity for a second look at ideas and facts. The translation became a new intellectual adventure.

Finally, I am grateful to Martin Baum for his willingness to consider the book for publication, and to the Oxford University Press team for their rapid and excellent work. My affection for Oxford is long-standing: at 13, I was a schoolboy at Magdalene College School; later, I was a visitor at Oxford University and Corpus Christi College.

Introduction

It is always advisable to perceive clearly our ignorance.
(*Charles Darwin [1]*)

We need to completely change, and perhaps even reverse, the way we think about making decisions. We are emerging from a century dominated by the power of reason. As the sine qua non of science, reason allowed us to discover the fundamental properties of matter and, armed with technology, to transplant the heart, symbol of love, from one chest to another. Reason brought us the moon, favoured muse of poets, and soon it will take us to Mars. Even now reason is enabling us to probe the brain—that extraordinary product of evolution—for the neural basis of the most sophisticated workings of cognition. It was reason that removed the demons believed to torment the brains of epileptic children and reason that vindicated the parents of children afflicted with disorders such as autism and schizophrenia—attributed until only recently to psychological trauma—by revealing their genetic origin. Reason underpins our conviction that the decisions our doctors make, like those made by our politicians, are the result of a logical analysis of observable phenomena.

But this rational thinking—arrived from Euphrates by way of Sumer, Jerusalem, Cairo, Athens, and Rome, this curious child of the East and West, of Arabic mathematicians and of astronomers from every continent who taught that we can predict the very movements of the planets—cannot be said to be a product of the Age of the Enlightenment. It is stiff and impersonal. It is indifferent to the soft fog of uncertainty; it shields itself from the wonders and vagaries of the imagination, and would have us believe that the world is amenable to reckoning, that the Vietnam War can be won by the Pentagon's computers.

Well aware of the limitations of reason (since Heisenberg showed that we cannot simultaneously know both the position and speed of a particle), physicists have turned to new theories that take uncertainty into account. Following on the calculation of probabilities and the theory of the Reverend Bayes, which links cause and effect, came theories of fractals, catastrophe, chaos, and complexity. In biology, too, uncertainty has become the companion of necessity. The principle of entropy helped to understand how neurotransmitters are released at the level of the synapses; and dynamic nonlinear systems and strange attractors have become indispensable conceptual tools for modelling processes as diverse as the neuronal encoding of movement in the brain and the evolution of a population of voles in Sweden.

Having predicted a backlash of emotion over reason, André Malraux wrote that the twenty-first century would 'be religious or [would] not be'. One could interpret that forecast as a recipe for a new kind of mystical thinking—since the human brain apparently still craves gods—but more tolerant, incorporating data from reason. In fact, we are witnessing a frightening return to the most obscure, sectarian, intolerant, and fanatical forms of religion amid a ruthless struggle for profit and global exploitation that is unprecedented in history. Such is the nature of the problems that have greeted the new century that the theories at our disposal are too crude to tackle them. How can we ever hope to solve them if we do not understand decision making?

We are convinced that decision making is the product of rational thinking, that it is unique to man and the structures situated in the frontal lobe of his brain, just as decision makers in big companies have their offices at the very top of sky scrapers. A classic model of this sort is that of Donald Norman and Tim Shallice [2], who postulated the existence of a 'supervisory' system regulating the flow of information in the systems that control action (see Fig. 0.1). Moreover, the dominance of formalist theories and the hegemony of linguists in the cognitive sciences have led us to believe that language joins logical reasoning as a sort of dynamic duo in the making of decisions.

Given this perspective, one can see the innovativeness of Antonio Damasio's elegant efforts to reintegrate emotion in the process of decision making, and, more recently, to reincarnate cognition as well, closely akin to some of my analyses and those of neurophysiologists such as Francisco Varela [3].

But we cannot stop there. Human progress always entails a change in point of view, and this book is intended as a radical reversal of perspective. Instead of looking at

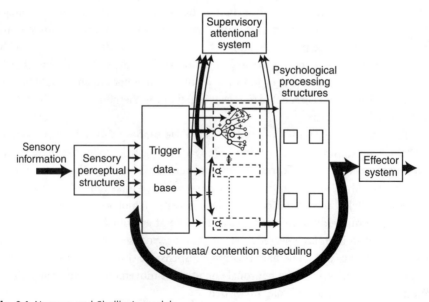

Fig. 0.1 Norman and Shallice's model.

decision making as a rational process that appeared recently in evolution owing to logical tools, this book proceeds from the opposite assumption (1) that decision making is probably *the* fundamental property of the nervous system and (2) that its origin is action. Action is not movement; it is the intention to interact with the world or with oneself as part of the world. Action always has a goal; it is always backed up by purpose. It thus becomes the organizer of perception, the organizer of the perceived world. Action is also embedded in a more general concept, the act. 'In the beginning was the Act', says Goethe's Faust [4].

In *The Brain's Sense of Movement*, I developed a theory of brain functioning founded on the idea that the brain is a simulator of action, a generator of hypotheses, and that anticipating and predicting the consequences of actions based on the remembered past is one of its basic properties. Current neurophysiology and cognitive psychology confirm these ideas already put forth by pre-Socratic thinkers and, closer to our time, among such great forerunners such as Nicolai Bernstein, Donald MacKay, J. J. Gibson and so on. The brain is thus essentially a comparator. It compares the state of the world with its hypotheses. It does not transform stimuli into motor responses or feelings. This activity of comparing is always linked to intention, to a 'project'—or plan—of action (in the sense of 'projection'). There is no mechanism of perception apart from action, no more than there are mechanisms of attention apart from the selections made continually by the brain.

Perceiving is deciding

I suggest, moreover, that perception is not only simulated action but also and essentially a *decision*. Perceiving means not only combining and weighting, it means *choosing* from among the variety of available sense data those pertinent to the action envisaged. Perceiving resolves ambiguity; thus, perceiving is deciding.

In contrast to the ideas maintained by some schools of cognitive psychology, I also suggest that decision making is not a process specific to humans, whose functioning depends on the prefrontal cortex [5] and mechanisms such as working memory, which enables him to hold in his mind facts and recollections [6]. I will argue that decision-making mechanisms are present in animals at various levels of their central nervous system.

We need to construct a *'hierarchical' and 'heterarchical' theory of decision making*, to understand how both parallel and serial [7] processing of information is organized. Deciding is about linking the present to the past and to the future; it is about organizing things. In this book, I will try to show what I mean by these ideas.

Formalist theories

Today, the literature boasts hundreds of articles about decision making in diagnostics and medical therapeutics, in economics, sports, the art of war and risk taking in accidentology, and so on. Examining this literature reveals a paradox: research suggests the best way to deliver medicine, to do (or not do) an operation, to launch a product, to introduce fiscal measures, and to referee a football game. But it says nothing about how the brain makes decisions. At most, one finds in economics allusions to the work (excellent, I might add) of Daniel Kahneman and Amos Tversky (see Chapter 1). More recently, Bernard Walliser [8] published a book on cognitive economy without providing a single reference to cognitive psychology and the process of decision making.

Cognitive psychology does have an extensive literature on the theory of decision making that relates to our subject. I will give you only a little taste of it here for purposes of contrasting these approaches with the cognitive neurobiology and physiology of decision making that we propose to construct. The need to bring the theory of decision making and the cognitive sciences closer together has often been stressed but rarely directly addressed. Cognitive theoreticians are believed to be quite prepared to adopt the Bayesian[1] principle of optimal interference, ignoring findings by researchers in decision-making theory, to wit, that people systematically violate the axioms of these optimal theories. For their part, theoreticians of decision making seem ready to adopt serial architectures of decision making for processing information, without taking into account work by cognitive scientists showing that humans perform substantial parallel processing. However, since the French edition of this book appeared, a new field called *neuroeconomics* has emerged which attempts to link theories of economics to the neural basis of decision making.

My criticism of the way economists represent decision making among consumers finds striking confirmation in this text by the Nobel prize winner in economics for 1994, Reinhard Stelten: 'Modern mainstream economic theory is largely based on an unrealistic picture of human decision making. Economic agents are portrayed as fully rational Bayesian maximizers of subjective utility. This view of economics is not based on empirical evidence, but rather on the simultaneous axiomization of utility and subjective probability. In the fundamental book of [Leonard] Savage (1954), the axioms are consistency requirements on actions, where actions are defined as mappings from states of the world to consequences. One can only admire the imposing structure built by Savage. It has a strong intellectual appeal as a concept of ideal rationality. However, it is wrong to assume that human beings conform to this ideal' [9].

---

[1] Bayesian probabilities are a combination of conditional probability (the probability of an event knowing that it occurs under certain conditions, which itself has a certain probability) and of total probability (the probability of an event that can occur under a multiplicity of conditions is the sum of probabilities of that event under each of the conditions influenced by their respective probabilities). Bayesian calculations enable us to revise our choices according to the information that we have about their consequences. For example, if a person chooses to drink coffee rather than hot chocolate because it is known that hot chocolate causes migraine with a probability p_m, the person might reconsider his choice when he learns that hot chocolate is also an antidepressant with a probability p_a. Bayesian probabilities are decision-making processes that take into account past experience.

Do not misunderstand me. I realize that major economists, such as Paul Samuelson, know full well the limitations of utility theory but think that it is the best of the alternatives until we have a theory of the brain.

Decision making is simulated action

Decision making, then, is not only about reasoning; it is also about action. It is never a purely intellectual process, a logical game for which you can write an equation. Making

Fig. 0.2 The *Judgement of Solomon,* by Poussin.

a decision involves thinking, of course; but all the while integrating elements of the past, it also contains in itself the act it is leading to. This act incorporates what I will term the 'acting body', which also fits well with Varela's concept of 'enaction' [10].

The physiology of memory must include that of forgetting; similarly, the physiology of decision making must also be a physiology of *not deciding*, that is, of indecision or inhibiting action. There are many such examples: a patient with Parkinson's who remains immobilized for want of dopamine; an airplane pilot who suddenly freezes, dumbfounded, when a truck appears in the takeoff lane; a driver who, like Buridan's ass, cannot decide whether to go left or right at a fork in the road and hits a tree in the middle; the strategist or politician who hesitates between two solutions. This paralysis is encouraged by the maxim, 'When in doubt, do nothing!' Subjects blocked in this way freeze so as not to persist in a previously adopted response. Likewise, psychologists have observed that during development, a child may fail to find a new solution because she is locked into one that corresponds to a preceding stage of her development and cannot replace the initial behaviour with a more sophisticated one.

Several fields of application of theories of decision making do not appear in this book: the problem of risk taking, for which readers are referred to the work of René Amalberti [11], questions of workplace security [12], which have inspired very complex theories, and the enormous problem of decision making in medicine, which occupies hundreds if not thousands of pages on the World Wide Web. I hope that this book will one day contribute to thinking in these very important fields of social science.

These many forms of decision making give an idea of the magnitude of the task. All the more so since decision making is also always, although to different degrees, related to emotion, as shown in the *Judgement of Solomon*, painted by Nicolas Poussin (Fig. 0.2).

References

1 C Darwin, *The expression of emotion in man and animals* (New York: Philosophical Library, 1955), p. 67.
2 Details of this model can be found in T Shallice, *Symptômes et modèles en neuropsychologie. Des schémas aux réseaux* (Paris: PUF, 1995), p. 421.
3 FJ Varela, E Thompson, and E Rosch, *The embodied mind: cognitive science and human experience* (Cambridge: MIT Press, 1993).
4 JW von Goethe, *Faust*, trans. W. Kaufman (New York: Anchor/Doubleday, 1963), line 12100, p. 502.
5 As shown brilliantly in the works of AR Damasio, *Descartes' error: emotion, reason, and the human brain* (New York: Putnam, 1994) and *The feeling of what happens: body and emotion in the making of consciousness* (New York: Harcourt Brace, 1999), the models developed by Stanislas Dehaene and Jean-Pierre Changeux [see J-P Changeux, *The physiology of truth: neuroscience and human knowledge*, trans. MB DeBevoise (Cambridge: Belknap Press, 2004)], as well as recent work in brain imaging, such as that by E Koechlin, GB Asso, P Pietrini, S Panzer and J Grafman, 'The role of the anterior prefrontal cortex in human cognition', *Nature* **399** (1999): 148–51.

6   The question of working memory has given rise to a substantial body of research over the past several years. Basic data on monkeys are the work of A Diamond and PS Goldman-Rakic, 'Comparison of human infants and rhesus monkeys on Piaget's AB task: evidence for dependence on dorsolateral prefrontal cortex', *Experimental Brain Research*, **74** (1989): 24–40; S Funahashi, C Bruce, and PS Goldman-Rakic, 'Mnemonic coding of visual space in the monkey's dorsolateral prefrontal cortex', *Journal of Neurophysiology*, **61** (1989): 1–19; S Park, S Holzman, and PS Goldman-Rakic, 'Spatial working memory deficits in the relatives of schizophrenic patients', *Archives of Neurology and Psychiatry*, **52** (1995): 821–828. The fundamental observation is that neuronal activity remains stable while the monkey is being asked to remember a piece of information that it will use later. According to Goldman-Rakic, specialized areas are devoted to maintaining working memory depending on the content (objects, space, and so on). An alternative theory proposed by Petrides hypothesizes a supervisor of activity that corresponds to particular functions (information processing, memory storage, and so on) independent of content. A final theory suggests that there actually exists an 'equipotentiality' among diverse regions of the brain [M Petrides, 'Lateral frontal cortical contribution to memory', *Seminars in the Neurosciences*, **8** (1996): 57–63].

7   E Koechlin, G Corrado, P Pietrini, and J Grafman, 'Dissociating the role of the medial and lateral anterior prefrontal cortex in human planning', *Proceedings of the National Academy of Sciences of the USA*, **97** (2000): 7651–6.

8   B Walliser, *Économie cognitive* (Paris: Odile Jacob, 2000).

9   G Gigerenzer and R Stelten, *Bounded rationality: the adaptive toolbox* (Cambridge: MIT Press, 2001), p. 13.

10 Varela, Thompson, and Rosch, *The embodied mind*.

11 R Amalberti, *La Conduite de systèmes à risques* (Paris: PUF, 1996).

12 DE Broadbent, *Decision and stress* (London: Academic Press, 1971); JM Hoc and R Amalberti, 'Diagnosis and decision making: some theoretical questions raised by applied research', *Current Psychology of Cognition*, **14** (1995): 73–101.

Part 1

Is decision making rational or irrational?

Chapter 1

The brain—gambler and logician

If you know exactly what you are going to do, why do it?
(*Popularly attributed to Picasso*)

Is the brain a calculator that uses probability theory to make decisions [1]? The basic principles of probability theory were first hammered out in an exchange of letters between Blaise Pascal and Pierre de Fermat on the subject of gambling. Pascal suggested that when in doubt, one should bet that God exists because that way there is everything to gain—life everlasting—and nothing to lose. After all, if God does not exist, neither does anything else. But the modern concept of probability is the brainchild of Jakob Bernoulli, who in 1713 set forth in *Ars conjectandi* a general theory for making rational decisions in uncertain circumstances. This theory was the basis for the classic interpretation of probability as a 'degree of belief', that is, reflecting *a state of human self-awareness as opposed to a state of the world* [2].

The term that really caught on and formed the basis of several modern theories is 'utility'. In 1728 Gabriel Cramer wrote, 'Mathematicians estimate money in proportion to its quantity, and men of good sense in proportion to the usage that they may make of it.' It is not the absolute value of money that matters, but the satisfaction it brings. Jakob's nephew, Daniel Bernoulli, took up the idea in 1738 and formalized it [3].

This perspective on the fundamentals of decision making takes us back to a theory of belief and of value. Indeed, people actually make decisions based on what they believe about value. Let us consider briefly what that means.

Utility, and theories of normative, descriptive, and prescriptive behaviour

The idea of utility was a tenacious one, because 200 years later, in 1944, John von Neumann and Oskar Morgenstern incorporated it into a 'utility function' to explain gambling in their epoch-making book *Theory of Games and Economic Behaviour* [4]. *Homo economicus* chooses to maximize the 'expected utility' of all outcomes. In other words, a person assesses, predicts, and gambles on the possible consequences of his choices, and maximizes the expected result. Here we encounter a model close to the one proposed by Richard Schmidt [5] that I talked about in *The Brain's Sense of Movement*, one that stresses the importance of predicting the consequences of action as a means of controlling that action as well as the choice of strategies.

In 1954, Leonard Savage extended von Neumann and Morgenstern's approach (which was based strictly on frequency) to include a subjective interpretation of probability. In *Foundations of Statistics*, he proposed to use a subject's choices to infer two functions that he called 'subjective probability function' and 'utility function', thus adding a new, more psychological dimension that he called 'personal probability'. Influenced by Bruno de Finetti and Frank Ramsey, who had interpreted probability as a degree of belief, Savage sought axioms that would bear directly on the rationality of actions rather than on their consequences. He also introduced a 'sure thing principle', according to which a preference for X over Y, whatever the state of the world, implies a preference of X over Y even if the state of the world is precisely known.[1]

These theories deeply influenced psychologists, who in the 1950s were inspired by them to develop their theories of decision making with contributions from the fields of risk taking, medical decision making (to correct decisions doctors make based on intuition) and so on. Current theories distinguish three principal, interdependent themes [6]: *normative* theories of optimization processes in decision making to determine how we should proceed; *descriptive* theories of the real processes based on which we make decisions; and, finally, *prescriptive* theories for developing workable models to make our choices more relevant.

The perfect decision maker: normative theories

The basic hypothesis of the normative school assumes that there is a perfectly rational decision maker who behaves according to very clearly formulated axioms. In contrast, descriptive theories look for patterns in the actual behaviour of decision makers. Here, theory emerges from observation, and not axiomatically, from principles established a priori. These theories, too many to summarize here, have given rise to countless variations. Some emphasize that deciding is gambling and that gambling can be studied empirically. Ward Edwards (1954) introduced psychologists to the theories of probability and decision making.[2] This research showed that every axiom of probability theory and every axiom of utility theory (not to mention their possible combinations) is largely contradicted by real subjects in well-controlled experimental situations, who nonetheless reproduce everyday choices. Life is a gamble: virtually all decisions are based on an insufficient number of sensory data, events, facts, and documents. We do not make decisions, whether motor or intellectual, based on a completely rational

---

[1] It is beyond the scope of this book to describe all the particulars of these theories. For example, the concept of a 'fair bet' is the core of the theory. It is an idealization comparable to Sadi Carnot's heat engine. It plays an essential normative role, but its cognitive legitimacy is debatable.

[2] He was, of course, not the first. Patrick Suppes and Donald Davidson, and many others (Friederick Mosteller and Philip Nogee as early as 1951, for example), had adopted the experimental method of choice among gambles to probe subjects' actual preferences. In 1947 Malcolm Preston and Philip Baratta even devised a curve of decisional probabilities.

analysis of the situation. Marrying, taking a job, and choosing a place to live all entail a gamble. And because decision making is a fundamental process of psychological life, it is suitable for scientific study and experimentation, in the laboratory, with subjects in gambling situations, as Antonio Damasio and neuropsychologists would later show.

Still, it wasn't until 1973 that two researchers, Daniel Kahneman and Amos Tversky (a student of Edwards), published their seminal paper in this field. First in 1979, then in 1993, they developed a complete cognitive theory (*Prospect Theory* and *Cumulative Prospect Theory*). Prospect theory does not counterpoise normative theories, but explains how and why real decision makers follow different laws. It underscores the value of normative theories for understanding these phenomena. For example, according to normative theory, if people have to make predictions and come up with a decision, they will predict a result that seems most representative of all the available facts [7]. In this way they make predictions that do not depend on the value of the facts or the probability that can be inferred from past consequences. People only consider the representative value of the information available to them, which explains many errors of judgement.

Errare humanum est. Perseverare diabolicum!

To err is human. To persist is folly. Yet humans really do make many errors of judgement [8]. Social pressure is one reason. It is often at the root of perfectly rational decisions. Nonetheless, the tendency to agree with others appears to be a powerful cause of collective error. It often leads groups to take more radical positions than those of its individual members. This amplification effect is substantial, especially if one-upmanship is involved. For example, in a holiday camp, children were divided into two groups for a contest. At the end, two friends who were put in separate groups had become enemies. Their common sense was altered by the exercise.

Another source is the power of hierarchy. For example, in England, an exploding motor caused an airplane to crash. The pilot made a fatal error: he shut off a motor that was working well, which caused the machine to fall onto the highway. The stewardess knew that it was the other motor that was on fire, but did not dare to tell the pilot out of respect for authority. These examples are so frequent that the French Army recently modified its regulations to allow soldiers to disobey if they believe that their commander is giving an unethical order or is not following basic procedure. It is nothing short of revolutionary to change a rule that a soldier's instruction manual requires him to obey 'unquestioningly'.

Emotion, which is known to guide decision making, is also a classic cause of error. It leads us to focus on immediate solutions and to neglect other possibilities. For example, when you have lost your wallet, you search frantically around you without imagining all the places where you might have lost it. Getting a grip on yourself allows you mentally to retrace your steps and to find the lost object.

Needing to be right is one of the most frequent causes of error. We are certain that the car we bought is the best, even though it took us forever to make the final choice. And once formulated, it is difficult to abandon a hypothesis. To illustrate this resistance, do the following experiment: In a series of three numbers, try to find the rule that governs their formation. Here is the first series: 2, 4, 6. Now choose yourself other series of numbers to test your hypothesis about the rule. The experiment shows that you prefer to choose series such as 22, 24, 26, that is, sets of numbers separated by 2 that seem to obey the most obvious rule in the first series rather than to try others. Because this rule is the most obvious, it can hide all other rules, such as 'all series of increasing numbers', which will give many more numbers than the preceding one. We do not understand the biological mechanisms for this persistence in fixed beliefs.

The ease or difficulty with which we access knowledge stored in memory is also an important factor in errors of judgement. For example, because words are organized in memory alphabetically, people think that there are more words that begin with a given letter than words that have the same letter in the third position.[3]

Errors in judgement can be linked to the way a problem is presented. Indeed, posing a problem this way or that can change or even reverse a person's preferences [9]. 'We use the term "decision frame" to refer to the decision maker's conception of the *acts* [italics mine], outcomes and contingencies associated with a particular choice.... Alternative frames for a decision problem may be compared to alternative perspectives on a visual scene' [10]. Note the analogy to perception of space.

The point here is that the 'frame' of the decision can, by changing, reverse a decision, whereas according to Tversky and Kahneman, two mountains should not change their relative height when viewed from a different perspective if they are accurately perceived. And yet it is obvious that a decision depends not only on content but also on context.

Disdain for objective facts in favour of opinion, however dubious, is also a significant cause of errors. In particular, we possess a repertoire of cognitive strategies that we mine at the risk of committing serious errors. For example, in a city, 85 per cent of the taxis are blue and 15 per cent are green [11]. An accident happens, and a witness remembers that a green taxi was involved. Statistical calculations show that actually, there is a greater chance that the taxi involved was blue. Despite the data, subjects falsely insist that the taxi was green. They are influenced by the witness, even though his testimony is not worth much. Generally speaking, use of statistics can avoid many errors of judgement, but the human mind does not trust statistics; it prefers personal opinion. In this book we will try to sort out some of the biological roots of this

---

[3] The data for English refer to words beginning with 'r' and words that have the letter 'r' in the third-from-last position. The actual ratio is 3 to 1, but everyone thinks the opposite. Even more striking is assessment of words ending with 'ing', and words that have the letter 'i' in the third-from-last position. In French, it works with 'nd' and 'n' in the penultimate position. See Massimo Piatelli-Palmarini's book *Inevitable Illusions*.

preference, at the same time beginning to sketch out a physiology of preference and analysing the neural basis of cognitive strategy.

Prospect theory

The idea that predicting the consequences of a decision can influence decision making was considered by a mathematical theory dubbed 'prospect' that proposes functions both for values and for subjective probabilities. A decision is based on a complex combination of these two functions (Fig. 1.1).

The value function is concave for gains and convex for losses, and its slope is greater for the latter than for the former. In other words, if one has big losses, naturally, one gives them a very negative value; the more one loses, the more desperate one becomes; if one wins big, again, naturally, one assigns the wins a very positive value but the satisfaction seems to level off the more one wins. In Chapter 8 we will see a splendid empirical verification concerning the neural basis of this prediction.

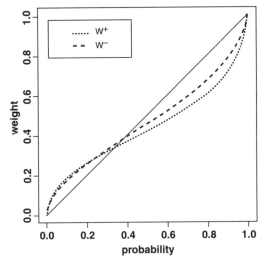

Fig. 1.1 Subjective probability. Curve linking the weighting factor to the subjective probability estimated by someone who is willing to take a risk involving gains and losses. The x-axis indicates the probability (objective probability, which the subject is given explicitly by the experimenter). The y-axis corresponds to the weight *this* particular probability exerts on decisions. The solid line represents an ideal linear relationship. A curve above the line indicates overestimation; a curve below the line indicates underestimation. The figure shows that we overestimate low probabilities and that we underestimate high ones. Curiously, the turning point is not 50 per cent but around 40 per cent (1/3 = ~0.37). The 0 and the 1 represent special cases, discontinuous from the rest. At values close to zero (very low probabilities), however, overestimates are enormous (100 times, 1000 times). After Tversky and Kahneman (1981)

The validity of utility theories was called into question by the analysis of 'nonconsequential' decision making [12]. In particular, when the future is fraught with alternatives—to pass an examination, or to fail it—then decisions violate the 'sure thing' principle established by normative theory [13].

Consider the following example of a decision with potentially appalling consequences, connected to incertitude and violating all the principles of rationality. During the Cuban missile crisis in the midst of the Cold War with the former Soviet Union, American strategists recommended that the United States fire a preventive first strike because the total number and characteristics of Soviet missiles were unknown. That's terrifying! The brain is clearly not a rational decision maker. It may be that failure to reason in a consequential manner is the fundamental difference between natural and artificial intelligence [14].

Economic man and the cognitive 'toolbox'

The second theme was inspired by Herbert Simon [15]. He proposed the idea that understanding the process of decision making requires knowledge of how cognitive, perceptual, and learning factors lead a human operator to deviate from the predictions of a perfect theoretical model constructed by 'economic man'. By economic man he meant a person who possesses a deep familiarity with his environment, a well-organized system of preferences, and the numerical ability to evaluate different solutions and their consequences and to optimize the relation between choices and preferences. According to Simon, the product of the interaction of our cognitive abilities with the complexities of the environment is a 'bounded rationality' [16]; that is, our decision-making behaviour reflects the limits of our systems of information processing.

This idea itself comes from observing the effect on cognitive ability in 'single-channel' experiments. Psychologists have observed that if a subject is given two tasks to perform at the same time, an effect of 'communicating vessels' is produced between them. If performance of one task improves, performance of the other declines. Do the experiment: Write your name several times on a piece of paper, then, after a few minutes, try to recite a poem or to describe a visual scene or solve a mathematical problem—a simple one—for example, counting backwards from 100. You will see your handwriting grow smaller and more disorganized, and you may even find yourself simply writing gibberish. This result is interpreted as the brain having only a single channel for cognitive operations, which also limits its capacity for short-term memory.

Simon suggested several cognitive strategies (averaging, adding, subtracting) to explain the behaviour of different economic agents. The most famous example is 'satisficing', which explains the behaviour of consumers who look for 'good enough' options in an uncertain environment where the search for solutions is costly. In these circumstances, people seem to stop at the first solution that satisfies one or more criteria, rather than hold out for the best solution whatever the cost. For example, when you are looking for a house in the country, you look at many of them. Then, all of a

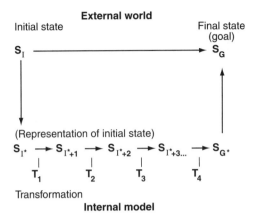

Fig. 1.2 Decision making as a sequence of operations. This model suggests that decision making is carried out by a progressive transformation from an initial state of knowledge (S_I) to a final state (S_G). The subject may, however, develop successive mental states $S_{I^*} > S_{I^*+1} > S_{I^*+2}$ and so on, thanks to the intermediate processes T_1, T_2, T_3. After Payne et al. (1995)

sudden, you choose one not because it meets all of your criteria but because it is the first that seems to satisfy some of them. It is a rational decision.

The brain appears to possess a 'toolbox' of cognitive strategies for linking the exterior world and internal models (Fig. 1.2) [17]. These include:

1 *The weighted additive procedure*: The decision maker considers each solution (e.g. the brand of automobile he wants to buy) according to its desirable features (power, comfort, reliability), to which he assigns a value and multiplies the value of each feature by its importance (e.g. he multiplies the value he assigns to a car's reliability by how important that feature is to him).

2 *The equal weight method*: Features still matter most, but the calculation is easier because the decision maker doesn't factor in the importance attached to each one. He simply adds up their values.

3 *Lexicographic strategy*: The decision maker chooses based only on the value of the most important feature. For example, in selecting a car, if he considers reliability to be most important, he will choose the most reliable. If he is looking for an ultralight portable computer, he will take the least heavy.

4 *Satisficing methods*: This strategy is based on the definition of the rejection threshold of a feature. The decision maker fixes minimum or maximum thresholds for all attributes—gas consumption, for example, in the case of a car. He examines each attribute one after the other. Whenever an attribute does not satisfy one of the criteria, he passes to the next.

5 *Elimination by aspects*: This approach combines the two preceding strategies.

The notion of an 'adaptive' toolbox [18] refers to the repertoire of strategies from which a decision maker can choose. This much-discussed idea is interesting because it assumes that the brain contains a limited number of mechanisms for decision making, which remain to be elucidated. In Chapter 8 we will see an example where we show through brain imaging how various brain structures are involved in different sorts of reasoning.

Cognitive algebra

Cognitive algebra is a third theme, associated with the monumental contributions of Norman Anderson (1981, 1991) and Kenneth Hammond (1966). They maintain that algebraic processes, expressed in the form of equations, constitute a good representation of fundamental cognitive processes. In particular, integrating information from multiple sources, which seems to be fundamental to the tasks of judgement and decision making, would be represented by weighted means or deviations.

The mathematical elegance of this approach contrasts with the piecemeal nature of previous descriptions [19]. Its disadvantage is that it is far removed from a workable biological theory of decision making.

Mental models: the surprising effectiveness of diagrams

The different theoretical streams mentioned above share the hypothesis of a *multiplicity* of cognitive strategies. But they differ in how agents predict and identify strategies. They also differ in the role of memory. They are all based on the limited capacity of working memory, that is, our short-term ability to remember concepts and facts, or words and numbers.

The distinction between the three themes—normative, descriptive, prescriptive—is not only a way of dividing up the vast literature that covers this field. Another theme is tied to the first three but nevertheless represents a stream of thought in itself. It derives from the work of Philip Johnson-Laird, who suggests that decision makers use 'mental models'.

Johnson-Laird's critique of normative theories is that they are not psychological theories but logical theories [20]. The formal character of the operations to which the brain submits according to these theories goes back to George Boole, who wrote in the middle of the nineteenth century on the subject of the calculus that the laws we must investigate are the most important laws governing the mental faculties, and that the mathematics we must construct is the mathematics of human intelligence. Jean Piaget fully agreed and asserted that reasoning is none other than propositional calculus itself.

Yet, according to Johnson-Laird, logic cannot determine the right solution to a problem among the infinite variety of possible solutions. Curiously, the theories that rely on logic to take into account deductive reasoning have largely ignored this problem. People make mistakes. Especially, they do not maximize utility. The major psychological discovery in reasoning and decision making is that people are not natural logicians, born statisticians, rational decision makers. The precise path of their thoughts and decisions is determined by complex, nonobvious processes that are still poorly understood.

Deduction in the brain does not depend on *syntactic* processes that follow formal rules but rather procedural semantics that involve mental models. These semantic procedures consist in constructing 'premise models', in formulating a small number of hypotheses based on these models and testing their validity by checking that no other premise model refutes them. If several models satisfy the question, a conclusion can be drawn by probabilistic reasoning.

Let us take an example of what Johnson-Laird considers a contribution of his theory to mental models. Formal logic offers means of testing whether the brain uses rules of inference but predicts that an inference is difficult if it calls for a long chain of steps or a rule that is difficult to access or to apply. For example, Johnson-Laird maintains that theories founded on formal rules of logical inference cannot explain the remarkable aid supplied by a diagram [21]. We know, too, that decisions that involve a disjunction (deciding whether a system is in state A or state B) are more difficult than decisions that involve a conjunction (A and B) [22], as in the following problem:

> Raphael is in Tacoma or Julia is in Atlanta, or both
> Julia is in Atlanta or Paul is in Philadelphia, or both
> What follows [23]?

Experiments show that using a diagram to aid this kind of reasoning saves time (around 15 to 30 s for this problem) and is more effective (around 30 per cent more valid conclusions). The experiments lead to three conclusions: 'First, difficulty in thinking increases with disjunctive possibilities. Second, this difficulty can be ameliorated by the use of an appropriate diagram that helps reasoners to make explicit all the possibilities. . . . Third, these two phenomena vindicate the theory of mental models because, unlike other theories of reasoning and other accounts of the effects of diagrams, it predicts both of them . . . Simon . . . reported anecdotally that engineers understood Supreme Court cases better when he represented them using circuit diagrams in which the switch positions corresponded to the yes/no decisions of the court' [24].

Rather than being based on formal rules of inference, the theory of mental models [25] postulates a more direct process that is an inverse reflection of the range of possibilities. People reason by constructing models based on information contained in basic knowledge and assumptions; by formulating conclusions that are valid for the models and subjected to other constraints such as parsimony; by seeking alternative models for which the conclusion does not hold; or looking for counterexamples. If no counterexample comes easily to mind, one infers that all is in order. If there are no other alternatives, the conclusion is considered to be valid. However, the mental processes in question must be 'calculable'; in other words, the brain can be studied using the metaphor of the computer.

The geometry of thought: a guide to divorce?

Johnson-Laird rightfully assumes that the brain constructs 'mental models'. But although he makes effective use of formal theories based on logic, he cannot explain the nature of these mental models. For a theory that is lucid but contradicted by experiment, he substitutes a vague concept. This is not the first time in science that we encounter vague but useful ideas! The concept of *internal model*, for example, is currently very popular among physiologists [26] and roboticists. But we don't really understand the neural basis for it.

The justification for Johnson-Laird's diagrams suggests an approach that I will come back to several times in this work (see, in particular, Chapter 8): the effectiveness of diagrams suggests that people like to *spatialize* complex problems, including reasoning. We know the power of the mathematical figure. It is sometimes forgotten in teaching mathematics, but should be restored. Remember also that space can aid memory [27]. I will come back to this point at the end of the book.

We get the sense that logical description, by nature serial, linear, successive, and somewhat like an egocentric description of a route—the succession of points of view that unfold as we walk—is, thanks to spatialization, being replaced by parallel, simultaneous description. This parallel description still makes it possible to create a path, just as a map of a city allows us to find new streets, to make connections, to compare itineraries, and to mentally manipulate connections between the elements.

The geometrization of thought by creation, as Alain Connes puts it, of a 'landscape' that the mind can grasp simultaneously from various perspectives lends fluidity to thinking and at the same time structures it like a fountain that both guides with its architecture and allows the water to find the best path [28]. In manipulating images, spatial reasoning enables the brain—a parallel processor if there is one—to call into play the new combinations and associations from which a solution will spring.

We encounter this spatial reasoning in a Scandinavian school of decision-making theory [29]. Starting with the observation that people can want one thing and choose another, and thus that their preferences are often dissociated from their actions, it separates *preference* from *action*. The key idea of the theory is that the decision maker structures the information he receives in a way to make it compatible with a possible decision and its consequences. He looks for what this school calls a 'dominant structure'. This theory can be summarized in four phases (see Fig. 1.3): preliminary, during which the important features are chosen for testing; the search for promising alternatives; the internal test of the best solution; the construction of the dominant (best) solution. The process culminates in a decision.

This study describes a painful decision-making process (divorce) that involves one or more shifts in perspective. These shifts are repeated until the individual reaches a stable situation where the points of view adopted lead to a decision (Fig. 1.4). Obviously, this process involves the search for a 'good structure' [30].

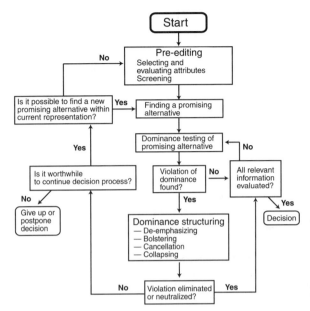

Fig. 1.3 Deciding to divorce: the succession of mental operations in seeking a dominant structure in a decision-making process. After Montgomery and Willén (1999)

Dominant structure theory has features in common with that of Johnson-Laird's mental model theory. But here, too, it is interesting to see that spatial reasoning is invoked explicitly by introducing the idea of adapting perspective that recalls the remarkable capacity of the brain for changing reference frames [31].

Compatibility, or how to light a burner

How many times have we cursed stove manufacturers because we cannot figure out which button to turn to light a gas burner or a heating element on a stove! To help us make this decision, all we have to do is to check for a certain compatibility between the disposition of the buttons and that of the burners. It seems, by the way, that this compatibility is very useful for making decisions or, more generally, when we have to connect a group of stimuli and motor responses.

An attempt was made to reconcile cognitive psychology and theories of decision making based on the *principle of compatibility* [32]. This principle states that when a stimulus and a response are represented mentally, the weight of one feature is increased if it is compatible with the required response. For example, in the area of visuomotor actions, subjects point towards a visual target more quickly if the stimulus is presented visually than if the instruction is given verbally. It is thus really the *spatial compatibility* of the stimulus and the response in the exterior world that play a role in deciding.

The following experiment will hopefully convince you: ask someone to turn on two lights located one to the left and the other to the right by pressing the switches for each light. The response is more rapid if the 'on' switch for the light on the left is on the left,

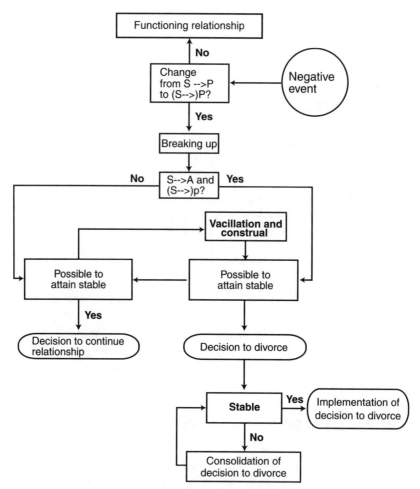

Fig. 1.4 The series of processes enabling a person to make the decision to divorce. The central hypothesis of this theory is that subjects seek a dominant structure. After Montgomery and Willén (1999)

and vice versa for the right. It is the compatibility of the switches and the lights, and not which hand is used, that matters. Indeed, if a subject is asked to cross his arms so that he responds with the right hand on the left button and so on, the result is the same. Compatibility does not depend on the anatomical contingency of the limbs with which the response is given.

Although psychologists are deeply interested in spatial compatibility, there are other kinds. Another is 'conceptual' compatibility. For example, a loud sound is more compatible with an intense light. Some stimulus properties are more *prominent* or more *striking* than others: the identity of a letter matters more than its size; the left-right axis stands out more than the top-bottom one.

These concepts of compatibility have been adapted to explain certain features of the decision-making process. For example, suppose that the members of a family are predicting what the weather will be tomorrow. The compatibility between the subject of the bet and the stakes influences the decision of the gambler. Defining wins in terms of money risks biasing the assessment of chance as a function of the assessment of the money and distracting from the calculation of probabilities. We say that the 'scale' on which the decision is made influences the process of decision making. All that is very intuitive, but it introduces bias and errors in decision making.

For example, subjects are asked to predict the performance of ten students in a history course, based on their performance in two other courses – philosophy, where they are graded A to D, and English, where they are graded 1 to 10. The subjects of the experiment are divided into two groups that must predict the grading in history, one between A and D and the other between 1 and 10. The result clearly shows a bias: the grading in history is influenced by the scale that was adopted by the two other fields.

Another phenomenon that can be explained by the theory of compatibility is the so-called reversal of preference. Subjects are presented gambles whose stakes have very different values. Gamble H offers a strong probability of winning a small sum (8 in 10 chance of winning 4 Euros). The other, gamble L, offers a weak probability of winning a substantial sum (1 in 9 chance of winning 50 Euros). Most subjects prefer gamble H. But on another occasion, subjects are offered to sell their 'chance' to one of the two gambles as one would sell a ticket at the races. Curiously, they sell gamble L at a higher price. This phenomenon has been observed not only in psychology laboratories but also at Las Vegas in professional gambling halls!

What explains this inversion of preference for the choice and evaluation of a reward? It appears that the main reason people assign greater value to the low-probability, higher win is the compatibility between the cost and the reward: because the result of the lottery is given in Euros, the rule of compatibility implies that big gains are valued more highly and desirably. Thus when value is the issue, L is preferred, but it is not chosen for the gamble. Here again, the scale of the response has a big influence on the decision.

Another example where the rational process of decision making is violated is in the opposition between choice and comparison (*matching*); knowing how to choose between two options of equal interest is an old philosophical problem.

For example, subjects are given pairs of presents composed of cash and coupons for books. In one pair, gift A contains both cash and coupons, but gift B contains only coupons. The subject is asked to say how much money he must add to A so that the two gifts have the same value. He chooses a sum. One week later, he is given both gifts with the combination he chose. It is observed whether he valued cash or coupons more highly. Whereas a week earlier he judged the two to be comparable, he no longer says they are equivalent but gives a different value. This bias is shown to be linked to a preference for one of the two elements. The choice between two parts of an alternative is guided by an effect of 'propensity'.

One is led to conclude that choice and decision making are both unstable and complex and that undoubtedly the current theories are not the best ones. Eldar Shafir writes: 'Are peoples' preferences better expressed when they are rejecting or appraising or when they are choosing? In contrast with classical analysis, which assumes that preferences are stable, one must conclude that they are constructed and that they depend on the methods that reveal them. Nevertheless, compatibility, a basic principle of human cognition, plays a significant role in the development of decision making' [33]. When we choose, we favour positive aspects; when we reject, we give greater importance to negative aspects. So an option that has many negative aspects and many positive aspects is both the most chosen and the most excluded.

History of a failure

The need to rethink theories of decision making was recently expressed by Samuel Leven and David Levine: 'The assumptions that decision makers, including consumers and producers, are always acting to optimize a measurable utility function have been increasingly challenged. In particular, there is wide recognition of the need for theories that include the dynamic effects of context on decision making, effects which are often not optimal or rational' [34].

For example, Coca-Cola decided to create a new product to compete against the classic soft drink. After many trials of a new product that was tested with many people and had been favourably received, the decision was made to launch it. It was a dismal failure. Even consumers who liked the new product clamoured for 'classic' Coke. The model is based on ideas developed by two groups of independent researchers. The first group constructed a theory called 'affective balance' [35] that consists of a neural network algorithm that calculates the 'affective' (in fact, utility disguised as emotion!) value of an anticipated event by comparing it with a different event. The architecture of this process of decision making is called a *gated dipole*, and its inspiration, at least terminologically, is very neurobiological. A schematic diagram [36] shows the organization of such a dipole and the structure of decision making proposed for the Coca-Cola problem (Fig. 1.5).

It uses a principle known in psychology as *opponent processing* that corresponds, for example, to reciprocal inhibition of two antagonist muscles. The gated dipole functions in the following way: A behaviour C is controlled by two antagonistic pathways, one positive and the other negative. The behaviour (in this case, 'to love the new Coke') will be triggered if the balance leans to the positive side. Two inputs influence the system. One is permanent and activates both pathways; the other (a new drink, for example) is temporary and excites the positive pathway. This first stage of the model projects onto a lower level that comprises synapses whose activity becomes depleted under certain conditions. A complicated mechanism causes frustration when 'familiar' Coke isn't available. A well-known phenomenon may be key to this effect: the subjective value of

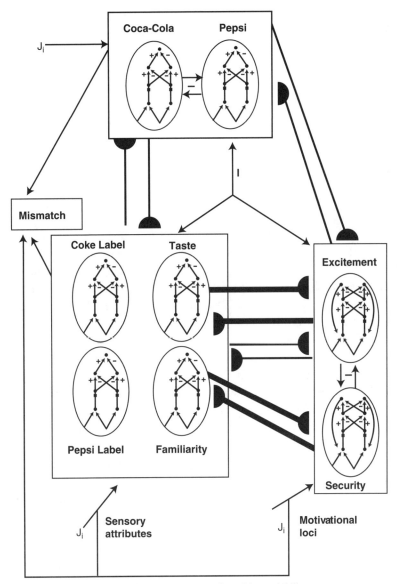

Fig. 1.5 Network simulation for finding a 'new soda'. This model illustrates an attempt to apply the theory of neuronal networks to the study of decision making concerning a new soft drink replacing or in addition to Coca-Cola. The basic elements of the model are the reciprocal inhibitory networks called dipoles (eight dipoles are shown inside the ellipses). These dipoles represent the effects of a sensory stimulus, and they can be conditioned through positive or negative reinforcement by modifiable connections represented by half circles. Two opposed motivations—excitement and security—are in competition (see the box to the right). Each sensory dipole has reciprocal connections that represent positive and negative motivations. The thick lines indicate strong connections. After Leven and Levine (1996)

B in the presence of A is not the same as the value of B when B is alone. This frustration was at the root of the failure that the Coca-Cola company encountered in launching its new product. Details of the model aside, it represents a primitive attempt to draw inspiration from biological mechanisms.

The authors of these models do not lack self-confidence: 'Neural network theory provides the only quantitative framework in which instinctive, rational, and affective processes can all be studied as interacting variables' [37]. But their certainty is at odds with the fact that no model is really satisfactory.

The struggle to unify the theories

It is sometimes difficult to find coherence in the study of decision making. Everything happens as though the fields of research were unconnected and their objectives different. Each school has its audience, conferences, and areas of application. Some groups favour Bayesian principles of inference, whereas others focus on demonstrating that humans systematically violate these principles, that they are hopeless gamblers, that they cannot reason their way out of a paper bag. The debate over whether humans are Bayesian processors continues and is by no means settled [38]. Still, the results of experiments validate many aspects of these theories.

It is also worth noting that several models rely on architectures of serial logical operations to represent information processing, whereas cognitive psychology experiments have shown that the human brain largely depends on parallel processing.

Following the distinction proposed by David Marr among the three levels of possible explanation of cerebral functioning (computation, algorithm, implementation), one might say that the theoreticians of decision making in areas of application such as economics prefer calculation, whereas cognitive psychologists prefer algorithms.

A clear sign of the current theoretical impasse is illustrated by Jonathon Baron's discussion of the two major models used by economists in theories of decision making: 'The years since 1953 have seen constant tension between the attackers and defenders of expected-utility theory as normative. Both camps have been engaged in efforts to develop better descriptive models of decision making. The attackers, who assume that people are generally rational, argue that better descriptive models will lead to better normative models. The defenders, who acknowledge the existence of irrational decision making, argue that the descriptive models will tell us where we fall short according to the normative model and will allow us to ask what, if anything, we can do about it. I take the view that our decisions are often irrational, and I shall defend utility theory as a normative model. I shall also point out, however, that there is room for various interpretations of utility theory as a normative model' [39].

Indeed, as Richard Lazarus said in critiquing economic theories of decision making, the problem we make in maximizing utility is that we have to know where our interest lies, and according to economists, we only know that when we have made a mistake. All the same, the connection between emotion and deciding, for example, is obvious.

Decisions are often made for reasons apart from personal interest. Other important human values come into play, such as religious or political ideas, sacrifice for children, loyalty, honesty, justice, compassion, reciprocity, and trust.

Up to now, none of these approaches has really taken into account the fact that decisions are made by a living brain. But the interest in 'neuroeconomics' is growing very rapidly, as is evident from recent work in brain imaging on interactions between people [40] and closer collaboration between psychology and economics. For years, the economist Richard Thaler has been publishing work on cognitive economics (with resistance from pure economists). There is also a field called experimental economics that links psychology and economics [41]. Somewhat more modestly, as a physiologist my contribution to this movement will be to examine the fundamental basis of decision making and to propose a neurobiology of decision making.

References

1. L Krüger, LJ Daston, and M Heidelberger, *The probabilistic revolution*, vol. 1, *Ideas and history* (Cambridge, MA: MIT Press, 1990). Also worth consulting is a recent introduction to theories of decision making: T Connolly, HR Arkes, and KR Hammond, eds, *Judgment and decision making: an interdisciplinary reader*, 2d edn (Cambridge: Cambridge University Press, 2000).

2. See also I Hacking, *The emergence of probability* (Cambridge: Cambridge University Press, 1975).

3. One should also cite the seminal contributions of Emile Borel, Bruno de Finetti, John Maynard Keynes, and Frank P Ramsey to the foundations of modern subjective probability. Savage spent several years in Rome studying with de Finetti. He always thought of himself as de Finetti's student.

4. J von Neumann and O Morgenstern, *Theory of games and economic behavior* (Princeton, NJ: Princeton University Press, 1944). In a forgotten article from 1928, von Neumann had already sketched the outline of utility theory: J von Neumann, 'Zur Theorie der Gesellschaftsspiele', *Mathematische Annalen*, **100** (1928): 295–320. On page 1 of his paper with Morgenstern, he makes clear that the theory was already almost twenty years old. The core idea is that if, and only if, a subject's choices (preferences) conform to a set of six axioms, does such a function exist, defined up to a linear transformation. Since, many attempts have been made to 'relax' one or another of these axioms (e.g. transitivity) and to show that a utility function still exists, but that its properties are less consistent.

5. RA Schmidt, 'A schema theory of discrete motor skill learning', *Psychological Review*, **82** (1975): 225–60.

6. R Hastie and N Pennington, 'Cognitive approaches to judgment and decision making', in J Busemeyer, R Hastie, and D Medin, eds, *Decision making from a cognitive perspective* (London: Academic Press, 1995), pp. 1–31.

7. D Kahneman and A Tversky, 'On the psychology of prediction', *Psychological Review*, **80** (1973): 237–51. See also D Kahneman and A Tversky, eds, *Choices, values, and frames* (Cambridge: Cambridge University Press, 2000).

8. R Amalberti, 'Évolution des concepts sur l'erreur humaine', *Médicine aéronautique et spatiale*, **34** (1995): 227–33; S Sutherland, *Irrationality: the enemy within* (London: Constable, 1992); M Piatelli-Palmarini, *Inevitable illusions: how mistakes of reason rule our minds*, trans. M Piatelli-Palmarini and K Botsford (New York: Wiley, 1994); RE Nisbett and L Ross, *Human inference: strategies and shortcomings of social judgment* (Englewood Cliffs: Prentice Hall, 1980); C Morel,

Les Décisions absurdes. Sociologie des erreurs radicales et persistentes (Paris: Gallimard, 2002); C Kerdellant, *Le Prix de l'incompétence: histoire des grandes erreurs de management* (Paris: Denoel, 2000); D Vaughan, *The Challenger launch decision* (Chicago, IL: Chicago University Press, 1997).

9 A Tversky and D Kahneman, 'The framing of decision and the psychology of choice', *Science*, **211** (1981): 453–8; Kahneman and Tversky, eds, *Choices, values, and frames*.

10 Tversky and Kahneman, 'The framing of decision', p. 453.

11 This case was covered in Piatelli-Palmarini, *Inevitable illusions*.

12 E Shafir and A Tversky, 'Thinking through uncertainty: nonconsequential reasoning and choice', *Cognitive Psychology*, **24** (1992): 449–74.

13 M Allais, 'Fondements d'une théorie positive des choix comportant un risque', in *Économétrie, Colloques internationaux du CNRS*, XL, 12–17 May 1952 (Paris: CNRS, 1953), pp. 127–40; 'Fondements d'une théorie positive des choix comportant un risque et critique des postulats et axioms de l'école américaine', ibid., pp. 257–332, produced by the French national printing office, 1955; 'Le comportement de l'homme rationnel devant le risque: critique des postulats et axiomes de l'école américaine', *Econometrica*, **21** (1953): 503–46; M Allais and O Hagen, *Expected utility hypotheses and the Allais paradox* (Dordrecht: Reidel, 1979).

14 Shafir and Tversky, 'Thinking through uncertainty'.

15 HA Simon, 'Theories of decision making in economics and behavioural science', *American economic review*, **49** (1959): 253–80.

16 G Gigerenzer and S Stelten, *Bounded rationality: the adaptive toolbox* (Cambridge, MA: MIT Press, 2001), p. 13.

17 JW Payne, JR Bettman, and EJ Johnson, *The adaptive decision maker* (Cambridge: Cambridge University Press, 1993).

18 G Gigerenzer and PM Todd, 'Fast and frugal heuristics: the adaptive toolbox', in G Gigerenzer and PM Todd, eds, *Simple heuristics that make us smart* (New York: Oxford University Press, 1999).

19 However, note also the elegance of the book by RD Luce, *Utility of gains and losses* (New York: Erlbaum, 2000).

20 PN Johnson-Laird and E Shafir, 'The interaction between reasoning and decision making: an introduction', *Cognition*, **49** (1993): 1–9.

21 MI Bauer and PN Johnson-Laird, 'How diagrams can improve reasoning', *Psychological science*, **4** (1993): 372–8.

22 JR Bruner, JJ Goodnow, and GA Austin, *A study of thinking* (New York: Wiley, 1956).

23 Bauer and Johnson-Laird, 'How diagrams can improve reasoning', p. 373.

24 Ibid., p. 378.

25 Johnson-Laird and Shafir, 'The interaction between reasoning and decision making', pp. 1–9; PN Johnson-Laird, 'Mental models and deduction', *Trends in cognitive science*, **5** (2001): 434–42.

26 DM Wolpert and M Kawato, 'Multiple paired forward and inverse models for motor control', *Neural network*, **11** (1998): 1317–29; J McIntyre, M Zago, F Lacquanitin, and A Berthoz, 'Does the brain model Newton's law?' *Nature Neuroscience*, **4** (2000): 693–4.

27 FA Yates, *The art of memory* (Chicago: University of Chicago Press, 1966).

28 A Connes's remark made during a colloquium in memory of Gilles Chatelet, Paris, 2001.

29 H Montgomery and H Willén, 'Decision making and action: the search for a good structure', in P Juslin and H Montgomery, eds, *Judgement and decision making* (Mahwah: LEA, 1999), pp. 147–73.

30 Ibid., p. 168.

31 A Berthoz, 'Reference frames for the perception and control of movement', in J Paillard, *Brain and space* (New York: Oxford University Press, 1990), pp. 80–111.

32 E Shafir, 'Compatibility in cognition and decision', in Busemeyer, Hastie, and Medin, *Decision making from a cognitive perspective*, pp. 247–74.
33 Ibid., p. 270.
34 SJ Leven and DS Levine, 'Multi-attribute decision making in context: a dynamic neural network methodology', *Cognitive science*, **20** (1996): 271–99, p. 272.
35 S Grossberg and W Gutowski, 'Neural dynamics of decision making under risk: affective balance and cognitive-emotional interactions', *Psychological Review*, **94** (1987): 300–18.
36 **Leven and Levine**, 'Multi-attribute decision making', p. 279.
37 Ibid., p. 291.
38 Y Weiss, EP Simoncelli, and EH Adelson, 'Motion illusions as optimal percepts', *Nature Neuroscience*, **5** (2000): 598–604.
39 J Baron, *Thinking and deciding* (Cambridge: Cambridge University Press, 2000), p. 226.
40 K McCabe, L Houser, L Ryan, VL Smith, and T Trouard, 'A functional imaging study of cooperation in two person reciprocal exchange', *Proceedings of the National Academy of Sciences of the USA*, **98** (2001): 11832–5.
41 See JH Kagel and AE Roth, eds, *Handbook of experimental economics* (Princeton, NJ: Princeton University Press, 1995).

Chapter 2

Decision making and emotion

> Although it is always linked to the body, emotion remains a part of consciousness because it is invariably an idea: the idea, which, for Spinoza, signifies awareness of a modification of the body. This permanent coexistence of affect and awareness should not surprise us since, according to Spinoza, the human mind which is 'idea of the body' is an effort to persist in being (a conatus) and thus is always aware of its effort.
> (*Robert Misrahi [1]*)

A theory of decision making cannot assume that subjects evolve in an indifferent world. The human brain, like that of animals, has different relationships with external objects depending on whether they are likely to aid or destroy it, whether they promise reward or punishment, satisfaction or pain. The world contains living creatures—prey and predators, partners and competitors—that can cause either happiness or misery. A physiology of perception must take emotions into account.

It is thus necessary to understand how the brain, in the enormous complexity of the material world, selects objects, guides action, directs attention, and specifies goals; and especially how it manages to choose among several behaviours to achieve a single objective, a process animal behaviourists call 'vicariance'. This flexibility of choice is what allowed organisms that appeared late in the course of evolution not to have to rely on poorly adapted reflexes that could do only one thing.

In this chapter I would like to show that emotions play a decisive role in several mechanisms of decision making: targeting objects in the world, letting the past guide future action, and flexibility in choosing behaviours. Twenty years ago, Jean-Didier Vincent concluded that our neurophysiology of perception was too reductionist and advocated a 'biology of emotions' [2]. According to Vincent, the major innovation of vertebrates was the capacity—thanks to the neuromodulator dopamine—to choose among several possible behaviours, and the key to that capacity is emotion.

The expression of emotion according to Darwin

Darwin concerned himself with 'the expressions and gestures involuntarily used by man and the lower animals, under the influence of various emotions and sensations' [3]. He criticized the lack of progress made since 1667, when the painter Charles Le

Brun, describing fright, affirmed, for example, that if 'the muscles and veins are swollen, it is only by the humours that the brain sends to those parts and that stir', a description Darwin thought typical of the 'specimens of the surprising nonsense which has been written on the subject' [4].

Darwin was referring to a rather broad repertoire of bodily and facial expressions published in 1872 and preceded by a treatise published in 1806 by Charles Bell, whose principal virtue, according to Darwin, was to have shown the close relation that exists between expressive movements and breathing, although Bell 'did not attempt to follow out his views as far as they might have been carried. He does not try to explain why different muscles are brought into action under different emotions; why, for instance, the inner ends of the eyebrows are raised, and the corners of the mouth depressed, by a person suffering from grief or anxiety' [5].

Darwin was also influenced by Jacques Louis Moreau's edition of Johann Kaspar Lavater's work on physiognomy, *L'Art de connaître les hommes* (How to understand people), but he took issue with Moreau's explanations. For example, Moreau explained that the purpose of wrinkling the eyebrows is to contract the face, 'to diminish oneself as if to offer less purchase and surface to formidable or troublesome impressions.' Rather than this falsely functional explanation, Darwin preferred, for example, the interpretations given in Thomas Burgess's *The Physiology or Mechanism of Blushing*, published in 1839; likewise, and especially, he appreciated the work of the French neurologist Guillaume Duchêne de Boulogne, who described with great skill in *Mechanisms of Human Physiology* the functional organization of the musculature, as well as Louis Pierre Gratiolet's treatise *De la physiologie des mouvements d'expression* (On the physiology of expressive movements, 1865). This last work accords motor expression an essential role: 'The senses, the imagination and thought itself, no matter how lofty or abstract one believes it to be, cannot be experienced without awakening a related feeling, and without this feeling translating itself directly, sympathetically, symbolically and metaphorically throughout the external organs' [6]. Darwin reproached Gratiolet for 'overlook[ing] inherited habit', that is, of not situating himself in an evolutionary perspective. But his commentary on the theory of Herbert Spencer was even more interesting. In his *Principles of Psychology* (1855), Spencer described how fear, 'when strong, expresses itself in cries, in efforts to hide or escape, in palpitations and tremblings; and these are just the manifestations that would accompany an actual experience of the evil feared.' Similarly, 'destructive passions are shown in a general tension of the muscular system, in gnashing of the teeth and protrusion of the claws, in dilated eyes and nostrils in growls; and these are weaker forms of the actions that accompany the killing of prey.' Darwin wrote admiringly, 'Here we have, as I believe, the true theory of a large number of expressions' [7]. He was also impressed by the idea that sensation, in emotive action, overflows into 'less habitual' routes.

Emotion is thus a simulation of action, an emulation of a hypothetical state or a reaction to a situation that may be pleasant or dreadful, and that circumvents the usual pathways of action.

Darwin's three principles

Darwin did not stop at borrowing or commentary. He noted first off that all reflective activity is expressed in the face. There is no thought without bodily accompaniment. Based on his observations, he described the expression of emotions in terms of three principles.

The principle of serviceable associated habits

Some actions are in general associated with mental states linked to satisfaction of our desires. When we induce one of these mental states, force of habit produces certain of the associated movements, which are judged expressive. Sometimes, we suppress a habitual movement and simultaneously make other little movements, themselves expressive. For example, when trying to decide, one person might scratch his head, another might rub her eyes, another chew his fingernails. Here Darwin intuited the role of selective inhibition, which I will come back to.

The principle of antithesis

Darwin constrasted the posture of a dog that is hostile and aggressive with the posture he adopts when he sees his owner. In the first case, he walks stiffly, head raised, his tail erect and rigid, ears forward and eyes fixed; in the second, his body sinks, flexes, tail lowered and ears bent backward, and so on. The two postures are opposed, 'antithetical'. So Darwin proposed a principle of antithesis. When a state of mind induces one combination of expressive movements, the opposite state induces an opposed series of movements, even if these are without usefulness or meaning. This principle is important in decision making because it shows the mechanisms that may be involved in the case of conflicting choices.

The principal actions of the nervous system independent of will and habit

Who has not experienced the agitated shaking that sometimes accompanies major decisions or emotions, and the many cardiovascular signs that appear to have nothing to do with what we are thinking about? Darwin formulated a third principle to take these into account. There are some actions that 'we recognize as expressive of certain states of the mind, are the direct result of the constitution of the nervous system, and have been from the first independent of the will, and, to a large extent, of habit' [8]. These actions include trembling, heart palpitations, and vascular effects that occur in stressful situations or after a failure or a bad decision and these struck Darwin by their strong and pervasive character. He thus proposed a principle *of nonspecific diffusion of emotional force*.

Other major principles have been proposed based on the observation of emotions of Wilhelm Wundt, the founder of experimental psychology. I will cite three of them. A first principle, equivalent to Darwin's third principle, holds that the intensity of muscular and vasomotor movements depends on the intensity of the emotion. The second is the 'association of analogous sensations'. If a person who reflects before making a decision is perplexed and scratches his head, coughs and rubs his eyes, he is expressing a slight physical discomfort and psychological confusion. Théodule Ribot, commenting on Wundt, wrote that these manifestations are 'a language turned aside from its primary signification, which in the order of gestures is the equivalent of metaphor' [9]. The third principle is the 'principle of the relation of movement to the perceptions of sense'. For example, an indignant man clenches his fists against an absent adversary, an expression of disdain or of contempt replicates the grimace of disgust and so on. Here again we encounter the idea of emotion as simulation, or rather emulation, of action.

The face of decision: eyebrows furrowed, mouth clamped shut

Let us return to Darwin and consider some of his observations. They will help us to understand that the bodily manifestations of emotions are not simply 'expressive' reactions. Imagine a person about to make an important decision. She wrinkles her eyebrows, closes her mouth, seals her lips; her look turns grave. A mental storm can be read on her face, and the contrary winds of opposing ideas are reflected in the bends and folds they imprint on her soft and tender skin. Why this gestural expression of mental effort? Darwin was especially interested in the fact that every act of thinking or mental effort, particularly when deciding, is accompanied by a wrinkling of the eyebrows. This movement is achieved by a 'corrugator' muscle that lowers the eyebrows and brings them together. Charles Bell thought this muscle the most remarkable of the human face because it reveals the 'energy of the mind'. Darwin noted its universal character: Australians, Malaysians, Hindus, the Kafirs of South Africa, all wrinkle their eyebrows. It is thus not a simple reflex; it signals an encounter with an obstacle. Darwin's observation about its universality was confirmed by the anthropological work of Paul Ekman [10]. Good naturalist that he was, Darwin remarked in his own children that the sign preceding a baby's anger is wrinkling of the eyebrows (Fig. 2.1).

Finally, Darwin thought that wrinkling was also associated with protecting vision by creating a sort of visor. Frans Cornelis Donders, a great Dutch physiologist who studied vision, believed that the contraction of the corrugator muscles facilitated near vision. That suggests to me that wrinkling allows our gaze to focus on the object of emotion when it is close by. I also have the impression that when we wrinkle our eyebrows, our arm and body muscles contract as they do before jumping forward. So actually an entire posture prepares for action and accompanies decision making. Besides, do we not say that we have 'jumped' at an opportunity when we have made a very quick decision? Have you noticed that you do not wrinkle your eyebrows when looking into the distance?

A closed mouth also often accompanies decision making. Darwin associated a hermetically sealed mouth with a strong character. Having a half-opened mouth is for him the sign of a weak character. When we make a decision, we shut our mouth and

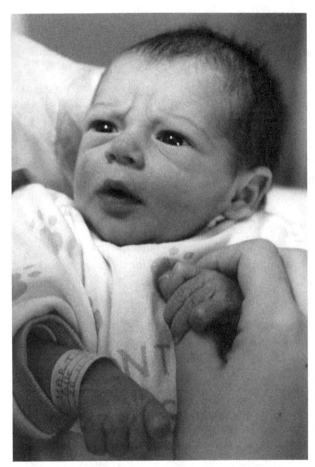

Fig. 2.1 Wrinkled eyebrows are a sign of discontent, but they are also often associated with decision making and with worry. Courtesy J. P. Martin

hold our breath. He remarked that the same is also true of a monkey executing a simple task. Indeed, says Darwin, every act, however insignificant, always requires a preliminary decision if it involves difficulty: 'The result would be a well-established habit, now perhaps inherited, of firmly closing the mouth at the commencement of and during any violent and prolonged exertion, or any delicate operation. Through the principle of association there would also be a strong tendency towards this same habit, as soon as the mind had resolved on any particular action or line of conduct, even before there was any bodily exertion, or if none were requisite. The habitual and firm closure of the mouth would thus come to show decision of character; and decision readily passes into obstinacy' [11]. Clearly, Darwin ranks decision making among character traits; today we would say that such-and-such a person has a 'determined character'. Note, by the way, Darwin's unexpected reference to the heritability of acquired characteristics.

For me, opening or closing the mouth constitutes a basic expression. Indeed, the mouth is the first space a baby explores; it's by the mouth and by gaze that the infant makes contact with the mother who nourishes him. Surprise, which leaves one open to the world, is always accompanied by opening the mouth. Closing the mouth stops the world from coming in; it interrupts the dialogue. Decision making requires withdrawal, concentration. It is hard to imagine a Buddha with an open mouth.

Thus, the state of profound reflection that accompanies making a difficult decision translates into facial expressions associated with the effort required by significant life events. Darwin believed these expressive gestures to be largely innate. But he went further, maintaining that we probably do not have just a repertory of expressions that accompany our deliberations; we have as well an innate capacity for immediately recognizing this same state in others' faces.

Darwin's analysis shows that our most complex decisions automatically activate actions belonging to an innate repertoire of expressions that through the play of habit and associations are linked to our mental mechanisms. That is why oriental theatre is more than a technique that uses the body to illustrate thoughts and emotions (Fig. 2.2).

We are beholden to Jerzy Grotowski and Ariane Mnouchkine for taking us back to it: 'We abandoned make-up, fake noses, pillow-stuffed bellies . . . We found that it was consummately theatrical for the actor to transform from type to type, character to

Fig. 2.2 Physical expression of emotion. After O. Aslan (ed.), *Le Corps en jeu* [The acting body], Paris, Éditions du CNRS (1993)

character, silhouette to silhouette—while the audience watched—in a poor manner, using only his own body and craft. The composition of a fixed facial expression by using the actor's own muscles and inner impulses achieves the effect of a strikingly theatrical transubstantiation, while the mask prepared by a make-up artist is only a trick' [12]. Western theatre readily tacks on the text bodily forms of convention that do not convey the suggestive power of real emotion.

The somatic anchor of emotion: the James–Lange theory

Despite these attempts to describe the expression of emotions and their connection to action, no physiological theory for it was proposed until the publication of an article by William James. In it he wrote, 'Our natural way of thinking about these standard emotions is that the mental perception of some fact excites the mental affection called the emotion, and that this latter state of mind gives rise to the bodily expression. My thesis on the contrary is that *the bodily changes follow directly the* PERCEPTION *of the exciting fact, and that our feeling of the same changes as they occur* IS *the emotion*' [13]. The circumstances thus induce changes in vegetative—cardiovascular, muscular, and so on—systems, which are then perceived as emotion.

Around the same time, the physiologist Carl Lange published similar ideas, stressing the role of the system that regulates our involuntary functions, in particular heart rate [14]. For this reason, the theory is called the 'James–Lange theory'. It was revolutionary because it reversed the order of cause. It gave priority to the sentient body even in the development of complex reactions. Like all so-called peripheralist theories that followed it, it was strongly criticized.

In his works [15], Walter Cannon advanced the following arguments against this theory:

1 Artificial production of visceral changes produces no emotion.
2 Because visceral organs have little feeling, it is hard to see how they could help in producing emotions.
3 Emotions are still observable in patients whose internal organs have been removed.
4 Visceral reactions are slow; they cannot explain the instantaneous character of certain emotions, for example, fear.

These criticisms are significant because they could be made today by detractors of Antonio Damasio's somatic marker theory. According to Cannon and Philip Bard, the emotional stimulus is first processed by the sensory thalamus (which is the portal to all sensations to the brain) [16]. Then it follows two pathways, the first, direct, involves the hypothalamus, which produces bodily reactions and sends messages simultaneously to the cortex and to the autonomic nervous system. A second pathway involves the cerebral cortex, which links both perceptual information transmitted by the thalamus and information about the internal state and reactions induced by the hypothalamus.

The notion of the limbic system

Helped along by advances in anatomy and physiology, over the course of years the processes that link emotion and the choice of associated behaviours began to look even more complex, and a third theoretical axis was proposed by James Papez [17]. This theory retained Cannon and Bard's idea of transmission of sensory information by the thalamus and of a double path or circuit, one pathway going through the hypothalamus via the hippocampus, and another passing through the anterior thalamus, the two reaching the cingulate cortex. This circuit, called the 'Papez circuit', has been the subject of numerous experiments but also of criticism in contemporary literature. It introduces the concept of *internal circuits* and not just circuits linking sensory stimuli to vegetative and motor responses (Fig. 2.3).

An important milestone in thinking concerning the mechanisms of emotions was reached owing to an experiment conducted by Heinrich Klüver and Paul Bucy [18], who removed the temporal lobe in a monkey. They then observed a 'psychic blindness' syndrome in the operated animal. It seemed to see objects clearly and to perceive its environment, but it could not assign meaning to them; it tried to eat any object it put in its mouth, and showed no reaction of fear or anger. The authors themselves attributed a central role to the hippocampus. Today their experiment must be reinterpreted in light of what we know about the role of occipitotemporal pathways in recognizing objects, but at the time, it gave rise to the so-called visceral brain theory of Paul MacLean [19] (Fig. 2.3). The experience and expression of emotion derive from the association of a great variety of external and internal stimuli whose messages are transmitted as nervous impulses to 'cerebral analysers'. For MacLean, the hippocampus—which he described as an emotional keyboard—was the central element of the mechanism that engenders and controls emotions. He included in his visceral brain the amygdala, the septum, and the prefrontal cortex, renaming it the limbic system, whose role consists in assuring the survival of the animal and the species in an integrated manner. The *global* character of the role of emotion remained (Fig. 2.4), but now the possibility emerged of an array of responses, like the play of organ pipes.

Finally, electrophysiologic analysis of the mechanisms of release of motor behaviours suggested a first hierarchical theory inspired by evolution. As MacLean saw it, the brain went through three stages over the course of evolution: *reptilian, paleomammalian*, and *neomammalian*. These three brains still exist in humans (Fig. 2.5).

1 The reptilian, or instinctive, and motor brain sends information to the thalamus and to the motor nucleus of the brain stem.

2 The emotional brain or limbic system has a number of pathways available that integrate information from several internal systems to *coordinate* emotional and motivational behaviour (action towards a goal).

3 The rational brain of mammals, or cortex, receives sensory signals from the thalamus and sends the bulk of these signals to the basal ganglia (in particular, the

THE NOTION OF THE LIMBIC SYSTEM | 31

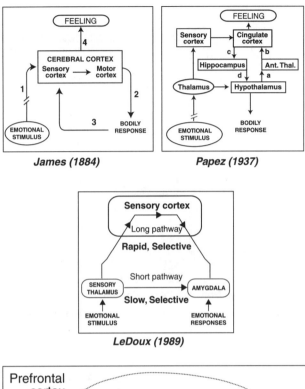

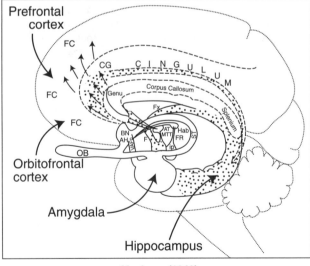

Fig. 2.3 Theories of emotions. Schematic representation of the theories of LeDoux, Papez, and MacLean. After MacLean (1949) and LeDoux (1996)

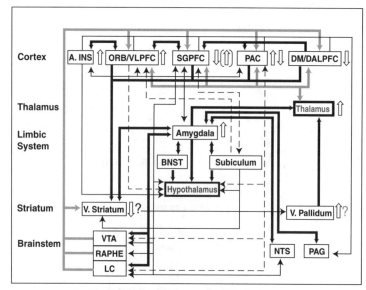

Manji, Drevets, Charneyin (2001)

Fig. 2.4 Circuits involved in emotions and mood disturbances. This schematic diagram summarizes many brain-imaging studies, observations in patients afflicted with mood disturbances (depression and so on), and it shows the regions believed to have structural abnormalities associated with these disturbances in patients with bipolar disorder. A.INS, anterior insula; ORB/VLPFC, orbitofrontal cortex and ventrolateral prefrontal cortex; SGPFC, subgeniculate prefrontal cortex (ventral and anterior to the corpus callosum); PAC, pregeniculate anterior cingulate cortex; DM/DALPFC, dorsomedial and dorsoanterolateral prefrontal cortex; VTA, ventral tegmental area; LC, locus coeruleus; NTS, nucleus tractus solitarius; PAG, periaqueductal grey matter. After Manji, Drevets, and Charneyin (2001)

caudal nucleus). The system of feelings and emotions signals internal states; the somatic and perceptual system signals the relation of the body to the external world; they converge on the reptilian system, which controls the motor programs and behaviours via the basal ganglia [20].

The amygdala and conditioned fear

Not all of MacLean's ideas withstood the test of time. For example, damage to the hippocampus and the mammillary bodies of the Papez circuit did not produce the expected abnormalities with regard to emotions. In contrast, Joseph LeDoux [21] established the role of the amygdala, which turned out to be critical for emotions and their relation to perception, such as for the rat in learning conditioned fear. For example, an electric shock is associated with a light, and subsequently it is observed that just the appearance of the light induces fear. Fear is an interesting model of behaviour that can be automatic and induced either by external stimuli or by thoughts. It can be triggered as easily by a snake as by a dream or by imagining the possible consequences of

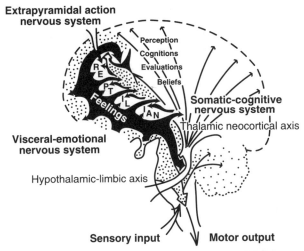

Fig. 2.5 Relations between emotions and behaviour. Schematic representation of the major axes of visceral information processing in the human brain: the hypothalamic and limbic axis of feelings, and the thalamo-neocortical axis of thought. The dorsal input is more concerned with information supplied by vision, hearing, touch, and is influenced only by the exterior world, whereas the more ventral portions concern input of a chemical nature (taste, smell, temperature, hormone level, energy, and hydration). The double flow of information converges towards the basal ganglia to induce behaviour in which the visceral and somatic processes are combined to ensure that behaviour is coherent. After Panksepp (1998)

an action. It exists in all species and very quickly (in less than 75 milliseconds) triggers stereotypic behaviours of immobility, flight, or aggression, a triad called the three Fs: 'fight, flight, freeze'. Fear can be conditioned by a single exposure, its memory having a profound influence on subsequent responses. These behaviours are characterized by their motor aspects, but equivalents can be described in the area of cognition: we say, 'When in doubt, do nothing!'; we speak of thoughts that 'escape' us; and we claim that the best verbal defence 'is attack'. The amygdala is involved in this short and rapid circuit (Fig. 2.3) with learning; but a second, slower mechanism, a long neural loop, that passes through the cerebral cortex enables evaluation of a situation and a choice of strategies, regulates the fast pathway and allows us, for example, to not be afraid if the creature slithering across the road turns out, after examination, to be a harmless grass snake and not a viper. This loop figures in anxiety and panic attacks, but also in the highest cognitive functions. I will describe these mechanisms more in detail in Chapters 9 and 11.

The amygdala receives information from many parts of the brain. When it detects danger, it sends messages to the hypothalamus. The hypothalamus activates the pituitary gland, which secretes adrenocorticotrophic hormone. This hormone signals the cortex to secrete steroid hormones which, in turn, modulate the activity of major centres involved in memory such as the hippocampus, or planning action such as the prefrontal cortex and so on. It is one of the mechanisms of stress induction, a response known for its association with having to make quick and difficult decisions (Fig. 2.4).

Psychological theories of emotions

In parallel with these physiological theories, philosophers and psychologists are interested in emotions and their connection with deliberation and decision making. There are more than 500 theories of emotion, not counting all the ones that have been proposed [22]. One might question the relevance of the very idea of emotion if it is so diffuse. Paul Griffiths puts it bluntly: 'The general concept of emotion has no role in any future psychology. It needs to be replaced by at least two more specific concepts. This does not necessarily imply that the emotion concept will disappear from everyday thought . . . Concepts like "spirituality" have no role in psychology but play an important role in other human social activities' [23].

Emotion and movement

The idea that emotion is preparation for action or at least associated with action also comes up in the writing of Théodule Ribot, a pioneer in experimental psychology [24]. He pointed out that 'emotion' contains the term 'motion', 'movement', which is fundamental. Emotion is movement. Its etymology is *e-movere*. '*Emotion is in the order of feeling the equivalent of perception in the intellectual order*, a complex synthetic state essentially made up of produced or arrested movements, of organic modifications (in circulation, respiration, etc.), of an agreeable or painful or mixed state of consciousness peculiar to each emotion. It is a phenomenon of sudden appearance of an limited duration; it is always related to the preservation of the individual or the species—directly as regards primitive emotions, indirectly as regards derived emotions' [25].

Anti-intellectual and antiformalist, Ribot was the philosophical counterpart of William James: 'An idea which is only an idea, a simple fact of knowledge, produces nothing and does nothing; it only acts if it is *felt*, and it is accompanied by an affective state, if it awakes tendencies, that is to say, motor elements. One may have thoroughly studied Kant's *Practical Reason*, have penetrated all its depths, covered it with glosses and luminous commentaries, without adding one iota to one's practical morality; that comes from another source, and it is one of the most unfortunate results of intellectualist influence in the psychology of the feelings that it has led us to ignore so evident a truth' [26]. Here is a professor at the Collège de France who suggested in 1930 what has been confirmed brilliantly today by Damasio! But at the time, his message fell on deaf ears.

Ribot essentially adopted the James–Lange thesis, but rejected its dualist character, as did the popular opinion James and Lange were fighting. The major difference is in the reversal of effects and causes: for some, emotion is a cause whose physical manifestations are effects. For others, physical manifestations are the cause of which emotion is the effect. According to Ribot, 'There would be a great advantage in eliminating from the question every notion of cause and effect, every relation of causality, and in substituting for the dualistic position a unitary or monistic one. The Aristotelian formula of matter and form seems to me to meet the case better, if we understand by "matter" the corporeal facts, and by "form" the corresponding psychical state . . . No state of consciousness can be dissociated from its physical conditions' [27].

Typologies of emotions and of decisions

Psychologists have attempted to divide emotions into many categories. Physiologists, assuming the existence of a basic repertoire of behaviours, have tended to distinguish a few fundamental emotions. Jaak Panksepp, for example, identifies four of them: *fear*, which leads to immobility, *flight*, or *aggression*; *panic*; *rage* or *anger*; and *curiosity*, which motivates exploratory behaviour [28] (Fig. 2.6).

Efforts to establish a typology of emotions are innumerable, and those that comprise less than six or seven so-called basic emotions are rare.

It is outside the scope of this book to review all of them, but the question is important, because if there really is a repertoire of emotions, one can also imagine a basic repertoire of decisions associated with those emotions.

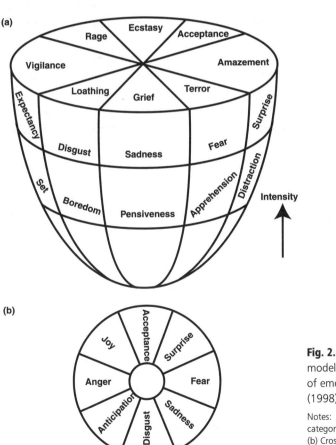

Fig. 2.6 Three-dimensional model of the Plutchik's types of emotion. After Strongman (1998)

Notes: (a) View of all the different categories of emotion.
(b) Cross-section of the model to show the architecture of its interior.

Theories inspired by phenomenology

The progress of physiology and the fascinating context of the James–Lange theory minimized the role of cognitive mechanisms actively involved in the regulation of emotions and decision making. The phenomenology of Edmund Husserl profoundly influenced the theory of emotions in Europe [29]. The idea that each individual constructs—assembles—an individual perception of the world and that the emotions he will experience will depend on this predisposition is one of the major contributions of phenomenology. As early as 1899, Carl Stumpf maintained that beliefs entail judgements that in turn provoke intentional mental states that excite emotions.

We have Husserl's disciple Jean-Paul Sartre to thank for his *Sketch for a Theory of the Emotions*, which opened many avenues of thinking that we will cross in our own examination of empirical observations concerning the neural basis of decision making. He begins with a critique of the James–Lange theory, adding to Cannon and Bard's critiques that 'even if the emotion objectively perceived presented itself as a physiological disorder, as a fact of consciousness, it is neither disorder nor chaos pure and simple, it has a meaning, it signifies something . . . ; it is an organized and describable structure' [30].

He also reproaches Pierre Janet for limiting himself to the untidy nature of emotions. He may have the great virtue of having reintegrated the psychic—today we would say 'cognitive'—dimension in the study of emotions, but he did so by attributing to it a weakened function: the defeatist attitude. All the same, he notes in Janet an 'underlying' theory according to which emotion has a certain purpose and a genuine 'behaviour', oriented as much to internal states and the subjects' own actions as towards the external world. For example, Janet wrote: 'A major failure of social behaviour is to react to *actions* and not simply to objects, as in primitive behaviour. People learn to react to the actions of their peers and then to their own actions. Our action becomes analogous to an external object that determines a reaction, and most likely these reactions to our own actions are an essential early component of consciousness. Indeed, feelings are made up of this group of reactions to our own actions, and in this way—conceiving of feelings as a regulatory system for action—they earn a place in the psychology of behaviour and can be explained similarly to other psychological phenomena' [31].

Sartre also mentions Henri Wallon, for whom emotion was a return to primitive behaviour, present in the infant, to which the adult reverts in difficult situations. Basically, Sartre tried to show that emotion is an intentional and organized behaviour, a way out that is particular to situations of conflict; he speaks of a 'special trick' [32]. He is also interested in Gestalt psychologists; in particular, he cites a long text of Paul Guillaume where Guillaume uses a spatial metaphor to describe the situation, for example, of an angry subject: he is imprisoned in a space with no exit and must surmount a barrier to get out. Emotion seems to be a 'weakening of the barriers between the real and the unreal' [33]. Anger appears as an evasion that restructures the world; it transforms the perception one has of the world to solve the problem posed by the world.

Understanding Sartre's thinking means turning one's attention for an instant to his reflections on consciousness. That is where most theories of emotion stumble. Sartre writes: 'Emotional consciousness is primarily consciousness *of* the world'; one is angry about something, one is afraid of something, one flees this animal or that foe. Emotion is the solution to a problem, but 'the subject who is seeking the solution of a practical problem is outside in the world, he is aware of the world at every moment throughout all his actions . . . And it is not necessary that the subject, between his failure in action and his anger, should turn back upon himself and interpose a reflective consciousness. There may be continuous passage from the non-reflective consciousness "instrumental world" (action) to the non-reflective consciousness "hateful world" (anger) . . . We tend too easily to believe that action involves a constant passing from the non-reflective to the reflective, from the world to oneself' [34].

According to Sartre, emotion constitutes neither a regression to infantile patterns nor a defeatist attitude, but a mechanism directly linking the world of impulsive action to the world perceived as beautiful or hateful, short-circuiting thinking and experience. Emotion thus ' [transforms] the world. When the paths before us become too difficult, or when we cannot see our way, we can no longer put up with such an exacting and difficult world. All ways are barred and nevertheless we must act. So then we try to change the world; that is, to live it as though the relations between things and their potentialities were not governed by deterministic processes but by magic' [35].

In *The Brain's Sense of Movement*, I suggested that *illusion is a solution* to a perceptual problem; now emotion, too, seems to be a solution that radically alters the world, like an illusion, to solve an impossible problem. The reference to magic is striking: for Sartre, all emotions 'are reducible to the constitution of a magic world, by making use of our bodies as instruments of incantation' [36]. One thinks of the tribes of New Guinea, described by Maurice Godelier, who draw shame on their faces.

For Sartre, emotion is thus a game of beliefs; it plays on the double status of the body as an object in the world and the direct experience of consciousness. It is thus a simulation of a state of the world that offers the means to overcome obstacles. I will come back to the fact that this theory endows emotion with predictive power: 'Thus', writes Sartre, 'in every emotion, a multitude of affective pretensions extends into the future and presents it in an emotional light' [37]. It is the predictive character of emotion that is so important in making decisions.

Cognitive theories and appraisal theories

During the deliberation that precedes decision making, we evaluate—we appraise—the elements involved. This activity of cognitive appraisal precedes judgement and is for some, Magda Arnold [38], for example, essential in the appearance of an emotion.

We automatically relate everything that we encounter to our intentions and our goals. Appraisal is a very familiar process that complements perception and produces in us the desire to do something.

The respective roles of cognitive factors and vegetative manifestations in emotions were presented in a simple way by Stanley Shachter in 1962 [39]. He developed a 'two-factor' theory of emotion that can be thought of as a cognitivist version of the James–Lange theory: when a situation awakens physiological functions, the brain evaluates and appraises this situation, which is labelled in some way with a type of emotion.

Psychologists also use the idea of frame of reference for decision making that emerged from the economists (see Chapter 1). For Lazarus, for example, appraising a situation has several goals. He writes: 'The premise of appraisal theory is that people (and infrahuman animals) are constantly evaluating relationships with the environment with respect to their implications for personal well-being . . . My concept of appraisal is that a person *negotiates* between two complementary frames of reference: first, wanting to view what is happening as realistically as possible and second, wanting to put the best possible light on events so as not to lose hope or sanguinity' [40]. This passage also evokes Sartre's theory of emotions: emotion colours the world so that it is perceived as acceptable.

This theory assumes processes at two levels. A preliminary appraisal is directly related to goals, hopes, commitments, and so on, and a secondary appraisal applies in situations of stress, when we must assess our *coping options* and decide which to choose and how to implement them. The questions vary according to the circumstances, but have to do with concrete problems: Do I need to act? What should I do? Is it doable? Which is the best option? Would it be better not to act? What are the consequences of acting or not acting? When should I act? Decisions concerning coping options are not etched in stone. They can be modified with the flow of events, but only up to a certain point. Emotion expressed or felt must be flexible and must depend on the context. The subject of emotions encapsulates the entire problem of decision making.

Before reacting in an emotional way and to choose the most appropriate reaction, the subject considers primary elements such as, for example, pertinence and coherence regarding the goal sought by the subject and his own desires (ego). Next come secondary elements: blame or approbation. The result of appraisal requires a decision, a judgement about something that can result in pain, benefit, and so on. This decision in turn requires establishing the fact that the action is well under the control of the agent or subject. Otherwise, he is not responsible for it. Expectations for the future and the consequences of possible choices also come into play, as well as coping ability. Lazarus stresses that appraisal is relative and not absolute. It is a 'warm' process and emotionally the contrary of cool logic: 'People in Western society, including many of its scholars, regard emotions as irrational. Typically, we regard emotions as a form of craziness and believe they fail to follow logical rules. We constantly pit emotion against reason . . . It is difficult to shake off more than 2000 years of Western habits of thought, with their roots in ancient Greece. During the Middle Ages, Plato's view of cognition and emotion was adopted by the Catholic church. The church fathers emphasized the antithesis between reason and emotion and the need for its parishioners, and people in general, to regulate their emotions and animal instincts by means of reason and an act of will. Despite the fact that such a struggle implies conflict between thought and reason, it is not wise to perpetuate an outlook in which emotion is reified as entirely independent of reason. The position taken by Aristotle . . . , which the church and Western civilization largely ignored, is a wiser but less familiar view—namely, that although conflict can take place between these two agencies of mind, emotion depends on reason. Appraisal theory is based on this Aristotelian view' [41]. The idea that there are appraisal mechanisms in the brain that control emotions was recently given a boost with respect to the possible role of the retrospinal cortex. I will return to it later on.

Which came first, cognition or emotion?

The idea that cognitive appraisal guides emotions was sharply criticized in the so-called Zajonc–Lazarus debate. At that time, affect was considered to be a postcognitive operation—'objects must be cognized before they can be evaluated'—in contrast to the theory of Wundt, which holds that 'the clear apperception of ideas in acts of cognition and recognition is always preceded by feelings' [42]. Robert Zajonc proposed to turn the order completely upside down, at least with regard to emotions that influence preferences. Objects need only be 'minimally' known. Zajonc makes an important distinction between thought and feeling. For him, feelings contain more energy and thoughts more information: 'Decisions are another area where thought and affect stand in tension to each other. It is generally believed that *all* decisions require some conscious or unconscious processing of pros and cons. Somehow we have come to believe, tautologically, to be sure, that if a decision has been made, then a cognitive process must have preceded it. Yet there is no evidence that this is indeed so. In fact, for most decisions, it is extremely difficult to demonstrate that there has actually been *any* prior cognitive process whatsoever' [43]. The problem is thus posed again between cognition and emotion, between liking or detesting and knowing, between the sentient body and reason.

Emotion has a fundamental role in *precategorizing* the stimuli that orient cognitive exploration. Affect is the greatest innovation of animals in comparison with plants. Zajonc takes as an example the flight of a rabbit confronted with a snake: the rabbit cannot take the time to analyse all the snake's characteristics. Zajonc senses that preference can result in a resonance between the global shape of objects and an individual's preperceptions [44].

This conception is close to one I will propose later of a resonance between the properties of the world and internal preperceptions. Zajonc bases part of his conviction on a known effect called the 'exposure effect'. When exposure to an object is repeated, subjects tend to increase their preference for the object. This effect is due in large part to emotion. For example, subjects repeatedly exposed to polygons use their subjective preference more than their recognition of the geometry of polygons to express their knowledge of them. Another example is face recognition. If, in memorizing a face, a person is told to judge it emotionally when the face appears, the memory of the face is better. Emotion is fundamental to memory, something that physiology confirms today.

Recent research has confirmed the importance of decisions based on emotions not consciously perceived by the subject [45]. Results show that one can love without being aware of it, and that the right and left cerebral cortex do not have the same role in differential processing of cognitive and emotional information. It even appears that to love and to know can be in conflict and can mutually suppress each other, which we will find in Chapter 11 in recent experimental findings concerning the anterior cingulate cortex.

The hierarchical nature of choice and the selection of emotional reactions shows up clearly in the debate between cognition and emotion. For Leventhal, for example, emotions are analysed and processed at several hierarchical levels (sensory motor, schematic, and conceptual), and it is at the two highest levels that complex interactions between emotion and cognition come into play. His approach is consistent with the so-called component-processing model, which describes emotion as an 'evolved . . . mechanism which enables an increasingly flexible adaptation to environmental contingencies owing to a decoupling of stimulus and response that creates a latency time for response optimization' [46].

Emotion as preparation for action

A decisive step was made in accepting the idea that emotion is preparation for action and not simply reaction. This opened the way for my hypothesis that emotion represents *anticipation of the future*—transformation of the world, as Sartre put it—and not just evaluation of the past. Nico Frijda writes: 'Emotions are action tendencies' [47], also referred to as 'action readiness'. That reminds me of the Scout expression, 'Always ready!' Action tendencies are 'not necessarily readiness for overt action'. Rather, says Frijda, they are 'states of readiness to execute a given kind of action . . . defined by its end result aimed at or achieved . . . [and] inferred from behavior . . . One action tendency is readiness for attacking, spitting, insulting . . . ; a different action tendency is readiness to approach and embrace' [48].

The computational approach

The metaphor of the computer brain has led several authors to propose what one might call both physiological and 'computational' models of emotions. This theme can be traced back to Ulric Neisser's famous book [49]. A novel feature of these theories consists in attributing two functions to an increase in vegetative activity [50]. One function is simply to prepare the organism physiologically to respond to the demands of the environment. It is the classic response incorporated into every theory since Darwin. But a second function consists in readying awareness to evaluate the situation, identify what triggered the activity, and *reorganize the plan of action*. This function corresponds to Joseph LeDoux's second loop. In 1987 and 1992 Keith Oatley and Johnson–Laird proposed a theory very close to Jean Mandler's [51]. They postulate that there is a series of basic emotions, joy, sadness, anxiety, anger, and disgust, and that their cognitive processes are 'modular' but must communicate with each other. Each module carries out a particular task, for example, searching memory for a name to put to a face.

These modules are organized hierarchically. Higher-order-level modules control lower-level modules, and at the very top, a central system coordinates everything. This recalls the central executive model proposed by Donald Norman and Tim Shallice (see the Introduction). In their model, subsystems are responsible for modules specialized in particular actions, but a supermodule coordinates the relations between the plans of action worked out by the brain and these functional submodules. This control and supervisory system, which can reorganize plans of action, contains within it a *model of the system itself as a whole*—a doppelgänger, a system double, the equivalent of a body schema (and the subject of Chapter 6). This global representation would have served the goal of preservation and survival, which is one of the essential functions of emotion. Communication between modules is divided into two different processes.

The first process is called 'propositional' or 'symbolic'. It deals with factual information about the world such as 'I met him at the corner of such-and-such a street.' The second process is not propositional and contains no information. It is essentially emotional and shifts the modules into a certain emotional 'state'. For example, suppose that you are a university student shopping at the supermarket. You are debating whether to buy tomatoes or leeks when, suddenly, you notice a young woman. Even before you remember her name, you start feeling very odd. After a moment, you understand why your discomfort has interrupted your mental activity and errands: it's your professor of English whose class you've been cutting for two months! These theories [52] are computational because each step has been simulated in models that are processable by calculation. But they are also appraisal theories.

Behavioural theories

These theories appeared with John Watson [53] (1929), who distinguished three fundamental types of emotions: fear, anger and hate, and love. In succession, Michael Millenson [54], Laurence Weiskranz [55], and recently Jeffrey Gray [56] have sought to explain emotions using the framework of reinforced conditioning dear to behaviourists. This framework involves classifying emotions linked to behaviours with which they are associated. For example, Gray distinguishes three distinct systems that contribute to the relation between reinforcing stimuli and response systems: an approach system, a behavioural inhibition system, and a fight or flight system. Edmund Rolls [57] is a brilliant representative of this school of thought and a creator of a 'neobehaviourist' stream. He has a few criticisms to make about the James–Lange theory that recall those of Cannon and Bard, mentioned above:

1. Peripheral manifestations, that is, all the so-called vegetative functions—heart rate, breathing rate, sweating, dilation of pupils, and so on, that are regulated by the autonomic nervous system—are too coarse to account for the subtlety of emotions.
2. Emotions are often evoked by imagery, in which case peripheral manifestations are weak.
3. Peripheral surgery does not suppress emotions.
4. Changing the vegetative system by injecting certain hormones produces no emotions.
5. Blocking vegetative mechanisms by drugs causes no observable change in emotions.
6. Finally, emotion does not always accompany success, but is experienced instead, for example, when one smiles at a friend.

In other words, these bodily responses, which can be very brief, often serve the needs of communication or action but do not produce feelings of emotion. I will return to Rolls's theories in Chapter 11.

What is worth remembering for our purposes is that these researchers have tried to link the neural basis of emotion to a basic behavioural repertoire. To investigate the neural underpinnings of choice among these behaviours, they propose neuronal mechanisms of association and conditional reinforcement. These data open the door to a neuronal neobehaviourism compatible with approaches that, today, stress the importance of Hebbian mechanisms of learning; that is, when input and output signals converge simultaneously on a synapse, the connections involved will be reinforced. We are currently seeing an efflorescence of new approaches to the study of reasoning under conditions of uncertainty inspired by what is called qualitative physics [58], a method that combines the determinism of physical calculations with notions of probability and that provides models whose biological plausibility will have to be tested in the future.

Decision and indecision: the ideas of William James

The ideas of William James on decision making are consistent with his theory of emotion. Since the former are less well known than the latter, I will review them here because they are suggestively very rich. In contrast with theories that consider emotions to be reactions to changes in the environment, William James tackles the problem of decision making via mechanisms of *voluntary initiation* of action and obstacles encountered in the desire to act. He proceeds in two stages. First, he proposes that the trigger of movement is associated with a representation of its consequences: ' "Idea of a movement" ... must precede it in order that it be voluntary ... The question is this: *Is the bare idea of a movement's sensible effects its sufficient mental cue, or must there be an additional mental antecedent, in the shape of a fiat, decision, consent, volitional mandate, or other synonymous phenomenon of consciousness, before the movement can follow?*' [59].

Based on the ideas of Rudolf Lotze [60] who had foreseen (like Darwin, I might add) that imagining a movement activates the same structure as carrying it out—which modern brain imaging has confirmed—James suggests that *awareness is always awareness of an action*. He then explains that deliberation appears when, for whatever reason, something stands in the way of a thought being executed. Faced with this obstacle, the brain seeks solutions through deliberation. The first act for James is indecision, where action is blocked by several possibilities that jostle and contradict each other and 'serve ... as an effective check upon the irrevocable discharge' [61]. In other words, one could summarize this mechanism as follows: 'I deliberate because I am blocked.' Deliberation is the product of a brake imposed on the impulse to act that links awareness and action. To decide is to conquer indecision.

James distinguishes five principal types of decision, this 'bracketing' of action, as Husserl might have said.

The first type is the one that James calls 'reasonable' decision making. The point of departure is a mental situation in which 'the arguments for and against a given course seem gradually and almost insensibly to settle themselves in the mind and to end by leaving a clear balance in favor of one alternative, which alternative we then adopt without effort or constraint ... But some day we wake with the sense that we see the thing rightly, that no new light will be thrown on the subject by farther delay, and that the matter had better be settled *now* ... The conclusive reason for the decision in these cases usually is the discovery that we can refer the case to a *class* upon which we are accustomed to act unhesitatingly in a certain stereotyped way' [62]. This idea of a ready-made repertoire of actions recapitulates what we have already seen in the case of emotions.

Deliberation brings one back to different ways of envisaging how to accomplish a planned act. James suggests that decision is possible when the brain can *simulate an action* that is already part of an acquired repertoire and whose consequences thus can be predicted. He continues: 'The moment we hit upon a conception which lets us apply some principle of action which is a fixed and stable part of our Ego, our state of doubt is at an end' [63].

It is interesting that he distinguishes the process of decision making and this particular state of indecision. James underscores that '*in action as in reasoning, then, the great thing is the quest of the right conception* ... A "reasonable" character ... does not decide about an action till he has calmly ascertained whether it be ministerial or detrimental to any one of these' [64]. James thus situates a decision at the end of a back-and-forth process between the intention to act and its goal, which constitutes the very foundation of brain operation, namely, predicting and choosing an action by estimating its consequences in advance.

In the second type of decision 'our feeling is to a certain extent that of letting ourselves drift with a certain indifferent acquiescence in a direction accidentally determined *from without*, with the conviction that, after all, we might as well stand by this course as by the other' [65]. This proposition puts one in mind of patients with schizophrenia. They feel that someone is deciding their actions for them. This can happen with basic actions, such as moving an arm, or more complex actions. The mechanism conceptualized by Christopher Frith [66] to explain this pathology is the following: Normally, when we decide to act (move) in some way, the brain, either before responding or at the same time, prepares the nerve centres involved in receiving pertinent sensory information. This is anticipation by extension of the 'corollary discharge' concept identified by Erich von Holst and Horst Mittelstaedt. It is fundamental in predicting the consequences of action, as I showed in *The Brain's Sense of Movement*. Here, it also enables attribution of a gesture to an agent. Patients with schizophrenia lack this mechanism which permits the brain to distinguish actions produced by the subject himself, such that when the action is triggered, they have the impression that the decision was made by someone else.

In the third type, *decisions brought about by internal circumstances*, decision making appears just as unintentional; but this time, it 'comes from within, and not from without. It often happens, when the absence of imperative principle is perplexing and suspense distracting, that we find ourselves acting, as it were, automatically, and as if by a spontaneous discharge of our nerves, in the direction of one of the horns of the dilemma. But so exciting is this sense of motion after our intolerable pent-up state, that we eagerly throw ourselves into it. "Forward now!" we inwardly cry, "though the heavens fall" ' [67]. James describes us as passive spectators of an 'extraneous force': making a decision appears to us to happen of its own accord as though by some miracle that is itself independent of our serene will.

Impulsive decisions of this sort give rise to a sort of 'cheering' that is 'too abrupt and tumultuous to occur often in humdrum and cool-blooded natures' but frequent 'in persons of strong emotional endowment and unstable or vacillating character. And in men of the world-shaking type, the Napoleons, Luthers, etc., in whom tenacious passion combines with ebullient activity, when by any chance the passion's outlet has been dammed by scruples or apprehensions' [68]. Such decision making clearly involves emotions. We are far from a formal theory of rational decision making as I described it in Chapter 1 of this book. Here, torrents of emotion carry away in their floods the edifices of reason! But this passion, this impetuous energy that annihilates reason is not a sign of stupidity. It evokes the following lines from Rimbaud's *Drunken Boat*:

> The tempest blessed my wakings on the sea.
> Light as a cork I danced upon the waves,

Eternal rollers of the deep sunk dead,
Nor missed at night the lanterns' idiot eyes!
...
I know the lightning-opened skies, waterspouts,
Eddies and surfs; I know the night,
And dawn arisen like a colony of doves,
And sometimes I have seen what men have thought they saw! [69]

A fourth type differs from the other three by the *mood that accompanies decision making*. Here we '*suddenly pass from the easy and careless to the sober and strenuous mood*, or possibly the other way. The whole scale of values of our motives and impulses then undergoes a change like that which a change of the observer's level produces on a view' [70]. Self-awareness is suddenly called into question. I will return later to ideas proposed by Janet on the importance of point of view in mechanisms of decision making.

The final type of decision making is characterized by the *direct influence of our will*. 'We feel, in deciding, as if we ourselves by our own wilful act inclined the beam; in the former case by adding our living effort to the weight of the logical reason which, taken alone, seems powerless to make the act discharge' [71]. James stresses that this awareness of slow and calm voluntary action constitutes a novel psychological characteristic, entirely absent from the four other types. About this 'feeling of effort', he notes simply that it has to do with 'peculiar sort of mental phenomenon', of an 'inward effort'. This type is different because in the others, 'at the moment of deciding on the triumphant alternative [the mind] dropped the other one wholly or nearly out of sight, whereas here both alternatives are steadily held in view' [72].

Decision making and point of view: the ideas of Pierre Janet

Decision making is not simply the result of rational deliberation or a unique emotional process. It can be completely changed by how we 'see' a situation. This idea is fundamental, because it determines our entire conception of the world. After all, one of the great scientific revolutions was the work of Copernicus, which changed how we thought about the movement of the Earth relative to the Sun! It is also generally accepted that decision makers are people who can look at a problem from a variety of vantages. To deliberate is to envisage several points of view (Fig. 2.7).

Think for a minute about the expression 'point of view'. Husserl spoke of 'views' that construct the perceived world. The expression 'point of view' is paradoxical, because it contains both the words 'point' and 'view'. To view is to 'point'. The expression thus means that one adopts a position and follows a direction at the same time. Everyday language also recognizes that our decisions depend on the 'perspective' by which we analyse facts. Judgement is spatialized—like many mental operations.

Janet talks about the importance of point of view in the context of a general study of the 'idea of position' [73]. He mentions that animals have a sense of spatial position and also a more symbolic sense of position in a social hierarchy. But he attributes

Fig. 2.7 Development of a decision often requires changing point of view. After *Traité de perspective* (Treatise on perspective), Paris, Éditions du Chêne (1976)

to humans a particular capacity to handle spatial shifts in perspective. From his experience as a clinician, he concludes that difficulty in managing these changes causes serious disturbances in patients, who lose the ability to navigate, for example, in their apartment or their town, or who are afflicted with allochiria, that is, an inability to make mental notations that enable one to know whether a statue is to the left or the right, depending on whether one is entering or leaving the hospital.

Janet gives the example of the Stinsteden illusion, introduced by Hermann von Helmholtz, which involves a deceptively simple decision-making task of perception. On a Dutch plain, in the evening, watching the wheels of a mill turn in the semidarkness that hides the building, the wheels may appear to be turning clockwise or counterclockwise depending on the point of view from which one imagines one is seeing the mill. This effect is similar to abrupt changes in perception like that posed by the Necker cube (draw a cube on a sheet of paper; your perception of its orientation in space will flip back and forth) or the roof illusion (draw a roof on a sheet of paper; its orientation will alternate) or many other illusions described by Gestalt theorists that show the influence of the brain on perception. For Janet, this problem of point of view is fundamental to all of intellectual life: 'Depending on our point of view, we see things one way or another. We also learn to change our point of view depending on the position we take or that we imagine taking' [74]. I have also argued that changing point of view is a fundamental mechanism of empathy [75].

A neurologist and clinician, Janet took an interest in this idea of point of view because it seemed to him to be characteristic of people. Indeed, according to Janet, if animals have a sense of position—for example, they know where they are in space—they can find their way home again. But he doubts that they can work out a concept of 'route', which requires imagining the same path from two opposite points of view: going and coming. This theme of shift in point of view recurs in this book. In Chapter 1 we encountered Johnson–Laird's analogy between rational thought and change in point of view. I will return to the problem later.

Decision making and emotion do appear to be intimately linked. What is needed is a physiology of preference, which I will outline in Chapter 11. Moreover, decision making does not simply mean calculating utility or gambling on likelihood. It is prediction experienced by a mind incarnated in a sentient body. It is the result of a hierarchy of many nested mechanisms that have now to be analysed. Examining the pathology of decision making will provide valuable examples of the extraordinary diversity of these mechanisms, which are the fruit of the slow work of evolution. They must be studied with patience and humility, leaving to those with concerns about time and profits the care of concisely formulating the complexity that is unique to living organisms.

References

1. **R Misrahi**, *Le Corps et l'esprit dans la philosophie de Spinoza* (Paris: Les Empêcheurs de penser en rond, 1992).
2. **J-D Vincent**, *The biology of emotions*, trans. J Hughes (Oxford: Blackwell, 1990).
3. **C Darwin**, *The expression of the emotions in man and animals* (Chicago, IL: University of Chicago Press, 1965), p. 27.
4. Ibid., p. 4.
5. Ibid., p. 3.
6. Ibid., p. 6.
7. Ibid., pp. 8–9.
8. Ibid., p. 66.
9. **T Ribot**, *The psychology of the emotions* (London: Walter Scott, 1897). Regarding the work of Wundt, cf. *Outlines of psychology* (Leipzig: Wilhelm Engelmann, 1897).
10. **P Ekman, ER Sorenson, and WV Friesen**, 'Pancultural elements in facial displays of emotions', *Science*, **164** (1969): 86–8; P Ekman, 'An argument for basic emotions', *Cognition and Emotion*, **6**

(1992): 169–200; P Ekman and J Davidson, *The nature of emotion* (Oxford: Oxford University Press, 1994).

11 Darwin, The *expression of emotions*, p. 236.

12 J Grotowski, *Towards a poor theatre* (London: Methuen), pp. 20–1.

13 W James, 'What is an emotion?' in *The works of William James* (Cambridge, MA: Harvard University Press, 1983), vol. 13, *Essays in psychology*, p. 170. First published in *Mind*, **9** (1884): 188–205.

14 C Lange, Über Gemuthsbewegungen (Leipzig: Theodor Thomas, 1887).

15 WB Cannon, 'The James–Lange theory of emotions. A critical examination and an alternative theory', *American Journal of Psychology*, **39** (1927): 106–124; **WB Cannon**, *Bodily changes in pain, hunger, fear, and rage* (New York: Appleton, 1929).

16 P Bard, 'The central representation of the sympathetic system', *Archives of Neurology and Psychiatry*, **22** (1929): 230–56.

17 JW Papez, 'A proposed mechanism of emotion', *Archives of Neurology and Psychiatry*, **79** (1937): 217–24.

18 H. Klüver and PC Bucy, ' "Psychic blindness" and other symptoms following bilateral temporal lobectomy in rhesus monkeys', *American Journal of Physiology*, **119** (1937): 352–3.

19 PD MacLean, 'Psychosomatic disease and the "visceral brain": recent developments bearing on the Papez theory of emotion', *Psychosomatic Medicine*, **11** (1949): 338–53.

20 J Panksepp, *Affective neuroscience: the foundations of human and animal emotions* (New York: Oxford University Press, 1998).

21 J LeDoux, *The emotional brain* (New York: Touchstone Books, 1996).

22 KT Strongman, *The psychology of emotion: theories of emotion in perspective* (Chichester: Wiley, 1996).

23 PE Griffiths, *What emotions really are: the problem of psychological categories* (Chicago: University of Chicago Press, 1997), p. 247.

24 A Berthoz, ed, *Leçons sur le corps, le cerveau et l'esprit* (Paris: Odile Jacob, 1999). This volume contains the inaugural lectures of a selection of professors at the Collège de France along with their biographies. See the chapter on Ribot, pp. 389–412.

25 Ribot, *The psychology of the emotions*, p. 12.

26 Ibid., p. 19.

27 Ibid., pp. 111–12. Ribot also critiques the theories of Darwin and specifies that in his opinion, the 'principle of the association of serviceable habits' is the most solid, and the 'principle of antithesis' ought to be abandoned.

28 Panksepp, *Affective neuroscience*.

29 N Depraz, 'Délimitations de l'émotion: approche d'une phénoménologie du coeur', *Revue de phénoménologie*, **7** (1999): 121–49.

30 J-P Sartre, *Sketch for a theory of the emotions*, trans. P Mairet (London: Routledge, 1962).

31 P Janet, *Les Débuts de l'intelligence* (Paris: Flammarion, 1935), p. 100.

32 Sartre, *Sketch*, p. 22.

33 Ibid., p. 25.

34 Ibid., pp. 34–5.

35 Ibid., pp. 39–40.

36 Ibid., p. 47.

37 Ibid., p. 54.

38 MB Arnold, *Emotion and personality*, vol. 1, *Psychological aspects*, vol. 2, *Physiological aspects* (New York: Columbia University Press, 1960).
39 S Shachter and JE Singer, 'Cognitive social and physiological determination of emotional states', *Psychological Review*, **69** (1962): 379–99.
40 RS Lazarus, 'Relational meaning and discrete emotions', in K Scherer, A Schorr, and T Johnstone, eds, *Appraisal processes in emotion* (Oxford: Oxford University Press, 2001), p. 41.
41 Ibid., pp. 59–60.
42 RB Zajonc, 'Feeling and thinking: preference needs no inference', *American Psychologist*, **35** (1980): 152.
43 Ibid., p. 155.
44 Ibid., p. 159.
45 TH Landis and LJ Christen, 'Dissociated hemispheric and stimulus effects upon affective choice and recognition', *Neuroscience*, **62** (1992): 81–7.
46 K Scherer, 'Toward a dynamic theory of emotion', unpublished manuscript on the Web, http://www.unige.ch/fapse/emotion/publications/pdf/tdte_1987.pdf, last accessed 16 May 2005.
47 NH Frijda, *The emotions* (Cambridge: Cambridge University Press, 1986), p. 70.
48 Ibid., 70–1, 76.
49 U Neisser, *Cognitive psychology* (New York: Appleton, 1967).
50 RR Cornelius, *The science of emotion* (Upper Saddle River: Prentice Hall, 1996).
51 G Mandler, *Mind and emotion* (New York: Wiley, 1975).
52 K Oatley and JM Jenkins, *Understanding emotions* (Oxford: Blackwell, 1996).
53 JB Watson, *Psychology: from the standpoint of a behaviorist* (Philadelphia: Lippincott, 1929).
54 JR Millenson, *Principles of behavioral analysis* (New York: Macmillan, 1967).
55 L Weiskranz, *Analysis of behavioral change* (New York: Harper and Row, 1968), pp. 50–90.
56 JA Gray, *The psychology of fear and stress* (London: Weidenfeld and Nicolson, 1971); JA Gray and N McNaughton, *The neuropsychology of anxiety*, 2d edn (Oxford: Oxford University Press, 2000).
57 ET Rolls, *The brain and emotion* (Oxford: Oxford University Press, 1999).
58 S Parsons, *Qualitative methods for reasoning under uncertainty* (Cambridge: MIT Press, 2001).
59 W James, *Psychology: the briefer course*, vol. 14, *The works of William James* (Cambridge, MA: Harvard University Press, 1983), p. 364.
60 H Lotze, *Medizinische Psychologie* (Leipzig: Weidmann, 1852).
61 James, *Psychology*, p. 369.
62 Ibid., p. 370.
63 Ibid.
64 Ibid., p. 370–1.
65 Ibid., p. 371.
66 C Frith, 'Neuropsychology of schizophrenia: what are the implications of intellectual and experiential abnormalities for the neurobiology of schizophrenia?' *British Medical Bulletin*, **52** (1996), 618–26.
67 James, *Psychology*, p. 371.
68 Ibid.
69 A Rimbaud, *A season in hell and the drunken boat*, trans. Louise Varèse (New York: New Directions, 1961), p. 93.
70 James, *Psychology*, p. 372.

71 Ibid., p. 372.
72 Ibid., p. 373.
73 Janet, *Les Débuts de l'intelligence*, p. 172.
74 Ibid., pp. 173–4.
75 A Berthoz and G Jorland, *L'Empathie* (Paris: Odile Jacob, 2004).

Chapter 3

The pathology of decision making

> It would be the pattern as a whole, rather than some bystanding area that observes it, that generates the experience and initiates the decisions.
> (*Marcel Kinsbourne [1]*)

Making a decision requires above all that perception faithfully reflect reality. Yet that is not always the case—far from it! Crossing the street one day, a woman friend of mine passed a man she thought was her husband and moved to make a familiar gesture. Realizing her mistake just in time, she checked her impulse. How was she able to avoid finding herself in an embarrassing situation of extending her hand, or her cheek, to a man she didn't know? How does the brain decide that a person is a friend, spouse, one's own child, or stranger? Who has not experienced similar confusion in recognizing a face?

How do we go from 'I see him' to 'I know him'? This ability to know another is not a question of language, because frequently we are able to recognize people without remembering their name. The decision made by our brain thus hinges on visual identification, on a sense of familiarity and on the recollection of common experiences—shared exploits or emotions and so on.

Disturbances in decision making occur in patients afflicted with lesions, psychiatric disorders, or addictions. These disturbances shed light on the neural basis of decision making. Indeed, identifying the brain areas involved shows that the process of decision making is not, for example, limited to the prefrontal cortex, but that it results from processes activating many areas of the brain.

My intent here is also to show that theories of decision making based on the idea of 'representation' or development of decisions based on formal rules must be supplemented, if not replaced, by a theory based on the concept of acts.

Recognizing a face or an object

Deciding whether a person encountered on the street is our mate requires that we identify the person but also that we assign to him or her certain features and associated memories. Brain lesions cause deficits in this process of identification and attribution of features. They are called agnosias, 'non-knowledge'. A person with agnosia can see visual shapes but cannot recognize them or say what they mean or are used for. (It has to do with 'gnosis', that is, high-level knowledge and not just raw sensations.) Yet not to err in deciding requires first of all that reality be accurately perceived, classified, and identified. How does one make the *right* decision when one's perception of the world is distorted, incomplete, or even illusory?

Agnosias [2] also include disturbances of association: the brain is basically an associator that combines, compares, links one percept with another, or with memory or with action that aids knowledge by changing point of view. Disorders of this capacity tell us something about the structures that perform these essential operations for decision making at whatever level.

To decide whether a face is familiar or whether a car in the parking lot is ours, at the very least we have to recognize the shape. There is a distinctive pathology related to shape recognition called 'apperceptive agnosia'. The term was invented by Wundt and taken up by Heinrich Lissauer in 1890 to refer to a first stage of conscious awareness of a sensory impression [3]. Lissauer distinguishes a second stage of shape perception that involves the association of other notions or of multisensory data. The inability to differentiate two very similar geometric forms is an important criterion in diagnosing apperceptive agnosia. We have all had this experience without having the disorder! Arriving in a foreign country, in Asia or Africa, for example, we have the initial impression that everyone looks the same.

Another disturbance of visual shape recognition is called 'associative visual agnosia'. This agnosia presents mainly in patients with lesions of the left hemisphere. It is a syndrome of disconnection of the right visual areas and the left language areas. It concerns the capacity for perceptual categorization [4]. Most of the time, perceptual decision making requires arranging several objects present in a visual scene. That is the case when we are in a strange town and have to decide whether we have already seen a place—a square, for example. Essentially it is the ability to put objects together again, to perceive a bicycle as a complete machine and not just an assemblage of parts (handlebars, pedals, and so on) (see Fig. 3.1).

I maintain that the capacity for making a perceptual decision is fundamentally linked to the capacity to construct a perceptual unity. The lack of such a faculty is called 'simultanagnosia'. The term was invented by Wolpert in 1924 to describe the difficulty in perceiving a whole as the sum of its parts. Patients cannot grasp more than one stimulus at a time. There are probably several sorts of simultanagnosias.

The first type is so-called integrative simultanagnosia. It manifests by a difficulty in integrating elements into a whole, for example, the parts of a bicycle (bilateral or unilateral inferomedial occipitotemporal lesion). The second is 'competitive' simultanagnosia: patients can integrate a single object on a homogeneous background, but they cannot simultaneously perceive several objects in the same scene. A sort of competition between the objects ensues. This disturbance relates to the neurological phenomenon called 'extinction', which in the patient manifests as the disappearance of a stimulus if a competing stimulus is presented at the same time. It is produced by

Fig. 3.1 Deciding requires a global view of a situation. This image is a test devised by Alfred Binet to examine whether patients with impaired ability to perceive a global situation can analyse the value of the 'simultaneous' presence of a group of elements (visual simultanagnosia). Here, patients do not grasp that the cyclist ('the telegram boy') is in trouble and is asking a passing driver for help. They perceive only the elements of the drawing (a hat, a wheel, a man, a handlebar, a car, and so on). After Grüsser and Landis (1991)

unilateral superior occipital lesions, distinct from the biparieto-occipital lesions of Balint-Holmes syndrome. Sometimes patients cannot direct their gaze towards the different parts of a scene because when they fix on one part, the rest of the image disappears. That is pseudo-simultanagnosia, which actually is part of the Balint-Holmes syndrome induced by biparieto-occipital lesions.

My father is an impostor!

What would you say if one evening on returning home your spouse or other member of your family said to you, 'You are an impostor! You are the spitting image of my husband (my wife, my son, my daughter). But you are just a double!' The sense of familiarity in the adult human sometimes fails. A very peculiar syndrome described in 1923 by the neurologist Jean Marie Joseph Capgras involves the ability of humans to recognize familiar people. Patients afflicted with the syndrome think that their family, friends, or acquaintances are impostors. They may even say that a person looks like their father but is not really their father. These patients generally are perfectly normal in many other respects. But their illusion is powerful. One patient, following a car accident, was convinced that his wife was dead and that the person living with him was an impostor.

The classic interpretation of Capgras syndrome is, obviously, Freudian. For example, a patient no longer recognizes his father. The Freudian interpretation may be that in infancy, the patient had a strong sexual attraction for his mother—a well-developed

Oedipus complex—which fed into an acute jealousy towards his father that was suppressed at adolescence. Following the unfortunate accident, the patient's jealousy reappeared: all of a sudden, he was sexually attracted to his mother and refused to recognize his father as his father.

But there is an alternative explanation. First, this form of decision making—identifying familiar features—may have appeared very early in evolution. The imprinting mechanism described by Konrad Lorenz and the ethologists enables baby animals to recognize their mother very early. It will not do to make a mistake and follow an animal that might eat you or lead you into a dangerous universe! Ethologists have shown us that the imprinting of the mother is an automatic mechanism that is produced at a critical stage of development. It is so automatic that if a chick or duckling is shown a leather bag at the time the imprinting mechanism kicks in, the animal will attach itself to the bag as if it were its mother! Because this mechanism arises in the first moments of an animal's life, it is probably the product of very old structures that do not require neurocomputation or the sort of complex formal logic carried out in the cortex.

A second possible explanation is disconnection of the corpus callosum, the bundle of nerve fibres connecting the two hemispheres. The disconnection prevents comparison of visual information with semantic information. Yet another theory blames the disorder on a memory deficit, a dissociation between the face seen and the one committed to memory. And finally, the deficit may be caused by a dissociation between the face, which is easily recognized, and its emotional significance, which is harder to retrieve. This last theory attempts to explain these 'delusions' [5] by functional or anatomical lesions. The systems that process faces are now better understood [6]. Some areas of the brain code specifically for faces, especially the temporal lobe and two particular areas within it known as TEO and TE (see Chapter 8). Among the major pathways involved, the first goes to the amygdala and the temporal lobe (the emotional system), and the second links the visual cortex, the superior temporal sulcus, the inferior parietal lobe, the cingulate cortex, and the hypothalamus. Prosopagnosia, the inability to recognize a face, is associated with lesions all along the first pathway, and Capgras syndrome with a lack of links with the second pathway, where the face of the father is identified for matching with the emotional value of the face coded in the limbic system [7]. However, that doesn't explain why it affects only people we know. Moreover, there are cases of Capgras that involve objects. Hence the need to remain open to a variety of interpretations.

For that matter, abnormalities involving face recognition are themselves very diverse. Patients with schizophrenia may see the faces of others transformed in a horrifying way—called 'paraprosopia'—becoming Dracula, for example.

Deliberation: a cortical function, or a subcortical one?

An important pathology of perceptual decision making and cognitive deliberation is the severing of pathways linking the two hemispheres, referred to as 'interhemispherical commisurotomy', which suppresses information transmitted by the corpus callosum. Many cognitive processes require relating the information processed by the two hemispheres. For example, the right hemisphere analyses the overall shape of objects and the left hemisphere their details. The right hemisphere is more concerned with space and the left with language. Generally speaking, as proposed, for instance, by Stephen Kosslyn, it seems that the right hemisphere mostly handles 'coordinated' types of information and the left handles 'category'-type information (see Chapter 8).

Although it is now clear that in humans there is a hemispheric specialization of cognitive functions and the capacity to represent one's own body [8], it is not possible to ascribe only to the cerebral cortex the essence of decision making during complex cognitive tasks. The neuropsychologist Justine Sergent [9] sparked a debate on this subject. She suggested that hemispherectomized patients could formulate judgements and even make decisions rapidly when cooperation between the two hemispheres was needed. For example, she visually presented a subject with two numbers, one to each hemisphere, and observed that the subject was still able to decide which was the greater. This decision requires a dialogue between the two hemispheres because each

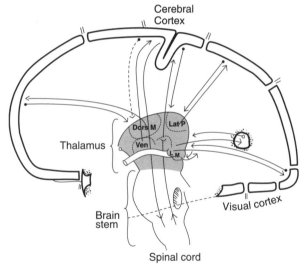

Fig. 3.2 Organization of the brain into loops and Penfield's centrencephalic theory. This schematic diagram shows the importance of the loops that connect the thalamus and the cerebral cortex (note that the arrows go both ways) via the basal ganglia (which do not appear in the diagram). Contemporary authors (see volume 1 of Llinás's *Vortex*) emphasize that communication between the areas of the cortex is more rapid through the thalamus than via cortico-cortical pathways. According to this idea, the unity of perception is ensured by synchronizing thalamic oscillators. The fact that the thalamus occupies a strategic position in the processes of decision making has been confirmed by recent work such as that of Komura and collaborators on retrospective and prospective coding of anticipated rewards in the sensory thalamus. After Penfield (1949)

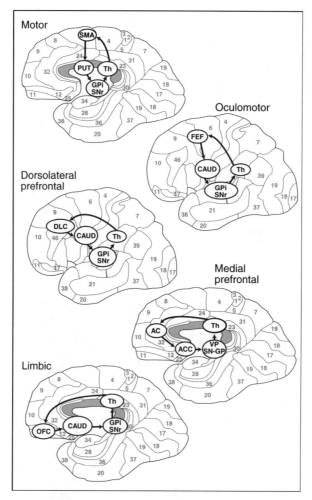

Fig. 3.3 Modular organization of cognitive and emotional behaviour: looped circuits linking the thalamus, the basal ganglia, and the cerebral cortex. Five major loops are shown here: motor, oculomotor, dorsolateral prefrontal, median prefrontal, and limbic. Some of these loops subdivide (see Chapter 11). Each has a corresponding (a) cortical area (SMA, supplementary motor area; FEF, frontal eye field; DLC, dorsolateral prefrontal cortex; AC, anterior cingulum; OFC, orbitofrontal cortex), (b) striatal area (PUT, putamen; CAUD, caudal nucleus; ACC, accumbens in the ventral striatum. In fact, each loop involves a different part of these structures; modularity is thus conserved at this level), (c) an area of the basal ganglia inside the internal globus pallidus (GPi) or the substantia nigra pars reticulata (SNr), (d) an area of the thalamus (Th). These are different parts of the thalamus that participate in different loops. Note that the reentrant projections going from the thalamus towards the cortex do indeed project onto the same area as the origin of the loop, but not to the same neurons. Moreover, these loops are transversally linked and thus interact with each other. After Alexander et al. (1986); modified with the help of L. Tremblay and E. Koechlin

receives one of two numbers. Moreover, most hemispherectomized patients are operated at a young age and undergo such extensive damage that the consensus is that the healthy hemisphere essentially takes over all major functions; this was notably shown for language by functional brain imaging.

Sergent's data are controversial because they suggest a much greater subcortical contribution to this kind of decision making than champions of the cortex would like to admit. The late Patricia Goldman-Rakic called such subcortical candidate structures as the claustrum, the pulvinar nucleus of the thalamus, and the basal ganglia 'third partners'. But earlier, Wilder Penfield had suggested that coordination between the cortical areas is carried out by thalamic relays and not by corticocortical pathways as shown in Fig. 3.2 [10]. Moreover, language areas have direct access to the basal ganglia and, via the pons, to the cerebellum.

Indeed, the brain is not organized solely according to corticocortical pathways. The most basic organization is radial, consisting in loops that link the thalamus (the brain centre that receives, for example, all sensory, visual, auditory, vestibular, tactile, olfactory information) to the cerebral cortex and that return to the thalamus via the basal ganglia (Fig. 3.3).

Whereas it was formerly thought that the brain receives sensory information, processes it, and responds in the form of action, it was subsequently discovered that it is in fact composed of a multitude of loops that Edelman [11] calls 'reentrant', permanent foci of neural activity that can function without input from the external world. Dreams are an example of this autonomous functioning of the brain, made possible by internal loops.

Among all these loops are *motor* loops that govern limb movements, whose output nucleus is the pallidum; an *oculomotor* loop that controls gaze (see *The Brain's Sense of Movement*), whose output nucleus is the substantia nigra; a so-called prefrontal loop, more cognitive, that subserves planning of action and its relationship with space; a temporal lobe loop where, for example, objects and faces are identified; and finally a limbic system loop involved in emotions that I will consider further in Chapter 11.

The power of the theory involving loops in pathology was reinforced by the recent discovery of abnormalities of these loops at the possible origin of disorders such as Parkinson's [12], obsessive-compulsive disorders, and other psychiatric and neurological disturbances. Rodolfo Llinás [13] suggests that dysregulation of molecular mechanisms at the level of neurons in the thalamus that control these low-frequency rhythms causes these various disturbances according to whether the group of neurons involved belongs to the motor or limbic thalamus and so on. For decision making, that means that its cognitive basis will have to be sought in these subtle mechanisms and not only by using a symbolic and formalist approach.

Hierarchy and heterarchy

The distinction is nontrivial: If decision making is the product of a few specialized structures in the brain (the visual cortex or the ventromedial prefrontal cortex), then what is needed is a *phrenology* of decision making. If, on the other hand, each decision involves many centres and deciding depends on a certain relational state among these different centres, then the physiology of decision making becomes more complex. We will need new tools and concepts to understand how the brain makes decisions. We will see that if each part of the brain is involved in the hierarchical and heterarchical[1] processes of decision making, the ultimate action requires that all decisions made *locally* be coordinated in a *global* fashion. One of the themes of this book is that emotion plays a special role in moulding the global character of decisions, as well as in the mental use of space.

What does pathology bring to the question? By and large, disorders of decision making have been observed in patients presenting with lesions of the frontal lobe. But these are not the only structures for which damage leads to impaired decision making.

Obsessional disorders

Two pathological forms of decision making are *obsession* and *compulsion*. Obsession presents in the form of persistent ideas, thoughts, images, or impulses that are perceived as involuntary, meaningless, or even repellant. The patient tries to suppress them. Compulsions are repetitive and apparently intentional behaviours, performed in a stereotyped fashion, with awareness of their irrepressible character coupled with the desire to resist them. The patient recognizes the uselessness of the action and takes no pleasure in it, although having done it brings a certain lessening of tension.

A combination of these behaviours is called obsessive-compulsive disorder [14]. The repertoire of behaviours includes rushes of affection towards others, fear of contamination, regrets over past actions, and washing, checking, and avoidance rituals.

Jean Esquirol first described the disorder in 1839, and Pierre Janet, in the still relevant *Obsessions et la psychasthénie* (Obsessions and psychasthenia) published in 1903, established a neural basis for it, whereas for more than half a century psychoanalysis had been dismissing every serious hypothesis concerning the biological basis of obsession in favour of purely psychological causes. Only the association of these symptoms with Parkinson's, then their resemblance to the motor signs of certain forms of epilepsy involving especially the frontal regions of the brain, steered research towards an organic basis of the disorder.

One of the symptoms of greatest relevance here is that of 'obsessive slowness'. It is characterized by a primary obsessive deliberateness which does not seem to be due to a slowing down of execution but to numerous meticulous checking rituals associated with difficulty in prioritizing actions. For example, Henry Head reported the case of a 25-year-old patient whose slowness resulted from difficulty in undertaking simple movements, passing from one sequence of movements to another, or carrying out

---

[1] A heterarchy is a system that combines both hierarchical mechanisms and parallel ones. It is a federation of hierarchies that may cooperate or compete.

several movements simultaneously. His barely commenced gestures were constantly disrupted by undesired movements and by his tendancy to be distracted by random stimuli.

These very particular disorders translate into symptoms of hesitation, rituals in even simple actions such as grasping a cup, and the breakdown of complex sequential actions such as teeth brushing or getting dressed into fragments of movements accomplished only after long reflection [15]. This type of disintegration has also been observed in Parkinson's as well as other neurological disorders such as Gilles de La Tourette syndrome.

Patients presenting with obsessive slowness show no abnormalities in pure spatial tests (memorizing drawings, arranging objects in the environment, and so on) but make mistakes if they are asked to change the rules for solving spatial problems. This difficulty in changing rules is associated with lesions of the frontal lobe and portions of the basal ganglia that are connected to it via the loops mentioned above. Yet it so happens that the basal ganglia play an essential role in deciding action. Analysis of the mechanisms of decision making thus must be rooted in those of control and choice of action and not be limited to a representational and logico-deductive theory of decision making whose abnormalities would be caused solely by the cerebral cortex.

Recent work shows that other areas may be crucial for these disorders. The orbitofrontal cortex is involved in the meaning attributed to the consequences of action, thereby subserving decision making, whereas the anterior cingulate cortex is particularly active in situations in which there are conflicting options and a high likelihood of making an error. The dorsolateral prefrontal cortex plays a critical role in the cognitive processing of relevant information. This cortical information is then integrated by the caudate nucleus, which controls behavioural programs. Failure anywhere along these networks can result in obsessive thoughts and behaviour. The fact that electrical stimulation of the caudate nucleus can provide relief from compulsive behaviour gives reason to believe that this neurophysiological interpretation of the disease may lead to significant improvements in patients that will alter their lives for the better. It also shows how modern integrative neuroscience can contribute alternative explanations to purely psychoanalytical theories of these disorders [16].

Perseveration

Many patients suffer from perseveration, that is, repeating the same action or pronouncing the same word and so on indefinitely. This disorder signals difficulty in disengaging from ongoing action and starting a new one. Perseveration may be due to the maintenance in memory of the word previously pronounced and not sufficiently blocked; its evocative threshold remains too low. Three types of perseveration have been proposed, distinguished by the level of complexity of the behaviour affected [17].

At the lowest level, a *continuous perseveration* consists in the uncontrolled repetition of elementary motor patterns such as those Alexander Luria described in 1965, for example, drawing series of loops. Luria confirmed the link between perseveration and lesions of the frontal lobe [18]. At an intermediary level, a *recurrent perseveration* consists in, for example, repeating answers produced beforehand, when the patient is subjected to a sequence of stimuli. The highest level is characterized by the *stuck in set perseveration*—that is, the inability to switch from one task to another. We will see that in fact this difficulty is not limited to cognitive tasks; it probably also involves the basal ganglia or at least the part of these structures that is linked to the prefrontal cortex.[2]

But these mechanisms are complex. For example, verbal perseveration can be due to a failure to suspend activity under way owing to internal processing of an action previously undertaken by the patient [19]. Perseveration thus would be due to impaired arbitration among various actions, and to the 'perpetual competition between past and future elements of verbal flow' [20]. Some of these ideas are similar to those I laid out in *The Brain's Sense of Movement* regarding Richard Schmidt's theories about the control of action.

The influence of the medial prefrontal cortex on the amygdala might also explain pathologies classified as perseveration [21]. We know that lesions of the lateral prefrontal cortex induce cognitive perseverations, that is, difficulty in changing action; for example, rules in a game of cards, in other words, adapting one's behaviour to changes in *context*. This capacity for selecting action is a function of projections of the prefrontal cortex both to the hippocampus and to the transitional structures involved in action such as the nucleus accumbens, as well as (directly) to the loops connecting the cortex to the basal nucleus, which contains the repertoire of actions. The medial regions of the cortex are involved in so-called emotional perserveration, that is, difficulty in changing and adapting emotions to changes in context. The association of these regions with the amygdala is critical in explaining this difficulty in patients. As a result, whereas animals placed before a fear-inducing pattern react in a stereotypical fashion, humans and other primates, owing to their developed prefrontal cortex, have the possibility of modulating their emotional responses.

---

[2] The prefrontal cortex can be delimited by the following regions, according to Brodmann's atlas: 8, 9, 10 (frontopolar cortex), 44, 46; area 32, called the cingulate gyrus, which plays a critical role, and the orbitofrontal cortex (areas 11, 47, 12); and finally three gyri that merge in the frontopolar gyrus. In the monkey, these areas have three types of connections. (1) Corticocortical: for example, 6-8-10 and 6-44-45-190. (2) Intrinsic: for example, 6-8-9-10. Area 32 is connected in a star formation with 6, 8, 9, and 10. (3) Extrinsic: the connections of the parietofrontal network are linked to the superior regions of the prefrontal cortex, and the regions of the temporal cortex (hippocampus) tend to connect with the inferior portions. Finally, very long loops link the prefrontal cortex with the basal ganglia, but also other basic loops for learning with the cerebellum, the thalamus, and so on. This cortex is thus a confluence of transversal and radial networks in the brain that make up a strategic zone for the entire organization of brain function. Worth noting is the absence of connections from area 10 with the posterior regions of the brain.

Impulsivity

Perseveration is manifested by ongoing repetition of the same behaviour, but there is another form of release from inhibition called 'impulsivity'. These patients cannot control their behaviour or resist acting. Their impulsive response to situations can lead to excessive risk taking and antisocial behaviour. Of course, normal people may also act this way to various degrees, and psychiatrists use the term 'borderline personality disorder' to denote deviations from normality. Here again tests point to potential dysfunction in the dorsolateral frontal or orbitofrontal cortex [22]. Impulsivity is also found in patients suffering from attentional disorders, for instance, attention deficit hyperactivity disorder (ADHD). Although ADHD is primarily a developmental disorder of childhood, it may persist into adulthood. It has been suggested that ADHD may be due in part to impairment of the monoaminergic system. However, other areas, such as the corticostriatal system and even the accumbens, cerebellum, and amygdala, may be involved (see Chapter 11) [23]. Impulsivity has also been found to be a component of some eating disorders, although the etiology here is very complex and involves a role of the serotonin (5-hydroxytryptamine) system, which along with dopamine and acetylcholine is one of the main neuromodulators [24].

Disturbances of decision making and lesions of the ventromedial frontal cortex

Patients with lesions of the frontal ventromedial cortex make mistakes in tasks of reward reversal. Subjects win points when a stimulus appears on the computer and lose points for another stimulus to which they were not supposed to respond [25]. After the patient has learned the difference between the responses, the reward is reversed.

The reward stimulus becomes the aversive stimulus. Patients often notice the reversal but are unable to respond to it. The patients' perseverance in touching the screen in response to the stimulus associated with the reward in the first battery of tests is very similar to the response of monkeys with lesions of the orbitofrontal cortex. And this difficulty of reversal is very strongly correlated with questions of sociability, that is, with the capacity of the subject to conform to social rules, not to use unacceptable language in public, and so on. These findings suggest that the abnormality in patients with orbitofrontal lesions is essentially emotional, if emotion is defined as a state induced by stimuli of reward and punishment.

These patients do respond normally to the Tower of London test. In this test, a subject is shown a stack of balls arranged in a certain order. The patient is asked to change the order of the balls, going through an intermediary step. The test requires the subject to mentally imagine the sequence of changes, a little like the game of Rubik's cube or chess. Patients with prefrontal lesions often have trouble with this test.

The inferior portion of the prefrontal lobe comprises several regions that are important in decision making. Damage to these areas leads to complex abnormalities. The historical account of a patient, Phineas Gage, in whom accidental lesions of this sort resulted in disturbed decision making and social behaviour was the starting point for a striking series of observations by Antonio Damasio's group regarding the ventromedial cortex [26]. The classical paradigm [27] was the following: Subjects had to choose cards from four packs. After each selection, they received a reward. Two packs gave large rewards without the patients being forewarned, but also significant losses requiring that money be given back. In contrast, the two other packs gave small wins but also negligible losses. The best way of winning was thus to choose cards from these last two packs. The significant result was that patients with lesions of the ventromedial cortex continued to draw cards from the riskiest packs and ended up losing the game. What is interesting about this test is that it models an important aspect of everyday decision making that consists in choosing among actions that bring different results but under conditions such that the laws or constraints that determine the win or loss are unknown. A fundamental uncertainty lies at the heart of every decision. The monetary game test developed by Damasio, Antoine Bechara, and their colleagues has become a classic and is now used in many studies under the name of the Iowa Gambling Task [28].

Loss of personal and social judgement, inadequate assessment of the consequences of action and, more generally, difficulty in deciding are the result of lesions of the orbital and ventromedial cortex. For example, a patient picks up his agenda to make an appointment, but thumbs through it endlessly without being able to stop at a date. Moreover, it has been shown that this difficulty does not stem solely from a deficit in the rational processing of information. Damasio's so-called somatic marker hypothesis assumes that a confluence of signals are produced in this region, some corresponding to rational formulation of decisions and others representing the emotional value of the consequences of a possible decision (Fig. 3.4). The marking occurs at several levels, some of which are conscious and others unconscious. The candidates are the ventromedial prefrontal cortex; the centres connected to mechanisms of emotion such as the amygdala, which can activate vegetative centres such as the viscera, the vascular bed, the endocrine system, nonspecific neurotransmission systems, and so on; and finally the somatosensory cortex and the insula. The insular cortex appears to be involved in risk-taking behaviour [29].

The novelty of this theory is that decision making does not necessarily require physical reactions to actually be expressed by the body itself. Faced with a situation that requires a decision, recollection of a memory activates implicit knowledge (dispositions), pertinent to the situation, in the associative regions of the cerebral cortex. This calling into memory of relevant associated facts enables a 'mental experiment' in a visualized form. Simultaneously, the regions involved in emotions linked to the ventromedial prefrontal cortex are activated, allowing reconstitution of a *learned group of emotion-facts*. The reactivation described above can be accomplished via a neuronal loop that includes the body. A change of 'somatic' state is thus induced which, in turn, activates the portion of the cerebral cortex that receives information from the body

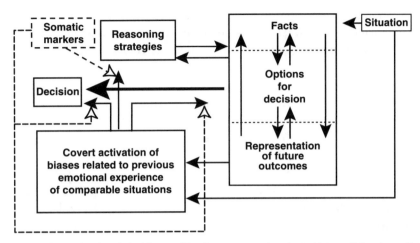

Fig. 3.4 Marking of rational decision making by past emotional experience. This schematic diagram summarizes some of Damasio and Bechara's somatic marker theory. A given situation simultaneously activates rational processes of decision making and the memory of past emotions. The memory of past emotion 'marks' and thus influences the final decision. After Bechara and Damasio (1997)

proper (the somatosensory cortex). This reactivation can also be relayed by another loop, the 'as-if body loop', that bypasses the body and conveys the somatic signals to cortical structures that then adopt appropriate patterns of activity.

Compared with the theory of William James, somatic marker theory is also novel in assuming that the body's own internal simulation mechanism can function as a genuine physical reaction. The brain uses what we call 'internal models' of the body, which enable scenarios of emotional expression to be played out. I described these at length in *The Brain's Sense of Movement* and will come back to them in the following chapters.

Note that the deficits singled out in this theory are absent in patients with lesions of the dorsolateral or dorsomedial prefrontal cortex. This suggests a dissociation between working memory and emotional memory that has been confirmed by experiment, as we are about to see now. It has also been suggested that a distinction should be made between 'emotional intelligence' and 'cognitive intelligence'. Patients with impaired somatic marker circuitry exhibited significant deficits in emotional intelligence, decision making, and social functioning, despite normal levels of cognitive intelligence [30].

Working memory and decision making: different mechanisms

When we make a decision, we temporarily commit to memory the information we need to be thinking about. So, to decide whether to holiday in the mountains or by the seaside, I need to keep in mind (in modern terms, in 'working memory') the advantages and disadvantages of both; I also have to recall the pleasures and annoyances of each solution to peg emotions to my memories; finally, I have to come to a decision. Working memory, emotion, and reasoning are thus major components of decision making. The question is whether they involve different regions of the brain.

There is a dissociation between working memory and decision making [31]. It can be revealed by comparing the performance of patients in two tasks, one involving working memory and the other a gambling task requiring decision making. These tasks were proposed to subjects presenting with damage to the ventromedial prefrontal cortex and dorsolateral/high mesial cortices, respectively. Subjects with an anterior ventromedial lesion exhibited abnormalities during the gambling game, but not the task of working memory, whereas a ventromedial posterior lesion caused problems in both tasks. Subjects with a right dorsolateral cortical lesion had trouble with the working memory task but not with the game. Subjects with a right dorsolateral cortical lesion showed no impairment in either task. These findings were the first to demonstrate a double dissociation between working memory and decision making.

From the marker hypothesis to the idea of emulation

Let us try to develop a little further the relation between cognition and emotion. First, I will try to establish that emotion is indeed a component of the mental simulation of action, but of a particular aspect of action that is neither the content of action nor its context, but a more global quality. Somatic signs, even those of an autonomous system, are signifiers. They *simulate* action, or at least its expected consequences. They do not express only the present or the value of past experiences, but predict the future.

Next, I will suggest a theory of the brain as *emulator* of action. We will see from it how emotion inclines the organism to internal creation of an emulated world (see also Llinás on this idea). The big difference between simulation and emulation is that simulation is a simple copy of a group of states, whereas emulation represents much more: it is internal creation and invention. It is the difference between illusion, which is based on reality, and hallucination, which involves creation of an imaginary reality—a waking dream. Emulation is close to Sartre's idea of emotion; it also transforms the world 'in a magical way' so that the world conforms to what it is possible for the brain to conceive or accept.

Another hypothesis involves the frontal cortex in inhibiting inappropriate behaviours [32]. Damage to the orbitofrontal cortex is known to cause loss of inhibitory control of the emotional component of behaviour, but it is especially the inhibitory function of inappropriate behaviour that is the issue here. This difficulty makes the subject a captive of the immediate situation owing to the emotional weight attached to it, which prevents him from evaluating a future plan of action. This essential role of the frontal cortex in suppressing behaviour will be taken up later on with respect to suspended action.

That brings us to a basic concept in understanding the neural basis of decision making, namely, inhibition of behaviour or undesired solutions. Decision making is not only choosing to do something, it is also choosing not to do it—suppressing irrelevant actions. Decision making is a competition, a dynamic process in which there are winners and losers, but in which the losers are never eliminated but stick around to try and win again, because suppressed actions are always potentially executable.

Disturbances of decision making and frontotemporal dementia syndrome

Some patients present with mental rigidity, no appreciation for subtleties of language such as irony, and ritual and stereotypic behaviours. They struggle with planning complex behaviours, including social interactions, and setting goals, eventually changing them or changing the rules (set shifting) as shown by the Wisconsin Card Sorting Test [33]. This test requires a subject to arrange cards according to a criterion, in increasing value, for example, and then to change the criterion. Some of the symptoms of these patients are similar to those of obsessive-compulsive-type disorders. Their visuospatial capacities are preserved. The deficit in decision making is thus not due to difficulty in analysing the situation on that score.

The disturbances of some of these patients can be revealed by another test. A subject was given a total of 100 points at the beginning of the experiment. He saw on the screen a series of red and blue boxes. He was told that the computer had arbitrarily hidden a yellow object inside one of the blue or red boxes. The subject had to decide, or rather to guess, whether the object was hidden in the blue box or the red box by pressing a button of the same colour. If the decision was the right one, he won points; if not, he lost points. The accumulated winnings were displayed in a box to the left (Fig. 3.5).

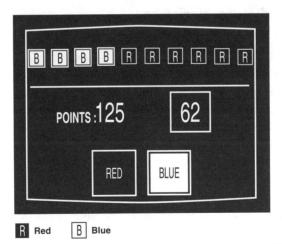

Fig. 3.5 Decision-making task with financial reward—a modified version of Damasio and Bechara's tasks. The subject sees blue (B) or red (R) boxes displayed on a screen, whose proportion changes from one session to another. The subject must press on one of the two squares marked 'blue' or 'red'. He sees the number of points he has obtained and the amount of accumulated winnings (see text). After Rahman et al. (1999)

Having seen the winnings, the subject could increase them by betting that his choice was correct. In other words, to the initial choice was added a betting task about the validity of the choice associated with a supplementary reward. The subject took a risk because the value shown to the right could be added if the bet was correct or subtracted if the bet was wrong. The supplementary winnings offered by the computer could be small but would be gradually increased by the automatic display ('increasing' situation). In a 'descending' situation, in contrast, the sum diminished. The amount of the supplemental bet was always a percentage of the subject's score, although the subject did not know that.

This test was constructed to evaluate three functions of decision making: (1) one could vary the proportion of blue and red boxes at the beginning of each trial and thus modify the effect of contingency weighting—for example, if there were 9 red boxes and 1 blue one, the probability of the object being in a blue box might be judged slight; (2) the supplementary bet evaluated the willingness of the subject to gamble his capital to win more; (3) the decreasing or increasing display said something about the subject's risk strategy.

The speed of decision making, its quality and adjustment to risk were evaluated. Patients with frontotemporal dementia estimate probability correctly, but are incapable of adequately adjusting the level of their bets. They act like genuine *risk takers*. Generally speaking, patients with frontotemporal dementia take a long time deliberating. Moreover, they have a limited capacity to suppress unprofitable strategies, but their principal difficulty really seems to be anticipating future consequences [34]. So the orbitofrontal cortex must be involved in predicting the consequences of actions.

Dopamine and motor decisions

Decision-making processes involve circuits, loops that interconnect several structures, in particular the thalamocortical system, including the thalamus, the basal ganglia, and the frontal cortex. Cognitive deficits linked to decision making in patients with Parkinson's or presenting with lesions of the frontal cortex suggest a role for dopamine. Indeed, Parkinson's disease, which manifests with postural rigidity, motor slowness, shaking, and cognitive and affective disturbances, is due to a lack of dopamine at the level of the basal ganglia. These patients have trouble *initiating* motor acts, for example, walking. This deficit is very curious, because it can be overcome by showing a subject a staircase, or having him take a first step. The question is genuinely one of difficulty in making a step voluntarily, in the absence of sensory stimuli.

That would seem to involve dopamine in the process of motor decision making. One way to verify this hypothesis is to study cognitive function in substance abuse patients, in particular those who have consumed large quantities of amphetamines. Chronic administration of this substance reduces the level of dopamine in the striatum and the prefrontal cortex in primates and rats. In humans, such a reduction was observed simultaneously in the striatum and the orbitofrontal cortex in amphetamine and cocaine addicts. These addicts were expected to show the same disturbances as patients with damage to the orbital prefrontal cortex. The same hypothesis was tested in opiate addicts and patients with frontal lesions. Finally, it is possible that serotonin is involved in cognitive processes linked to decision making, because the structures that produce serotonin or that use it are sensitive to amphetamine. Other kinds of addictions have also revealed deficits of decision making [35], including tobacco smoking [36] and cocaine addiction [37].

Amphetamine and opiate abusers take longer to predict the consequences of their decisions [38]. This accrued time to deliberation might stem from impaired neuromodulation of the circuits comprising the ventral regions of the prefrontal cortex, the ventral striatum, the amygdala, and perhaps also the nucleus accumbens. Amphetamine abusers also share with patients with prefrontal lesions a tendency to choose the least likely outcomes.

What conclusions can we draw from this examination of the pathology of decision making? First, recall that my purpose is neither to be exhaustive nor to describe the detailed physiopathology and etiology of these disorders. That is for doctors to do, and I do not pretend to be an expert in this area. But the results of careful experimental work, which, for that matter, is still being carried out in multidisciplinary teams, can help us to make a few useful observations.

First, it is clear that the brain uses very varied mechanisms to come to decisions. Some mechanisms concern the neural basis of movement and the acting body. Others will have to be sought at the level of perception itself and its interpretation as a function of intention and context. Finally, others involve selection and inhibition of actions among all those available to us both through the repertoire bequeathed us by evolution and those we have learned. We have also seen the close exchange between emotion and decision making. It is clear that the cerebral cortex is not the only part of the brain involved in decision making, even in complex deliberations; subcortical structures such as the basal ganglia, themselves tightly linked with the cerebral cortex, play an essential role, and, within them, neuropharmacological mechanisms of neuromodulation. The chapters that follow will examine in succession these different elements that constitute the physical and cognitive basis of decision making.

References

1 M Kinsbourne, 'The material basis of mind', in LM Vaina, ed., *Matters of intelligence* (Dordrecht: D. Reidel, 1987), p. 425.
2 O-J Grüsser and T Landis, *Visual agnosias and other disturbances of visual perception and cognition* (Basingstoke: Macmillan, 1991).
3 H Lissauer, 'A case of visual agnosia with a contribution to theory', trans. M. Jackson, *Cognitive Neuropsychology*, 5 (1988): 157–92, in S. Yantis (ed.), *Key readings in visual perception: essential readings* (Philadelphia, PA: Taylor and Francis, 2001), p. 286, original work published in 1890.
4 Tests specific for this pathology are so-called associative tests: for example, a patient is asked to compare actual objects with photos of different items of the same class. 'Optic aphasia' presents as a visual agnosia, but the patient may demonstrate his comprehension by miming the action of the object.

5 HD Ellis and AW Young, 'Accounting for delusional misidentifications', *British Journal of Psychiatry*, **157** (1990): 239–48.

6 JV Haxby, EA Hoffman, and MI Gobbini, 'The distributed human neural system for face perception', *Trends in Cognitive Science*, **4** (2000): 223–33.

7 HD Ellis and MB Lewis, 'Capgras delusion: a window on face recognition', *Trends in Cognitive Science*, **5** (2001): 149–56.

8 T Nagumo and A Yamadori, 'Callosal disconnection syndrome and knowledge of the body: a case of left hand isolation from the body schema with names', *Journal of Neurology, Neurosurgery, and Psychiatry*, **59** (1995): 548–51.

9 J Sergent, 'Processing of spatial relations within and between the disconnected cerebral hemispheres', *Brain*, **114** (1991): 1025–43.

10 W Penfield and T Rasmussen, *The cerebral cortex of man: a clinical study of localization of function* (New York: Macmillan, 1957).

11 GM Edelman and G Tonini, *Consciousness: how matter becomes imagination* (London: Penguin, 2001).

12 P Cesaro, P Damier, JP Nguyen, and H Ollat, *La Maladie de Parkinson*, 2d edn. (n.p.: Monographies de l'ANPP, 2003).

13 RR Llinás, *I of the vortex: from neurons to self* (Cambridge: MIT Press, 2001).

14 J Cottraux, D Gérard, L Cinotti, J-C Froment, M-P Deiber, D Le Bars, G Galy, P Millet, C Labbé, F Lavenne, M Bouvard, and F Mauguière, 'A controlled positron emission tomography study of obsessive and neutral auditory stimulation in obsessive-compulsive disorder with checking rituals', *Psychiatry Research*, **60** (1996): 101–12.

15 H Head, 'The neurology of obsessional slowness', *Brain*, **114** (1991): 2203–33.

16 B Aouizerate, D Guehl, E Cuny, A Rougier, B Bioulac, J Tignol, and P Burbaud, 'Pathophysiology of obsessive-compulsive disorder: a necessary link between phenomenology, neuropsychology, imagery, and physiology', *Progress in Neurobiology*, 72 (2004): 195–221.

17 J Sandson and ML Albert, 'Varieties of perseveration', *Neuropsychologia*, 22 (1984): 715–32.

18 AR Luria, 'Two kinds of motor perseveration in massive injury of the frontal lobe', *Brain*, **88** (1965): 110.

19 L Cohen and S Dehaene, 'Competition between past and present. Assessment and interpretation of verbal perseverations', *Brain*, **121** (1998): 1641–59.

20 GS Dells, MF Schwartz, N Martin, and SEM Gagnonga, 'Lexical access in aphasic and nonaphasic speakers', *Psychological Reviews*, **104** (1997): 801–38.

21 MA Morgan and JE LeDoux, 'Differential contribution of dorsal and ventral medial prefrontal cortex to the acquisition and extinction of conditioned fear in rats', *Behavioral Neuroscience*, **109** (1995): 681–8.

22 J Dowson, E Bazanis, R Rogers, A Prevost, P Taylor, C Meux, C Staley, D Nevison-Andrews, C Taylor, T Robbins, and B Sahakian, 'Impulsivity in patients with borderline personality disorder', *Comprehensive Psychiatry*, **45** (2004): 29–36.

23 JA King, J Tenney, V Rossi, L Colamussi, and S Burdick, 'Neural substrates underlying impulsivity', *Annals of the New York Academy of Science*, **1008** (2003): 160–109; M Ernst, AS Kimes, ED London, JA Matochik, D Eldreth, S Tata, C Contoreggi, M Leff, and K Bolla, 'Neural substrates of decision making in adults with attention deficit hyperactivity disorder', *American Journal of Psychiatry*, **160** (2003): 1061–70; RA Chambers, JR Taylor, and MN Potenza, 'Developmental neurocircuitry of motivation in adolescence: a critical period of addiction vulnerability', *American Journal of Psychiatry*, **160** (2003): 1041–52; CA Winstanley, DE Theobald,

RN Cardinal, and TW Robbins, 'Contrasting roles of basolateral amygdala and orbitofrontal cortex in impulsive choice', *Journal of Neuroscience*, **24** (2004): 4718–22.

24 H Steiger, 'Eating disorders and the serotonin connection: state, trait, and developmental effects', *Journal of Psychiatry and Neuroscience*, **29** (2004): 20–9.

25 ET Rolls, J Hornak, D Wade, and J McGrath, 'Emotion-related learning in patients with social and emotional changes associated with frontal lobe damage', *Journal of Neurology, Neurosurgery, and Psychiatry*, **57** (1994): 1518–24.

26 AR Damasio, *Descartes' error: emotion, reason, and the human brain* (New York: Putnam, 1994); *The feeling of what happens: body and emotion in the making of consciousness* (New York: Harcourt Brace, 1999); MA Cato, DC Delis, TJ Abildskov, and E Bigler, 'Assessing the elusive cognitive deficits associated with ventromedial prefrontal damage: a case of a modern-day Phineas Gage', *Journal of the International Neuropsychological Society*, **10** (2004): 453–65.

27 A Bechara, AR Damasio, H Damasio, and AR Lee, 'Different contributions of the human amygdala and ventromedial prefrontal cortex to decision making', *Journal of Neuroscience*, **19** (1999): 5473–81.

28 LM Ritter, JH Meador-Woodruff, and GW Dalack, 'Neurocognitive measures of prefrontal cortical dysfunction in decision making', *Schizophrenia Research*, **1** (2004): 65–73. Patients with schizophrenia also show prefrontal cortical dysfunction in decision making. For instance, in a comparison of a group of schizophrenic patients versus a healthy group using two different neurocognitive tasks—the Iowa Gambling Task (thought to involve the ventromedial prefrontal cortex) and the Wisconsin Card Sorting Test (thought to involve the dorsolateral prefrontal cortex)—patients with schizophrenia perform worse on the gambling task with respect to total monetary gain and loss.

29 MP Paulus, C Rogalsky, A Simmons, JS Feinstein, and MB Stein, 'Increased activation in the right insula during risk-taking decision making is related to harm avoidance and neuroticism', *Neuroimage*, **19** (2003): 1439–48.

30 R Bar-On, D Tranel, NL Denburg, and A Bechara, 'Exploring the neurological substrate of emotional and social intelligence', *Brain*, **126** (2003): 1790–1800.

31 A Bechara, H Damasio, D Tranel, and SW Anderson, 'Dissociation of working memory from decision making within the human prefrontal cortex', *Journal of Neuroscience*, **18** (1998): 428–37.

32 KC Plaisted and BJ Sahakian, 'Dementia of frontal type: living in the here and now', *Aging Mental Health*, **1** (1997): 293–5.

33 S Rahman, BJ Sahakian, JR Hodges, RD Rogers, and TW Robbins, 'Specific cognitive deficits in mild frontal variant frontotemporal dementia', *Brain*, **122** (1999): 1469–93.

34 G Schoenbaum, A Chiba, and M Gallagher, 'Orbitofrontal cortex and amygdala encode expected outcomes during learning', *Nature Neuroscience*, **1** (1998): 155–9.

35 O Kelemen, A Mattyassy, and S. Keri, 'The role of ventromedial prefrontal cortex in addiction disorders', *Ideggyogyaszati szemle*, **20** (2004): 4–10; A Verdejo, F Aguilar de Arcos, and M Perez Garcia, 'Alterations in the decision making processes linked to the ventromedial prefrontal cortex in drug-abusing patients', *Revista Neurologica*, **38** (2004): 601–6.

36 E Rotheram-Fuller, S Shoptaw, SM Berman, and ED London, 'Impaired performance in a test of decision-making by opiate-dependent tobacco smokers', *Drug and Alcohol Dependence*, **73** (2004): 79–86.

37 KI Bolla, DA Eldreth, ED London, KA Kiehl, M Mouratidis, C Contoreggi, JA Matochik, V Kurian, JL Cadet, AS Kimes, FR Funderburk, and M Ernst, 'Orbitofrontal cortex dysfunction in abstinent cocaine abusers performing a decision-making task', *Neuroimage*, **19** (2003): 1085–94.

38 RD Rogers, BJ Everitt, A Baldacchino, AJ Blackshaw, R Swainson, K Wynne, NB Baker, J Hunter, T Carthy, E Booker, M London, JF Deakin, BJ Sahakian, and TW Robbins, 'Dissociable deficits in the decision-making cognition of chronic amphetamine abusers, opiate abusers, patients with focal damage to prefrontal cortex and trytophan-depleted normal volunteers: evidence for monoaminergic mechanisms', *Neuropsychopharmacology*, **20** (1999): 322–39.

Part 2

Decision making with my second self

Chapter 4

Fight or flight

> The object of the idea constituting the human mind is the body, or, a certain actually existing mode of extension, and nothing else.
> (*Spinoza [1]*)

One spring morning around 8:45, in front of the Lycée Montaigne, which was built simultaneously along the lines of a train station, a hospital, and a prison, the street was deserted aside from a few hurried latecomers. For a few minutes, a smooth flow of students invaded the building, but now it had managed to swallow them up, aided by the bells. A silent confrontation resumed between the school windows, which are always shut, and the Luxembourg Gardens, which invited wandering beside the apple trees, whose extended branches appeared to mimic the orderly rows of benches on which the sequestered students listened to their teachers. Motionless, alone on the street, a young girl was looking at the school, her bookbag slung across her shoulder. I was impressed by her stillness. She was turned halfway towards the building and halfway towards the street, her face frozen in a mournful mask. Her gaze was directed towards the door of the school as she deliberated silently, letting out a few big sighs, visibly torn between the desire to escape and the obligation to enter into the 'dungeon'. I stopped a distance away, fascinated by her hesitation, and sympathized with the dilemma that her indecision was causing this child. Finally, sighing even more deeply than before, she headed—with obvious reluctance—towards the school! How had she managed to suppress such a strong desire to flee?

We make decisions constantly: from the neuron that fires 'all or nothing', the breath that suddenly stops and starts, the step that hesitates, to the smile that stiffens before broadening. Decision making is apparent in gestures that are barely sketched before they suddenly change purpose or direction, in the look that alters its focus depending on the interest of the subject, in anger overcome and desire contained. Decision making is found, too, in the herd of buffalo that in the morning follow the direction of their leader; the school of fish that abruptly change course or the flock of storks taking wing; in the group of birds that opt to spend the night in one tree rather than another.

But while mindful of the variety of types of decision making that William James attempted to distinguish, let me suggest a *general definition of decision making* that we will consider here: an *act* by which the brain, faced with several solutions for identifying an object, guiding movement, or resolving a problem, *cuts* in favour of one solution over another. Decision making is judgement: what better illustration than the

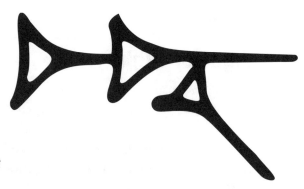

Fig. 4.1 Assyrian form of an Etruscan character for the word 'DI' dating from the tenth to the seventeenth century BC. It literally means 'to cut' in the sense of to judge or to decide in the case of a legal trial.

judgement of Solomon! I deliberately used the word 'cut' in the sense of 'choose', because it is the word the Sumerians used 3000 years ago to represent deciding in cuneiform writing (Fig. 4.1).

Let us consider a few examples to show that decision making can be found at all levels of processing linking perception and action.

The Mauthner cell, or deciding to flee

The first example is the Mauthner cell. This giant neuron is responsible for the flight response in fish (Fig. 4.2). It is the extraordinarily subtle and complex seat of multisensory integration [2]. Far from being a simple relay controlling a flight response, the Mauthner cell detects danger signals. Indeed, its true function is to make the decision to flee as a function of the context and pattern both of external signals and the internal state of the animal.

The flight response exists in many animal species, from anemones, annelids, insects, crayfish, fish, rats, cats, and monkeys to humans. It is undeniably one of the most important components of the behavioural repertoire of natural selection because it makes it possible to escape a predator!

It shows up very differently in diverse species, so much so that its neuronal substrate was once believed to be fundamentally different. Yet although over the course of evolution many variations appeared and its complexity increased, the general principles of the cell can still be studied in a simple model like the fish. Moreover, the flight response shares many features in common with another defence mechanism: the rat's response to noise.

To a fish swimming along minding its own business, a stone dropped into the aquarium activates the Mauthner cell, this fantastic giant neuron: its cellular body or 'soma' can reach up to 50 to 80 μ and its dendrites up to 500 μ. It is located in the reticular formation of the brain stem and descends towards the spinal cord to activate the muscles of the body and produce the life-saving flexion that will propel the fish to the side. The speed of reaction is incredible: 6 to 10 ms (10 thousandths of a second)! The neuron has only to produce a single action potential for the fish to bend its body

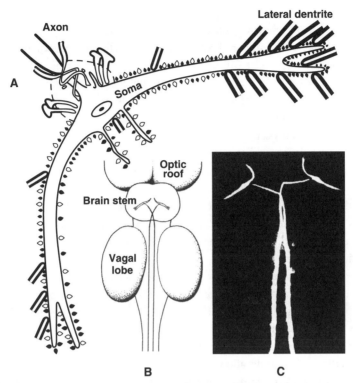

Fig. 4.2 The Mauthner cell: a neuron from the brain of a goldfish, which makes a decision (flee) when sensory information signals 'danger'. **(A)** Schematic representation of the Mauthner cell in the fish. Distribution of the synaptic endings of afferent nerves. This giant neuron can reach 0.5 to 0.6 mm from one end of the dendrites to the other. **(B)** Location of the soma and the descending axon in the spinal chord of the two Mauthner cells on each side of the median line. The neuron is situated in the brain stem, under the cerebellum.
(C) Photograph of two Mauthner cells. After Faber et al. (1989) and Korn et al. (1990)

to escape a predator with an acceleration that can reach 50 metres per second (5g). We are talking about a response whose power and astonishing speed are the animal's one chance to survive.

For us, what is remarkable is that the decision to flee is made very quickly (15 ms). The animal cannot afford to make a mistake. It has to determine the position of the predator and the best direction to flee at the risk otherwise of hurling itself towards the enemy. Neurobiologists have shown that 'it is clear that central processing is critical for decision-making in the escape response' [3].

Perhaps the reason this network forms the basis of many of the most evolved mechanisms of decision making is that the flight response is not an isolated one. Not only does the fish have to decide to flee at the right time and in the right direction, but at the same time it has to suppress other behaviours that could conflict with flight or impede it. Given its vital role, a flight response must have priority over whatever the fish is doing once the behavioural decision has been made.

This reaction is not a simple reflex; the Mauthner cell is also activated in capturing prey. Moreover, firing can be followed by a correction to the trajectory, and finally, activation depends on several complex contextual factors. Similarly, in the sea slug *Pleurobracheus*, the impulse of a single neuron simultaneously causes swimming and suppression of feeding. The same cell can be involved in several movements with different functions. This capacity for 'switching', changing, choosing can be studied in the Mauthner cell. It is one of the basic elements of decision making.

Suppression of behaviour is governed by a very strong inhibitory mechanism. When the fish is swimming, activation of the Mauthner cell produces a reconfiguration of the motor and sensory circuits that control the animal's movement. Inhibitory interneurons block the swimming centres and leave the field open to the action of the Mauthner cell. The neuronal component of the flight response is carried out automatically by reorganizing the wiring.

This mechanism has an amazing capacity for adapting to multiple situations. For example, the sensitivity threshold of the Mauthner cell can be modified through a device known as 'long-term potentiation' that steadily increases its susceptibility to the inhibitory neurotransmitter glycine. The response of the cell accordingly diminishes when a dangerous stimulus appears. It isn't a matter of simple habituation but of a mechanism that, for example, allows the fish not to flee when it finds itself in turbulent waters [4]. It suggests the roots of our ability not to lose our cool in a traffic jam or when threatened by an adversary.

The neuron does not only carry out multimodal integration; it compares sensory inputs with a reference state (internal model) of what signifies 'danger'. If the fish encounters an obstacle, it has to avoid it; but if a predator then appears in the direction the fish has chosen to get around the obstacle, it must execute two very rapid movements in opposing directions that involve two Mauthner cells in succession. Observations show that fish can manage this feat in about 35 ms for both movements and, thus, survive. Footballers would be happy to achieve that kind of performance! Mechanisms of control of synaptic noise involving probabilistic processes are at the root of this remarkable flexibility [5].

The flight response, one of the major determining survival responses over the course of evolution, is thus a classic example of a process of decision making in a fundamental component of the central nervous system. It is in fact often accompanied by a decision either to stay or to flee. In many species of animal the fear response is more immobility 'freezing'—than flight. These choices are in no way automatic. They can become highly developed, since the social status of a crayfish in its group can determine whether serotonin is a positive or negative modulator of its flight response. For the dominant crayfish, serotonin increases the reaction, and in subordinates, it diminishes it [6]. It isn't only in humans that social aspects affect the decision to remain or to flee in the face of adversity! Perhaps serotonin plays a role in encouraging a captain not to abandon a sinking ship.

Approach or flee? That is the question

The decision whether to approach a prey or to flee it is a fundamental question. We will be talking about the toad, but is it not also a decisive choice that men have to make on encountering an attractive woman (and vice versa)? or that a naïve mushroom enthusiast makes when lured by the shimmering colours of a poisonous mushroom? Even on the floor of the stock exchange, deciding not to buy a hot stock results from this choice between opposites.

I touched on the behaviour of the toad in *The Brain's Sense of Movement*. There the issue was to show that evolution had found ways to simplify complex tasks. But this mechanism is also the neural manifestation of a subtle process of decision making. When the toad sees a moving object (Fig. 4.3), it has several options: *ignore* it; *stay put* and let it pass by; *pounce* on it if it decides it's an earthworm; *flee* if it decides it's a predatory bird. The toad can decide between the last two behaviours in the following way: if the object is long and moves in the direction of its greatest length, it could be an earthworm; if the object is large and expanding, it could be a bird approaching. Two specialized neuronal networks enable the toad to trigger the two contrary behaviours of approach and flight [7].

The mechanism is located in the toad's mesencephalon and tectum. Retinal messages are conveyed towards two neural pathways that make it possible to distinguish between the two types of sensory patterns. The first projects onto the optic tectum and the second towards the thalamus. Following these first relays in each pathway, an intermediary

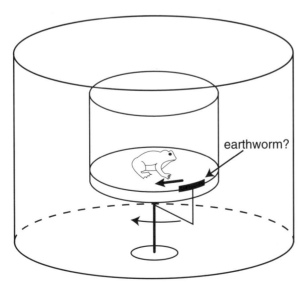

Fig. 4.3 Approach or flee? A toad must decide whether the object that is moving is prey (an earthworm) or predator. Here, because the object is long and moving in the direction of its greatest length, a network of specialized neurons will trigger an approach behaviour in the toad and at the same time suppress the opposite behaviour (to flee). After Glagow and Ewert (1999)

stage is sensitive to a variety of influences. For example, the response of the second neuron is suppressed by the other pathway so that the animal isn't approaching and fleeing at the same time. The two behaviours are thus mutually exclusive thanks to a simple reciprocal inhibitory mechanism. The decision to approach is taken at several levels; first at the periphery by the toad's visual system's properties of spatial filtration, which classify patterns of stimuli into 'prey' and 'predator', then at the intermediate stage where reciprocal inhibition is carried out. The execution stage produces the behaviour, whose intensity or direction, for example, can be varied.

The neurons of the first pathway in the optic tectum only fire when shown a pattern of the type 'earthworm' and stop firing in the presence of the pattern 'anti-earthworm'. The two networks interact in such a way that if one is activated, the other is inhibited. So this mechanism comprises first a visual means of identifying the shape of the moving animal and the nature of its movement; typically, that is the function of the tectum. Next, a network of neurons activates a motor synergy that enables either approach or flight. Other mechanisms can abort initiation of these two actions and induce immobilization. This paralysis is also important in allowing the animal to 'consider' the situation or to deliberate in his toad way. In the chapter on emotion we will pick up the topic of short loops for triggering simple behaviours and longer loops that control those behaviours.

Very subtle mechanisms of habituation enable the toad—when repeatedly shown lures—to identify the particular pattern of stimuli that corresponds to the lure and that which deceived the toad into approaching it. Reward is already a factor here to allow the animal not to respond anymore. This role of reward in decision making is also found in mammals, primates, and humans. We will briefly explore those mechanisms later on.

Dopamine, an important neuromodulator involved in learning functions, plays a determining role in regulating approach behaviour in the frog. Dopaminergic effects on various structures modulate mechanisms of prey capture in Anuri in terms of 'hunt' or 'wait' strategies. Hunting prey involves locomotion, orientation, and capture. In this strategy, successful capture depends on the mobility of the animal, at the risk of becoming the victim of a predator itself. Another strategy, waiting in a safe place, requires a low capture threshold to increase the chance of catching a prey on the run, but introduces the risk of accidental capture of nonprey. Some species of toad tend to adopt hunting strategies (*Bufo bufo*), whereas other species tend to adopt waiting strategies (*Rana esculenta*). These different behaviours appear to depend on different levels of dopamine in the neuronal networks involved. Such adjustments are probably not determined, which allows each individual to choose a combination of hunting and waiting behaviours as a function of external and internal conditions [8].

Another interesting example of the mechanism of decision making present even in the frog is that of *detour*. Taking a detour is often considered to be a cognitive task requiring complex development involving the frontal cortex. Even the frog is capable of detouring in a situation where it must avoid an obstacle. It knows how to get around a wall that it cannot hop over.

The mysteries of the colliculus

It is winter. In the courtyard of an apartment building in Paris, a skinny, homeless cat crouches, famished. It hears a scratching noise: Is it a man coming home, or a dog, its enemy—in which case it must flee—or is it a sparrow, a welcome meal? Deciding quickly whether a moving object is predator or prey is one of the most important behaviours for survival even in so-called higher species. Let us see how the decision to approach or flee is organized in mammals, whose decision-making behaviour is certainly more highly developed than that of the toad.

The superior colliculus—in mammals the equivalent of the tectum in toads and birds—is a brain structure involved in movements of orientation towards prey. Its neurons form a genuine visual map of the world. When an object appears, its image is projected onto the neurons of the superficial layers of the superior colliculus. They cooperate with the neurons of the inferior colliculus, which receive messages regarding the sound clues associated with the movement of the object. For example, if a cat sees a bird land on a branch, the image of the flying object is projected onto the superior colliculus, and its song activates the neurons of the inferior colliculus. The fact of receiving information from more than one modality (e.g. hearing at the same time as seeing) increases the probability of the animal heading in the direction of the target [9].

Retinal mapping ('retinotopy') of the visual world is thus retained; a directional movement is induced via activation of the motor neurons of the eyes and head, even the trunk. These orientation responses have the characteristics of a reflex. However, Walter Hess showed that they can be the object of selection, depending on the animal's context [10]. He proposed a model (Fig. 4.4) for a selector of movement, a simple prototype of a decision-making mechanism. He started with the fact that the eye is controlled by six muscles, and that a movement of gaze orientation towards a target coming into view requires rapid implementation of a specific combination of these muscles. But this combination depends on the initial position of the eye at the instant the target appears. So the brain must be able to choose which muscles to activate based on this initial position. A suitable *synergy* has to be chosen. What is important in the model shown in the figure is the idea that the choice is produced by a selecting mechanism superimposed on the motor apparatus.

But the superior colliculus can also be involved in producing movements other than those of orientation towards a visual or acoustic target, associated with predatory behaviour. Indeed, it may be involved in movements of *defence* and *avoidance*. Lesions of the superior colliculus can destroy the impulse to flee or to freeze that very often accompanies fear. A wild rat with a damaged superior colliculus will not react to the sudden presence of a human. On the other hand, electrical stimulation of its colliculus

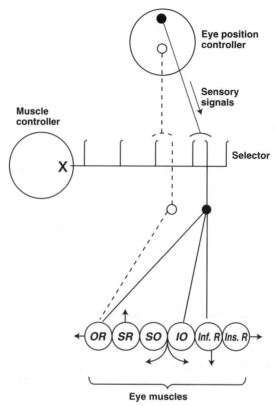

Fig. 4.4 A selector of action. This scheme shows a mechanism of gaze control. It acts by selectively transmitting stimuli, by interposing an 'organ of selection', a primitive mechanism of nearly automatic decision making. The sensory signals coming from the retina can potentially trigger many different configurations of activity of the different eye muscles. These muscles are indicated by circles. OR, outside right; SR, superior right; SO, superior oblique; IO, inferior oblique; Inf.R., inferior right; Ins.R, inside right. The dashed line shows the path of the impulses when the eye is in a resting position; the solid line corresponds to an elevated position of the eye. Input is redistributed based on contraction of the superior right muscle. The main effect is that together all these muscles appear to act as a functional unit that reorganizes its activity depending on context. After Hess (1981)

can induce numerous defensive behaviours, such as immobility or motions of flight. The effect is distancing—the suggestion of flight—as opposed to an orientation response.

A repertoire of defensive movements (retreating, lying flat, or running away) can thus be induced by stimulation. At least some of these responses are due to neurons in the colliculus borrowing pathways that project to the same side as the part of the colliculus involved in contrast to the neuronal pathways that contribute to orientation responses. The neurons concerned with movements of orientation and those involved in immobilization or distancing responses are different. We currently lack a functional neurophysiology similar to the one developed for the toad to explain how the two major types of responses mutually exclude each other at the level of the colliculus, in particular through the intermediary of the intersecting, partly inhibitory pathways that connect the two colliculi.

Involvement of the colliculus in complex processes of decision making is also illustrated by the fact that it contributes to *cortical arousal*: when an animal is dozing, its cortical activity is characterized by low-frequency waves and moves to a higher frequency when it wakes up. It seems that activating the neurons of the superior colliculus can desynchronize the cortical electroencephalogram via at least three distinct ascending pathways (remember that, during sleep, neuronal activity is synchronized and that waking is accompanied by a destructuring of the rhythm):

1. Signals concerning the *consequences of movements of orientation* may, through bifurcation of the axons of the tectoreticulospinal neurons, reach the thalamus and influence the slow rhythms of neuronal activity that are found in the thalamocortical loops. Recent data on slow rhythms suggest that they play an essential role in body awareness and that they are disrupted in major illnesses such as obsessional disorders (see Chapter 3), schizophrenia, and so on, with profound consequences for cognitive activity, in particular the ability to make decisions.
2. Signals indicating *danger* can be addressed by the intermediary of the basal nucleus of Meynert towards relays in the cuneiform and peripeduncular regions.
3. Signals indicating *an unidentified event* could also be sent to the basal nucleus of Meynert via a third pathway, as yet unknown.

All these ascending projections suggest that the decision to approach or flee or direct oneself towards one target or another is actually made by a complex system in which the colliculus is involved both in delivering perceptual messages but also as the place where descending signals are combined, as a selector of action.

Unfortunately, no group anywhere in the world is currently pursuing this research because the cerebral cortex seems to be more interesting to physiologists than the colliculus. A neurophysiology of selection and decision making has yet to be invented.

Looking, a fundamental decision?

Gaze is a fundamental mode of active exploration of the world, but it also has tremendous social meaning. For instance, among certain South American tribes, a young woman agrees to marry a young man simply by looking at him. To look is therefore to make a decision. Vision is not only 'the brain's way of touching', as Maurice Merleau-Ponty has said, it is *deciding by looking*. At a restaurant, when the room is full and you are waiting impatiently to be served, you will notice that the waiters work very hard not to catch your eye. They pass by with exasperating nonchalance. They know that if you get their attention, they will have to serve you. In some societies, looking directly at people is impolite; not looking at them is thus the utmost in civility. To look is both to win another and to give oneself to the other. Brain imaging has shown that as soon as we establish 'eye contact' with another person, the amygdala, one of the most important brain areas for emotion, responds vigorously. Societies where women must hide their faces recognize that there is more to an exchange of gazes than simple eye movements. It is also known that deficits in social communication, in autistic patients for instance, are always associated with gaze avoidance.

The control of gaze movements and, in particular, the jumps—saccades—that the eyes make are an infant's first feat of exploration. Still tiny, barely born, a baby turns his gaze to his mother and can be attracted by moving objects. The saccade is the first movement that allows the infant to explore the world around him. Already, looking is deciding, since to free himself of the almost 'magnetic' visual link he has with his mother, the baby must disengage to look elsewhere. Shifting one's gaze also means shifting one's attention [11]; it means choosing from the external world what one wishes to introduce into one's own internal world.

The organization of gaze shifts is one of the most basic models of the process of decision making. However, the brain's mechanisms for gaze control range from automatic to cognitive and emotionally driven. Summarizing the findings to date on this question is beyond the scope of this chapter; rather, the reader may wish to consult several recent reviews [12].

Gaze is subserved by a repertoire of movements that enable stabilization of the image of the world on the retina (via the vestibulo-ocular and optokinetic reflexes), maintenance of the image on the fovea (via pursuit) and gaze orientation, either reflex or voluntary (via saccades). I introduced the elements of this repertoire in *The Brain's Sense of Movement*. Here I will discuss only saccades, the jumps of the eye made by the reader's gaze while reading this book or looking around. Saccades are fundamental movements in our voluntary exploration of the environment. They call into play nearly all the structures of the brain involved in deciding, and as such they constitute one of the best models for studying the neurobiology of decision making.

Gaze shifts can be either single or organized in sequences. Even single gaze shifts decompose into several steps. It is well known that for voluntary gaze shifts, an initial movement brings the direction of gaze close to the target of interest and the brain then produces a 'corrective saccade' which adjusts the direction of gaze precisely. This elementary sequence is under the control of the cerebellum. But even the launching of the first saccade can be decomposed into two steps probably related to the examination of the visual world followed by a decision phase [13]. This decision depends largely on the expectations of the viewer. A model called 'LATER' has been proposed to explain the variability of saccadic reaction times and to relate saccadic performance to the expectations of the subject and important contextual elements, such as the prior probability of events, the urgency of the response, and observer uncertainty. A variety of models are now available for simulating saccadic decision making at this behavioural level [14].

Scan paths

Generally, in looking around, we use a sequence of saccades. Pioneers such as Alfred Yarbus in Moscow and Lawrence Stark in Berkeley described the succession of changes in gaze made for exploring the environment. They called them 'scan paths' and stressed the highly cognitive basis of the construction of these motor sequences and their dependence on the context and content of the visual scene, but also intentions and even mood. In fact, the brain has several ways of organizing scan paths; some processes are truly sequential in time, but others [15] are parallel. Some models use iconic scene representations derived from spatiochromatic filters at multiple scales, which treat the process of guiding the eye to the target independently, and decision processes that deal with information at the target site do rather a good job of simulating the scan paths [16].

But the decision to look is subject to many biases that are not only cognitive but can also have a systematic origin. For instance, we usually look at the left part of a figure or a painting or a face. This perceptual tendency is probably due to a hemispheric bias, the right hemisphere being more involved in processing faces [17]. A similar tendency is to be expected when the scene is charged with emotional value, because the right hemisphere is also dominant in processing emotional stimuli. An opposite (rightward) bias may occur when we are searching for a categorical aspect of a scene versus a more global (metric aspect) and so on.

Another example of the sophisticated organization of gaze saccadic sequences is the phenomenon called 'inhibition of return'. It has been shown that we never or rarely look back to where we have just looked when scanning a scene: the brain constrains our choices according to rules that have appeared throughout phylogeny, probably because they were useful. Another interesting demonstration of the highly predictive nature of gaze control is the fact that information acquired during fixation between two gaze changes is not necessarily used immediately in guiding the next saccade but may be stored to be used for a subsequent saccade [18].

The neural mechanisms for scan path generation are obviously complex. They involve a hierarchy of mechanisms, and I have proposed that inhibition plays a fundamental role in the gating of decisions at various levels of the neural network for gaze control [19]. Figure 4.5 gives a highly schematic description of these processes.

A first level in this hierarchy, already present at birth before a baby's cerebral cortex is completely developed, is that of orientating reflexes towards visual or auditory targets and the mother's capture of the baby's gaze. We can call this the 'reactive-orienting' subpart of the gaze system. Gaze orientation is achieved, for instance, when jiggling a toy in front of a baby, by a first very rapid neural loop that connects the retina, the superior colliculus, and the centres of the brain stem that produce the saccades. Fixation on the mother's gaze is probably triggered by innate mechanisms of face recognition that may involve the amygdala [20], as is suggested by findings of brain

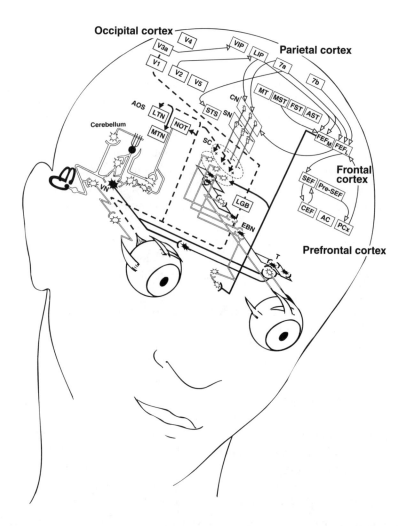

Fig. 4.5 The decision to look: a choice between different subsystems of the oculomotor repertoire. This schematic diagram summarizes a few of the brain systems that are involved in the control of gaze movements. (These partial diagrams were published in *The Brain's Sense of Movement* in 2000.) The scheme includes the vestibulo-ocular systems that together with the optokinetic pathways ensure gaze stabilization. It also includes the system of ocular pursuit as well as the saccadic system. The choice of appropriate ocular movement and the organization of gaze movements depending on the task or memory is made at several levels of this complex system. VN, vestibular nuclei that are relays for the vestibulo-ocular reflex but also the optokinetic nystagmus; AOS, accessory optical system, which is involved in the optokinetic nystagmus; MTN and LTN, medial and lateral terminal nuclei; NOT, nucleus of the optic tract. These nuclei control the different horizontal and vertical elements of the optokinetic nystagmus and project them towards the vestibular nuclei and the cerebellum. A short circuit involving the superior colliculus (SC) projects onto the brain stem and induces

imaging, whose functions we will consider in Chapter 11. These two mechanisms (orienting to a peripheral target and fixation) are often in competition and call into play neural decisions (fixating or disengaging from fixation to attend to another target) to which the system of so-called fixation neurons in the colliculus and the brain stem contributes under the supervision of the frontal and prefrontal cortex.

A crucial structure for these reactive saccades to static or moving targets is the superior colliculus. The superficial layers of the colliculus constitute an internal map of visual space. It is 'retinotopic', meaning that its organization is similar to the retina. It is a sensory map. In the deeper layers of the colliculus, neurons are arranged to constitute a 'motor map' because these neurons project to the brainstem and can trigger saccades. This 'motor map' is in register with the sensory map of the superficial layers. The neurons of the motor map are under a constant inhibitory barrage from the substantia nigra portion of the basal ganglia in the oculomotor loop.

So when a target of interest appears at the periphery of the visual field, the collicular neurons are activated. But they can only trigger a saccade to this target (the brain can only decide to look at the target) if the neurons of the substantia nigra are themselves inhibited. A cascade of inhibitory mechanisms participate in the decision when and where to make a saccade [21], and these mechanisms, which are under the control of the cortex, are mediated by the descending inhibition–disinhibition process.

The brain is even more sophisticated. It not only contains mechanisms for selecting the time and direction of a saccade but also anticipates the future position of the target. In the superior colliculus, mechanisms of anticipation help to orient gaze towards

discharge of excitatory burst neurons (EBN), which trigger orientation of gaze by saccades. A long pathway, beginning with the first-order visual relays in the thalamus (LGB, lateral geniculate body), projects images of the world onto the cortical visual areas V1, V2, V3, V4, V5; VIP, LIP, ventral and lateral inferior parietal cortex; STS, indicates a projection towards the superior temporal sulcus that is involved although not shown in this scheme. Gaze movements are also processed in the temporal cortex, for example, in the hippocampal formation. MT, MST, FST, and AST indicate temporal areas involved in ocular pursuit that project onto the pontine nuclei (not shown here), which activate the eyes via the brain stem and the cerebellum. FEF_M and FEF_L, frontal and lateral eye fields that are the final cortical stations of ocular saccades and pursuit, but are also involved in gaze fixation. SEF and pre-SEF are supplementary eye fields, the equivalent of the supplementary motor area and the pre-SMA for the limbs; CEF, AC, cingulate eye field and anterior cingulum, involved in intentional aspects and gaze guidance according to the affective and cognitive features of the task. PCx, prefrontal cortex, where decisions are made to direct gaze towards one point or another in space and also the point of convergence for information coming from the dorsal and ventral pathways of the visual system, but also other sensory information, motivational aspects, and memories. Also shown is the inhibitory cascade that in the striatum participates in the decision to look (CN, caudal nucleus, and SN, substantia nigra). These last two structures belong to the oculomotor loop shown in Fig. 3.3. After Berthoz (2000)

the future position of a moving target, such as we do when we want to catch a ball. We have to look at the place where the ball will be in a fraction of a second. This mechanism exists in all animals. The frog cannot catch a fly if it send its tongue to where the fly is at a given instant; it has to send it in advance. A tennis player cannot hit at the direction of a ball; she has to anticipate its future position at the time of impact with the racket. The brain looks 'one step ahead', and thus the decision to shift gaze is a predictive mechanism [22].

The loop for gaze shift decisions is a short one. But as we just saw, this fast loop is very complex and very clever. I have known 'blind' people who could play table tennis and volleyball even though they could not read. They had a type of subcortical vision involving the superior colliculus and could catch a ball very efficiently. A way to show the involvement of a brain structure in decision making is to have the subject (or the animal) make what is called a 'countermanding gaze shift with a saccade'. That is, instead of responding to the appearance of a target by a saccade towards the target, the subject is asked to hold his (or its) response (or, in the case of the so-called antisaccade, to make an eye movement contrary to the saccade needed to reach the target). It has been shown in monkeys that superior colliculus neurons discharge during countermanding tasks [23]. It has even been suggested that neuronal activity in the superior colliculus is related to the 'likelihood' of the possible targets; the collicular map would therefore be an area where 'the decision to select a new goal would occur each time the null hypothesis, represented by neurons in the rostral superior colliculus, was rejected in favor of an alternative represented elsewhere in the colliculus' [24]. Special neurons in the superior colliculus called 'buildup' neurons are thought to play a role in the process of reaching a decision [25].

A second level of organization of saccades, particularly developed in mammals and primates and in humans, involves a longer loop. Visual information is processed through the thalamus to the parietal cortex, where retinal data are mixed with other information from the body as well as reciprocal descending influences from the frontal and prefrontal cortex [26]. The identity of visual forms in the environment is processed in the ventral pathway and spatial aspects in the dorsal stream. These two streams converge in the frontal and prefrontal cortex, from which information is fed back to the parietal cortex.

At present there is no complete, coherent explanation of the decision process, and it will probably turn out that there are several processes that lead to any decision given the task and context. However, a few recent experiments shed some light on fragments of these operations. First, we know that the frontal eye fields are probably the most important cortical centre for the generation of saccades. It was thought that they were a sort of final motor centre for eye movements, and that is probably true. They play a role equivalent to that of the motor cortex for limb movements. It now seems, though, that their function is more subtle. For instance, the frontal eye fields may serve as a saccade probability map or 'saliency map' [27], receiving a combination of bottom-up and top-down influences for target selection under the control of attention. They may also be involved in imagined shifts of attention, as predicted by the 'premotor theory of attention'.

For instance, we have studied the decision to look to the left or to the right in a simple self-produced saccade task. Recording brain activity by functional magnetic resonance imaging suggests that in the frontal lobe, preparation of a saccade involves the frontal eye fields whether the free choice of saccade direction involved the additional activation of the pre-supplementary eye field, the anterior cingulate cortex, or the dorsolateral prefrontal cortex. It has also been suggested that the dorsolateral prefrontal cortex, which is known to participate in working memory—the short-term memory of a recent event that we want to keep in mind—may be important in inhibiting unwanted reflexive saccades, maintaining stored information to guide a subsequent gaze shift, or facilitating mechanisms of anticipation by providing memory of previous saccades and so on [28]. These neurons also seem to encode the attributes of remembered stimuli in a way that enables adaptive decision making [29].

Recent interest in prefrontal and frontal structures in saccadic decisions should not, however, preclude a crucial role for other brain centres in the decision process. Most probably the decision to look will not be made in one brain area but will result from a competitive process involving large but definable networks of areas, each contributing its own properties and 'personality'. For instance, subcortical structures such as the caudate nucleus, the premotor cortex, the putamen [30], and the thalamus [31] may be involved through long basal ganglia-thalamo-cortical loops in prospective decision-making mechanisms. Some operations probably also involve other structures of the temporal cortex, such as the hippocampal formation.

Saccades can be used to study the neural basis of decision making during perceptual tasks requiring decisions about what is contained in the visual scene. This field, best known as 'visual searching', uses eye movements to understand the successive steps taken by the brain from the initial detection of visual targets or visual motion cues to perceptual decisions. A very successful paradigm consists in showing the subject or the monkey a visual scene in which there are clouds of moving dots with a certain percentage moving in one direction and another percentage moving in the opposite direction. The animal or the human subject is then requested not only to respond manually by pressing a button to indicate the direction perceived but also to make a saccade in the direction of the perceived motion [32]. It seems that in the brain, decisions are progressively elaborated from the parietal cortex, which accumulates sensory signals relevant to the selection of a target for eye movement, to the prefrontal cortex, which actually carries out the decision by mechanisms which are still not certain. The parietal cortex is not only involved in these simple perceptual tasks for decision making; it has also been shown to be involved in foraging tasks and therefore is part of the network dealing with competing actions and evaluation of their reward value [33].

One has to go into the prefrontal and frontal cortex to find neurons that really appear to be linked to the decision to look here or there and that integrate sensory data, intentions to act as well as desired objectives or rewards along with their emotional significance [34]. In itself the problem of the neural basis of emotional influence on gaze control would be worth a long commentary. But it would be impossible in this short essay to review all the modern theories of saccade generation, for instance, the recent extensive use of Bayesian concepts of probability, which are directly relevant to a potential theory of gaze and decision making.

A final word about an interesting concept: likelihood. Which mechanisms allow the brain to decide to make a saccade in one direction or the other? It seems that the cortical neurons evaluate not probabilities but *likelihood*, that is, preferences based on a comparison between two or more hypotheses. In other words, the brain does not calculate an absolute value for each possibility, but makes comparisons between solutions. Recently, Joshua Gold and Michael Shadlen [35] (Chapter 9, note 12) proposed the idea that preference based on an estimation of likelihood results from competition between neurons and 'antineurons' whose activity represents two alternative hypotheses. Two ideas are essential here: the first is that the brain does not process absolute values, but errors or differences; the second is that competition between excitation and inhibition plays a fundamental role in decision making (since the activity of antineurons in the mathematical models proposed is negative and thus equivalent to competing excitation and inhibition). It is interesting to note that the most recent models of competition recapitulate somewhat models of fight-or-flight competition that we have examined in the toad. The most highly developed functions of the brain do seem to be based on principles that are at core rather simple, but are reproduced in neuronal 'contexts' that are very diverse. In other words, like a Bach fugue, the brain's apparently complex activity consists of many, subtle variations based on a repertoire of themes whose elements, though limited, are nonetheless amenable to combinations, as in the *Well-Tempered Clavier*.

Deciding means choosing

One could argue that even though these mechanisms are complex, they do not really belong to what is ordinarily considered to be a truly conscious decision, the result of rational deliberation. Nevertheless, I still think it is worth seeking among those mechanisms some of the neural principles that underlie complex decisions.

These principles teach us first of all that decision making that selects one behaviour over another does not consist in constructing a behaviour adapted to a given situation from scratch. We have access to a repertoire of well-defined behaviours. So the decision is first of all to select already present behaviours. Of course, these behaviours can be learned or belong to the genetic repertoire of a species.

Not only are these behaviours developed in the nervous system through a complex neuronal machinery of the motor system, they are already reflected, inscribed as it were in the sensory analysers at the earliest levels. Their formula could be: 'Looking means having decided.' Indeed, the brain projects onto the world its preperceptions and hypotheses. It is essentially a comparator of predictions and reality. The Russian school of physiology deliberately used the term 'analyser' rather than sensory 'receptor'. In so doing, the Russians emphasized that the brain doesn't merely passively take in data from the physical world [36]: the senses are already organized according to possible behaviours. In the 1970s, I and others visited the laboratory of Alexander Batuev at Leningrad (today, St Petersburg). He proudly showed us a discovery he had made: the neurons of one of the brain centres of the auditory system responded much more to natural stimuli of predators or members of the same species than to combinations of sounds. This observation, revolutionary at the time, supported a more ecological theory of central nervous system organization. Today we can go further and say that the rationale for this type of encoding is that it already contains the behaviour that will be adopted in the presence of this stimulus. Whence the idea that perception itself is decision making. To perceive is to classify according to a repertoire of possible actions.

In this book I would like to extend the analysis I made in *The Brain's Sense of Movement*. There, I maintained that action influences perception at its source. Here, we must yield to the evidence that action is itself written in the workings of sensory receptors, at least in the early relays, because these choose, filter, and organize, for example, visual data, based on a repertoire of possible actions.

I think the future will show that modulation of the primary visual relays by what is called 'attention' will in fact prove to be modulation determined not only by attention, that is, the portion of the visual field that draws our interest, but by the reason for acting that accompanies this interest. We have yet to understand the organization of visual processing based on the goals of action.

When I show an audience a film of a toad choosing between approaching and fleeing, people burst out laughing to see the toad attentively observe a little bit of wood moving. The animal appears to be reflecting. Either it produces a quick little movement of its mouth that accompanies capture, or it remains indifferent, or sometimes it recoils as if in disgust. The toad really does give the impression of reflecting, thinking, deliberating, and suddenly deciding.

Theoreticians have tackled the problem in reverse. They begin with the idea that decision making is a general, abstract process that obeys formal, disembodied rules, and then they try to find all the exceptions to these general and formal rules, and end by declaring that decision making depends on the task, context, frame of reference, body, and so on. And yet observation of nature and animals teaches us straight off that decisions depend on the actions that are possible for a given species in a given context. A theory of decision making must begin so: The world is constructed ('constituted', phenomenologists would say) by the animal according to its goals and in a way that best allows it to survive in a given environment (*Umwelt*).

From this modest attempt at analysis, we extract yet another theoretical proposition, namely, that the concept of utility must be replaced or augmented with that of affordance ('doability'). The animal decides based not only on what it judges useful but what it is actually able to do—how it can act—to survive. The concept of patterning, both of motor programs but also the receptors that are interrogated by the brain, is fundamental here. I have proposed it in another form: substitution of one motor system for another in the case of a lesion or dysfunction [37]. I think that a real 'decision' takes place, a choice that does not involve cognitive deliberation but that probably contains the germ of the basis for more cognitive decision making. It is within the mechanisms of deciding to act that one must seek the basis of mechanisms for thinking that decides. Decision making itself is an act.

Inhibition (again and always)

These examples lead me to stress the fundamental importance of inhibition in the process of decision making. In the classical models of the nervous system, inhibition is important in coordinating movements. We have seen that even in the fish, it is needed to keep two opposing behaviours from both happening at the same time. We know that this reciprocal inhibition of contrary behaviours is also present in the system of gaze control; in the vestibular nuclei, inhibitory interneurons block the vestibulo-ocular reflex. This reflex turns the eyes opposite the movement of the head to stabilize gaze in space; but when we want to orient our gaze towards an object of interest, the eyes turn in the same direction as the head. Just as with fight or flight, the choice between stabilization and orientation is regulated by inhibition.

Inhibition also enables *temporal* selection of the moment when we want to direct our gaze towards a point of interest, as it does in the spatial selection of targets. It is present at all levels of the nervous system, and we will see later on that this feature of inhibition blocks behaviour when we decide not to do something. To decide is thus to block; deciding means not only choosing between several solutions but also blocking undesired behaviours. Remember that in the nervous system, the main mode of action of all the major structures in the choice of behaviours is inhibitory. This is true for the cerebellum, for the basal ganglia, and for the prefrontal and frontal cortex.

The simple inhibitory mechanism that, together with visual analysers that have already specified a response, enables the toad to choose between two behaviours probably became widespread for precisely that reason. The fundamental innovation of evolution was made once for all, namely, decision making could be accomplished by an inhibitory mechanism in competition with excitatory mechanisms. You can begin to see how, through the addition of superimposed loops, evolution managed to construct ever more complex mechanisms of decision making.

Neuromodulation and reconfiguring networks

Further evidence for the existence of fundamental mechanisms of decision making in the nervous system is supplied by the selectors of motor strategies in functions as essential as locomotion. Sten Grillner, in Sweden, dissected neuron by neuron the mechanisms of the network that produces swimming in the lamprey [38]. And yet, what Gillner really discovered is the general model for how locomotor generators work. Like swimming, locomotion results from activation of a network situated in the spinal cord that (after an appropriate sequence) produces a neural rhythm that translates to the muscles of the limbs. This coordinated rhythm is produced by a very small, very particular network of neurons that both excite and inhibit. Until recently, the challenge was to understand how this network enables an animal to decide whether it would go at a walk, a trot, or a gallop. How can the nervous system produce movements whose coordination is so different?

One solution would have been to have several circuits whose anatomical connections and neural organization compose the parts of the motor repertoire. But evolution chose another, much more elegant and economical solution: specialized neurons in the spinal cord release neuromediators that alter the properties of transmission at the level of the synaptic junctions of the same network. This neuromodulatory action translates in turn into an actual *functional reconfiguration* of the network like that demonstrated by the pioneering work of the physiologist Maurice Moulin in the stomatogastric ganglia. Some neurons are facilitated, others blocked, others ever so subtly change their membrane properties, which modifies or curtails their dynamic properties; finally, still others, little in demand in one mode, are recruited by another and so on. The decision to walk, trot, or gallop is thus made by these neurons, which reconfigure the circuit.

You might be tempted to say: 'Careful! misleading use of language! Decision making is done upstream of these neurons because they must themselves be excited for the network to be reconfigured.' But for us what matters is to establish that decision making does not involve complicated neurocomputations. An automatic mechanism makes it possible to simplify change in such a way that the decision can be induced simply by activating a few neurons. We saw above that deciding to approach or flee is possible thanks to reciprocal inhibition of two competing circuits; here we see that the decision to walk or to gallop involves reconfiguring of the same circuit. Decision making is indeed a fundamental property of the nervous system.

References

1 Spinoza, *Ethics*, ed. and trans. GHR Parkinson (Oxford: Oxford University Press, 2000), p. 124.
2 D Faber, J Fetco, and H Korn, 'Neuronal networks underlying the escape response of the goldfish', *Annals of the New York Academy of Sciences*, **563** (1989): 1–33.
3 H Korn and D Faber, 'Escape behaviour: brainstem and spinal cord circuitry and function', *Current Opinion in Neurobiology*, **6** (1996): 826–32, p. 828.
4 Y Oda, K Kawasaki, M Morita, H Korn, and H Matsui, 'Inhibitory long term potentiation underlies the auditory conditioning of goldfish escape behaviour', *Nature*, **394** (1998): 182–5.
5 K Hatta and H Korn, 'Tonic inhibition alternates in paired neurons that set direction of fish escape reactions', *Proceedings of the National Academy of Sciences of the USA*, **96** (1999): 1290–1295.

6 SR Yeh, RA Fricke, and DH Edward, 'The effect of social experience on serotonergic modulation of the escape circuit of the crayfish', *Science*, 271 (1996): 366–9.

7 H Buxbaum-Conradi and J-P Ewert, 'Responses of single neurons in the toad's caudal ventral striatum to moving visual stimuli and test of their efferent projection by extracellular antidromic stimulation/recording techniques', *Brain, Behavior, and Evolution*, 54 (1999): 338–54.

8 M Glagow and J-P Ewert, 'Apomorphine alters prey catching patterns in the common toad: behavioral experiments and 14C-2-deoxyglucose brain mapping studies', *Brain, Behavior, and Evolution*, 54 (1999): 223–42

9 TJ Anastasio, PE Patton, and K Belkacem-Boussaid, 'Using Bayes' rule to model multisensory enhancement in the superior colliculus', *Neural Computation*, 12 (2000): 1165–87.

10 WR Hess, *Biological order and brain organization* (Berlin: Springer, 1981), fig. 15.12.

11 L Chelazzi, M Biscaldi, M Corbetta, A Peru, G Tassinari, and G Berlucchi, 'Oculomotor activity and visual spatial attention', *Behavioural Brain Research*, 71 (1995): 81–8; M Corbetta, E Akbudak, TE Conturo, AZ Snyder, JM Ollinger, HA Drury, MR Linenweber, SE Petersen, ME Raichle, DC Van Essen, and GL Shulman, 'A common network of functional areas for attention and eye movements', *Neuron*, 21 (1998): 761–73; M Corbetta, FM Miezin, GL Shulman and SE Petersen, 'A PET study of visuospatial attention', *Journal of Neuroscience*, 13 (1993): 1202–26; M Corbetta, GL Shulman, FM Miezin, and SE Petersen, 'Superior parietal cortex activation during spatial attention shifts and visual feature conjunction', *Science*, 270 (1995): 802–5.

12 C Araujo, E Kowler, and M Pavel, 'Eye movements during visual search: the costs of choosing the optimal path', *Vision Research*, 41 (2001): 3613–25; PW Glimcher, 'The neurobiology of visual-saccadic decision making', *Annual Review of Neuroscience*, 26 (2003): 133–79.

13 BA Reddi, 'Decision making: the two stages of neuronal judgement', *Current Biology*, 11 (2001): R603–6.

14 BR Beutter, MP Eckstein, and LS Stone, 'Saccadic and perceptual performance in visual search tasks. I. Contrast detection and discrimination', *Journal of the Optical Society of America, A, Optics, Image Science and Vision*, 20 (2003): 1341–55; K Viikki, E Isotalo, M Juhola, and I Pyykko, 'Using decision tree induction to model oculomotor data', *Scandinavian Audiology, Supplementum*, 52 (2001): 103–5; RH Carpenter, 'Contrast, probability and saccadic latency; evidence for independence of detection and decision', *Current Biology*, 14 (2004): 1576–80; JC Leach and RH Carpenter, 'Saccadic choice with asynchronous targets: evidence for independent randomisation', *Vision Research*, 41 (2001): 3437–45; BA Reddi, KN Asrress, and RH Carpenter, 'Accuracy, information and response time in a saccadic decision task', *Journal of Neurophysiology*, 90 (2003): 3538–46; BA Reddi and RH Carpenter, 'The influence of urgency on decision time', *Nature Neuroscience*, 3 (2000): 827–30.

15 PH Schiller and J Kendall, 'Temporal factors in target selection with saccadic eye movements', *Experimental Brain Research*, 154 (2004): 154–9; K Kurata and H Aizawa, 'Influences of motor instructions on the reaction times of saccadic eye movements', *Neuroscience Research*, 48 (2004): 447–55.

16 RP Rao, GJ Zelinsky, MM Hayhoe, and DH Ballard, 'Eye movements in iconic visual search', *Vision Research*, 42 (2002): 1447–63.

17 S Butler, ID Gilchrist, DM Burt, DI Perrett, E Jones, and M Harvey, 'Are the perceptual biases found in chimeric face processing reflected in eye-movement patterns?', *Neuropsychologia*, 43 (2005): 52–9. See also O Grüsser and T Landis, *Visual Agnosias* (London: Macmillan, 1991).

18 D Melcher and E Kowler, 'Visual scene memory and the guidance of saccadic eye movements', *Vision Research*, **41** (2001): 3597–11.

19 A Berthoz, 'The role of inhibition in the hierarchical gating of executed and imagined movements', *Brain Research, Cognitive Brain Research*, **3** (1996): 101–13.

20 AJ Calder, AD Lawrence, J Keane, SK Scott, AM Owen, I Christoffels, and AW Young, 'Reading the mind from eye gaze', *Neuropsychologia* **40** (2002): 1129–38; R Kawashima, M Sugiura, T Kato, A Nakamura, K Hatano, K Ito, H Fukuda, S Kojima, and K Nakamura, 'The human amygdala plays an important role in gaze monitoring. A PET study', *Brain* **122** (1999): 779–83; N George, J Driver, and R Dolan, 'Seen gaze direction activates fusiform gyrus and its coupling with other brain areas during face processing', *Neuroimage* **13** (2001): 1102–112.

21 GD Horwitz, AP Batista, and WT Newsome, 'Representation of an abstract perceptual decision in macaque superior colliculus', *Journal of Neurophysiology*, **91** (2004): 2281–96.

22 A Caspi, BR Beutter, and MP Eckstein, 'The time course of visual information accrual guiding eye movement decisions', *Proceedings of the National Academy of Sciences of the USA*, **101** (2004): 13086–90.

23 M Pare and DP Hanes, 'Controlled movement processing: superior colliculus activity associated with countermanded saccades', *Journal of Neuroscience*, **23** (2003): 6480–89.

24 RJ Krauzlis, D Liston, and CD Carello, 'Target selection and the superior colliculus: goals, choices, and hypotheses', *Vision Research*, **44** (2004): 1445–51.

25 R Ratcliff, A Cherian, and MJ Segraves, 'A comparison of macaque behavior and superior colliculus neuronal activity to predictions from models of two-choice decisions', *Neurophysiology*, **90** (2003): 1392–407.

26 RP Hasegawa, M Matsumoto, and AJ Mikami, 'Search target selection in monkey prefrontal cortex', *Neurophysiology*, **84** (2000): 1692–6.

27 KG Thompson, NP Bichot, and TR Sato, 'Frontal eye field activity before visual search errors reveals the integration of bottom-up and top-down salience', *Neurophysiology*, **93** (2005): 337–51.

28 C Pierrot-Deseilligny, RM Muri, CJ Ploner, B Gaymard, S Demeret, and S Rivaud-Pechoux, 'Decisional role of the dorsolateral prefrontal cortex in ocular motor behaviour', *Brain*, **126** (2003): 1460–73.

29 C Constantinidis, MN Franowicz, and PS Goldman-Rakic, 'The sensory nature of mnemonic representation in the primate prefrontal cortex', *Nature Neuroscience*, **4** (2001): 311–16.

30 BR Postle and M D'Esposito, 'Spatial working memory activity of the caudate nucleus is sensitive to frame of reference', *Cognitive, Affective, and Behavioral Neuroscience*, **3** (2003): 133–44.

31 MT Wyder, DP Massoglia, and TR Stanford, 'Contextual modulation of central thalamic delay-period activity: representation of visual and saccadic goals', *Journal of Neurophysiology*, **91** (2004): 2628–48.

32 JI Gold and MN Shadlen, 'The influence of behavioral context on the representation of a perceptual decision in developing oculomotor commands', *Journal of Neuroscience*, **23** (2003): 632–51; JI Gold and MN Shadlen, 'Representation of a perceptual decision in developing oculomotor commands', *Nature*, **23** (2000): 390–94; MN Shadlen and WT Newsome, 'Neural basis of a perceptual decision in the parietal cortex (area LIP) of the rhesus monkey', *Journal of Neurophysiology*, **86** (2001): 1916–36.

33 LP Sugrue, GS Corrado, and WT Newsome, 'Matching behavior and the representation of value in the parietal cortex', *Science*, **304** (2004): 1782–7; comment, *Science*, **304** (2004): 1753–4.

34 **AN McCoy and ML Platt**, 'Expectations and outcomes: decision-making in the primate brain', *Journal of Comparative Physiology, A, Neuroethology, Sensory, Neural, and Behavioral Physiology*, 12 October 2004 [Epub ahead of print]; J Diniz Filho and TB Ludermir, 'Modeling a particular decision process by using a modulatory activation function', *International Journal of Neural Systems*, 13 (2003): 111–18.

35 **Gold and Shadlen**, 'Influence of behavioral context'; Gold and Shadlen, 'Representation of a perceptual decision'.

36 **V Gurfinkel, EE Debreva, and YS Levik**, 'The role of internal models in the position, perception, and planning of arm movement', *Fiziologiya Cheloveka*, 12 (1986): 769–76.

37 **A Berthoz**, 'The role of gaze in compensation of vestibular dysfunction: the gaze substitution hypothesis', *Progress in Brain Research*, 76 (1988): 411–20.

38 **S Grillner, P Wallén, L Brodin, and A Lansner**, 'Neuronal network generating locomotor behavior in lamprey: circuitry, transmitters, membrane properties, and simulation. *Annual Review of Neuroscience*, 14 (1991): 169–99.

Chapter 5

Walking and balance

If, on a beautiful summer afternoon, it pleases you to take a leisurely stroll through the Luxembourg Gardens, reading the poems of Verlaine, it is unlikely you will be making many strategic cognitive decisions. Still, watch out for the child who suddenly comes looking for his ball between your dreamer's legs and trips you! But if you play for a French football, rugby, or basketball team, if you are a mountain hiker, a police officer chasing a thief, a roofer on the top of a five-storey building, it's a completely different matter. You can choose to go quickly or slowly, to walk or run, to jump from one foot to the other, or with both feet together. When Ronaldo, the Brasilian striker for the 2002 World Cup, scored the two victory goals, it wasn't because of his muscles but his ability to make incredibly quick decisions about changing position and direction based on what his opponent was doing. This flexibility of the motor repertoire requires very rapid decision making that sometimes spells life or death.

Old people falling: a cognitive decision-making problem

Keeping your balance while walking and coordinating posture and movements is only possible thanks to complex mechanisms that select postural responses from a repertoire that is both innate and learned. Far from being able to reduce them to a series of reflexes, analysis of these mechanisms reveals their great *flexibility* depending on context. The brain can make changes of strategy equivalent to genuine decisions because it has to choose among several equivalent possibilities. Suppose, for example, that you are in a bus at a bus stop. The conductor, a little churlish that day, takes off rather quickly and you lose your balance. Your brain can then choose at least three responses: correcting the backwards rocking of your body by contracting the flexor muscles in your calf, which will steady you; taking a step forward; or grabbing onto a support with your hands.

How is the decision made to bring into play completely different groups of muscles (what we call 'synergies')? How is the pertinent sensory information selected [1]? Here again, speed is essential. Everything happens in 100 ms, so there is no time to reflect. And yet, to make the right decision, the brain must take many factors into account: the force and speed of the perturbation, the possibility of catching yourself, the presence of other passengers, what's in the bags you are carrying, your sense of dignity, the memory of past falls.

Likewise, when a person is subjected to a perturbation of increasing intensity, he can chose to maintain his balance without moving by proportionally increasing the contraction of his muscles. This ability to step up the effort of resistance requires predicting the force needed, but also a decision to employ this strategy.

The need to reorganize motor strategies appears when, for example, we want to reach an object that is just beyond our grasp. This 'reaching beyond reach' [2] requires very rapid reorganization of balance and movement, and is often the cause of falls in old people and children.

In elderly people, falls are a major cause of injury and even death, and are often attributed to muscular weakness or problems related to bones. In fact, falls are also—and perhaps even especially—related to an inability to rapidly make decisions that trigger appropriate corrective movements. Falls often occur in circumstances where integration signals provided by vision, muscle receptors, the environment, and so on, are contradictory, such that the brain is disoriented. But here the question is one of spatial orientation, a highly cognitive process. So falls are also a failure of decision making.

A hierarchy and heterarchy of nested levels

Getting a sense for the neural basis of this functional flexibility requires a bit of physiology. Let me very briefly describe several nested levels of hierarchical control that allow the brain to make its strategic decisions.

In bipedal walking [3] (Fig. 5.1), forward progress is made by cyclic movements comprising stance, a phase in which propulsion is handled by the extensors, and a swing phase (40 per cent of the cycle) with activation of the flexors. This cycle is produced by a spinal locomotor pattern generator. Work with animals has taught us the neuronal organization of this generator. It is made up of a complex network of neurons that control walking, running, jumping, and so on. Suppose that during the rhythmic activity of walking, an obstacle appears. I have to decide which strategy to adopt. Should I ignore the obstacle and continue walking? Should I stop? Relieve the weight on my hurt foot? Jump, in case another obstacle appears alongside the first one? Should I change rhythm and start running to get away from a dangerous area? The nervous system really must decide very quickly. It can employ several strategies to do that.

I would like to propose that the complex problem of control was solved by a hierarchy of levels of control that each exerts on the *internal models of the level that precedes it*.

A first level is located in the spinal cord, which contains the internal models that simulate the dynamics and maybe even the geometry of the limbs. Implementation of these internal models is handled by mechanisms that simplify control. When we reach for an object with our hand, the angles of the arm, the forearm, and the trunk are linked in a very precise relationship. It therefore suffices to regulate a single variable to control the movement. This law of phase relation also applies to locomotion. The three angles of the calf, the leg, and the pelvis are also very tightly associated.

A second level in the cerebellum, associated with the brain stem, contains the most global [4] internal models and controls the operation of models in the spinal cord thanks

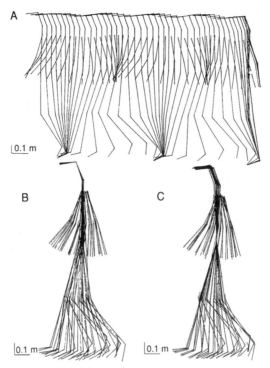

Fig. 5.1 Man walking unimpeded. **(A)** Markers were placed on various points of the body, and a videocamera connected to a computer enabled reconstruction of the movements of the markers. **(B)** The computer plotted these movements relative to a reference point on the ear. Note the stabilization of the head in rotation, which suggests that coordination of walking starts from the head, which forms a stabilized platform for the visual and vestibular sensors. **(C)** The computer superimposed the plots with reference to a point on the subject's trunk. After Berthoz and Pozzo (1994)

to very high level variables called 'composite' variables [5]. For example, if you want to control the *position* of your hand, it turns out to be simpler to use a mix of position, speed, and acceleration. This combination seems more complex, but it makes it possible to treat nonlinear problems in a linear way. The use of such composite variables keeps the brain from having to worry about the details of the force exerted by each muscle and to have global control of the movement of a limb, which, obviously, simplifies decision making for implementation of this or that group of muscles. Remember, too, that control of movement is not continuous. It is driven by oscillating mechanisms that give it an essentially discrete character [6]. This intermittent character of motor control suggests the possibility of a level of microdecision making, but that is all that we know about it.

A third and fourth level, which correspond to the basal ganglia and the cortex, control the internal models located in the cerebellum and the spinal cord. In this way, a global internal model—a model of the acting body—begins to emerge.

First internal model in the spinal cord?

The generator system for walking produces a motor command that it sends to a network of neurons that copy the properties of the limbs. These networks contain basic generators of so-called primitive motor synergies, in other words, everything that is needed to make a movement towards a goal and to regulate its accomplishment. Incidentally, Russian researchers in the 1960s showed that cutting a frog's spinal cord at the level of the bulb (interrupting the commands coming from the brain) did not prevent its leg from reaching a point in space even when the initial position of the limb was changed. So the spinal cord is much more than a simple relay or centre of coordination. It contains networks of neurons that enable internal simulation of movement. In other words, if a signal for movement is sent to this network, on exiting, the signal will be modified as if it had been transmitted through a complex unit composed of the muscles of the arms, the skeleton, and so on. This network constitutes an internal model of the biomechanical and dynamic properties of the limbs. At the output of the network, signals predict, internally, how the state of the system should be if the prediction matches a fixed objective. If the sensory (afferent) messages detected by the muscle, joint, or skin receptors signal that the body is in a state different from the one predicted or signal the presence of a dangerous or painful stimulus, then comparison of the output of the internal model and the sensory inputs produces an *error*.

During walking, it may also be the comparison of actual behaviour with expected behaviour based on data stored in the course of learning by the network that causes changes in strategy or alterations adapted to different portions of the cycle.

So an 'optimal' system like this one needs *three* components to work: first, a control system that produces the motor program; second, an afferent message that gives the actual state of the system; and finally, a circuit that enables internal simulation—an internal model [7].

It would be tedious and beyond the scope of this book to summarize the many mechanisms for selecting action at the level of the spinal cord. Still, I will cite one example of a mechanism of selective suppression of sensory information even before it reaches the spinal cord. A sort of anticipation by suppressing information: a real biological censor!

A gateway for filtering sensory data

You are walking barefoot on the beach, admiring the thrust of the waves, psyche soothed by the thousand delights offered you by the wind, the light, and perhaps a pleasant companion! Suddenly, a sharp pain in your foot alerts you that a small piece of metal hidden in the sand has injured you. Very quickly, your brain interrupts your step or raises your hurt foot; it reorganizes movement to respond to this unexpected event.

A laboratory investigation of the mechanisms of reorganization in response to such a perturbation can be made by placing a subject on a treadmill to study his muscular motor activity and movement during a slightly painful electrical stimulation. The first thing you observe is that painful stimulation induces a response that involves a group of muscles that constitute a 'synergy'. In other words, the response is organized from the outset as muscles acting together; it is not a simple reflex affecting one or two muscles.

Moreover, the response is flexible; it depends on the phase of the locomotor cycle during which the pain is felt. That is how the nervous system more or less opens the door to reception of pain signals. These mechanisms have been analysed [8]. For example, a flexor muscle—the biceps femoris to be precise—is only activated when pain is felt during a small part of the cycle. For the extensors, the sequence of time is very different. It can be explained by a circuit in which the central locomotor pattern generator, while activating the muscles to respond to a pain stimulus, also allows intermittent access to muscles of different synergies. This decision to change synergy as a function of the point in the cycle is made at the segmentation level, in the spinal cord. It persists even when communication between the spinal cord and the higher centres is interrupted. There is a local, decentralized decision that chooses the best motor strategy to respond to a stimulus.

A remarkable mechanism, 'presynaptic inhibition', contributes to this selection (see Fig. 5.2). It blocks sensory information even before it arrives at the spinal cord. We call it 'gateway' by analogy with the devices that deny passengers access to the platform without a ticket. The brain has not processed this information. The decision is taken to close the door, not to let it enter.

Consider an example. The soleus muscle is responsible for extending the foot. It is activated during the propulsive phase of walking. When it is stretched, the neuromuscular bundles that measure its length and speed of stretch are activated and in turn activate the soleus muscle's own motoneurons via a reflex pathway in the spinal cord. This autofacilitation can be useful because it reinforces the generator activity of walking. But it isn't useful at every phase of walking. Sometimes, the activity of the soleus muscle has to be suppressed. To do that, the excitability of the reflex circuit is modified during walking such that there is facilitation only during the part of the cycle where information is useful. This is where presynaptic inhibition comes in. Its influence is exerted just prior to the synapse between the sensory messages of the bundles and the motoneurons, *preventing the sensory signals from getting to the spinal cord*. This modulation, and possibly this inhibition, which varies according to the phase of the cycle of walking, plays a key role in the selection of pertinent sensory information. Furthermore, anticipation in the spinal cord is handled by very subtle circuits under cortical control [9].

The mechanism that in the spinal cord produces the locomotor cycle is thus much more than a simple rhythm generator. It controls changes of motor synergies. It is a foreman who can organize the activity of his teams. It is a local decision maker.

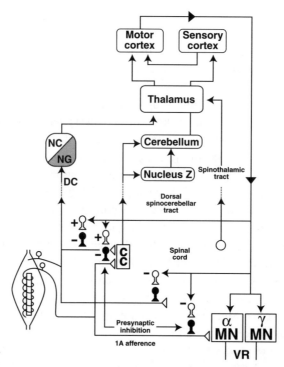

Fig. 5.2 Selection of sensory information at its source; mechanism of presynaptic inhibition. Information about the length and speed of muscle stretch is measured by the neuromuscular bundles. This information is transmitted to the spinal cord and influences the motoneurons, which control the muscles. Moreover, this information travels up through the spinal cord by way of ascending pathways (spinocerebellar, spinothalamic, and so on). Depending on the intentions of the animal and the context of action, a command descends from the cerebral cortex and activates the inhibitory interneurons in the spinal cord that suppress the transmission of sensory information before it reaches the motoneurons or the other circuits in the spinal cord. This is known as 'presynaptic' inhibition, that is, it acts before the synapse that links the sensory nerve with the motoneuron. This scheme shows that presynaptic inhibition can exert itself in at least two places: at the level of the projection of the afferent sensory fibres on the motoneurons, but also at the level of transmission of sensory information towards the cerebellum. The brain can also select information that regulates activity in the reflex circuits, as well as activity that reaches higher levels of control (the cerebellum, for example). NC, nucleus cuneatus; NG, nucleus gracilis; DC, dorsal column; CC, Clarke's column; MN, motoneuron; VR, ventral roots. Courtesy of P. Rudomin (2002)

Second internal model: the cerebellum

The spinal cord cannot handle all necessary decisions and adaptations alone. The neurons that reconfigure the spinal networks must themselves be directed. Indeed, the problem is not only one of choosing an appropriate strategy and stepping up response according to the intensity of the disturbance. To guide walking and maintain balance, it is also important to regulate the timing of the motor reactions, that is, to coordinate the temporal sequence of gestures that accompany movement very precisely and to associate anticipatory synergies to the postural response.

'Companyyyyyyyyy, halt!' What young man over the years has not heard this order hundreds of times? It is true that, today, because military service is not obligatory in France and elsewhere, only career soldiers know the crack of the 'C' that prepares the stop, then the long wait at the sound of 'nyyyyyyyyy', which goes on forever, then, this preparatory command finally ended, the 'halt!' that comes like a knife chop announcing the obligation to stop at once. The aim of this long 'nyyyyyyyyy' is to enable the soldier both to stop immediately and, by some miracle of brain organization, to land in exactly the right phase of his step so he neither moves nor falls. But that's not all. Actually, the 'nyyyyyyyyy' shouted by a familiar warrant officer always has the same duration. A soldier can thereby anticipate the order and act as if he had made the decision to stop himself. In the course of training, the soldier would have progressively internalized the order from without. The cry to halt thus becomes a simple confirmation of the fact that now really is the time. This switch of the order from external to internal initiative reveals one of the great secrets of how the brain works: it prefers to be its own decision maker.

How do you stop walking if not by creating a motor synergy of deceleration after the signal 'STOP' and activating a braking program? Here again, a special, very complex synergy has to be activated. In doing the experiment, one observes that the first muscle to activate is the soleus muscle (recall that it is the muscle in closest contact with the ground) and the extensors of the hips (which take part in other synergies). Also to be decided is the best moment to brake! But the system is capable of picking the right time, which depends on the pace. Research has shown that the same synergy is activated but that the relative contribution of the opposing muscles changes according to speed. So using a different pattern of the same muscles solves a complex problem. The cerebellum appears to be crucial in this reorganization of sequencing within a synergy. For that, the cerebellum must be informed not just of the state of the muscles but also of the state of the spinal cord generators and of its internal models. The cerebellum plays an essential role in the *temporal coordination of movements*, as was recently demonstrated in a study of brain imaging [10].

The cerebellum is indeed informed of what is going on in the spinal cord. An ascending pathway called the 'ventral spinocerebellar tract' tells the cerebellum about the internal state of the spinal cord. Moreover, the 'spino-olivo-cerebellar' pathway reports motor errors to the cerebellum. This is not the place to go into the physiology of the cerebellum. Remember only that an internal model in the spinal cord enables the cerebellum not to have to control an exterior object itself (e.g. a leg) but only the walk generator and errors that arise between the generator's commands and the movements actually made. Even at this level, control of movement and mechanisms of thought have a certain equivalence by virtue of their not operating directly on the exterior world but on internal models [11].

A third and fourth level: control by the basal ganglia and the cortex

The first level of decision making at the level of the spinal cord offers a choice of several ways of walking or regulating posture to be able to react quickly in the event of an obstacle. A second level involves the cerebellum and enables alternating synergies and especially adapting the rhythm and coordination of postural and locomotor movements to the demands of the environment and the task. These levels are still not capable of triggering such changes *voluntarily*. It takes a third level to achieve this intentional flexibility. But here again, the brain predicts and detects errors by constructing and using internal models from the lower levels.

Let us take an example. A subject is walking on a treadmill. He is asked to pull on a handle when he feels like it. This action could potentially destabilize him. How does the brain decide which strategy to adopt? Research shows that it spontaneously chooses to pull on the handle in the phase called 'double support' when the walker is sustained by both feet, that is, when he is most stable. That happens at two very specific moments in the locomotor cycle of walking but with very different synergies. In contrast to simple adaptation of the motor programs to the sorts of perturbations we analysed above, the system must be aware of the voluntary motor system at the same time that it is adjusting to the generator activity of walking. Here, a loop connects the neocerebellum to the cortex by the red nucleus via the thalamus, which acts on the motor area of the cortex (area 4) but also on the premotor area (area 6), which sends information about anticipated movement to the cerebellum as well as to areas 6, 7, and 8, which, via the olive, inform the cerebellum of the potential error. It is the same three elements again, but *the object controlled is the cerebellum*.

The principle that each level is aware only of the lower level and only receives errors already processed by the internal models (which compare a desired state with an actual one) is also found at the higher level. In a way, the *centres do not need to be in contact with the real world*. Note here the elements of a modern formulation of the idea that the brain contains one or more independent models of the physical body. Yet the word 'model' is too graphic, too representational; it may be better to use the concept of 'schema', that is, a mechanism for simulating action. Under these conditions, the perception of a phantom limb or the projection of a 'double' onto the world no longer seems strange.

Cognitive control of locomotor trajectories

I recently proposed that, actually, locomotor trajectory formation and guidance is a process which obeys laws similar to those of arm trajectory formation. Mathematician Henri Poincaré said: 'To localize an object simply means to represent to oneself the movements that would be necessary to reach it' [12]. You can reach a nearby point with your hand, but to get to a distant point, you have to walk. The decisions we make when we execute a locomotor trajectory are highly cognitive. The validity of this analogy between the gesture of the arm and that of walking, illustrated by dancing, is supported by the fact that the same rules seem to apply, for instance, to walking and to moving the arms: a dissociation between the control of distance and direction [13], the validity of the law that relates the curve of the trajectory and its geometry (the two-thirds power law) [14], the laws of phase coordination that link the angles of the upper limbs during a movement of the arms and the lower limbs during walking [15].

In other words, the relationship between walking and balance requires hierarchical global and local mechanisms of decision making that in my opinion were the prototypes for the neural basis of even the most cognitive decisions. Indeed, sometimes we say that a person is 'on the way to a decision', that someone is 'running into disaster', that we have 'halted our deliberations', that judgement 'has been suspended'.

Postural decision making and Parkinson's disease

The contribution of cognitive factors to decision making in the control of balance and walking is also revealed by the changes observed in major neurodegenerative diseases such as Parkinson's. For example, patients with Parkinson's are known to have difficulty initiating walking. For the patient to commence walking, they often need the help of visual cues. They present a 'postural inflexibility'. If such a patient is placed on a platform that moves sharply backwards or forwards, even if he is warned, he cannot *step up* his effort according to the intensity of the perturbation. Similarly, the patient has difficulty *alternating motor strategies* if he is asked simply to go along with the perturbation or to resist it.

In Chapter 3, we examined the abnormalities caused by the reduction in dopamine in Parkinson's disease, in particular in the basal ganglia. These structures are known to contain networks of neurons that constitute a repertoire of stereotyped motor behaviours. This pathology suggests that the basal ganglia play an important role in decision making. They allow automation of motor routines, of preselecting and controlling their execution and consequently conserving attentional resources. Indeed, if the cerebral cortex was continuously involved in executing movements, we would never be able to turn our attention to new things or even to think. The performance of daily motions, such as the sequence of gestures required to make coffee in the morning, thus appears to be entrusted to circuits in the basal ganglia [16].

But this role is cognitive as well as motor. The disease does of course manifest through 'akinesia', that is, the inability to move, accompanied by rigidity, difficulty in initiating walking, and shaking. But it is also accompanied by cognitive disturbances, such as mental slowness, depression, and anxiety, disturbances of verbal fluency and memory, and visuospatial abnormalities. Current knowledge concerning these cognitive deficits is often contradictory, yet it clearly indicates involvement of the basal ganglia. As for memory, the disturbances are similar to those observed in lesions of the frontal cortex, for example, in tasks involving the ability to spontaneously generate efficient coding strategies for memorizing a problem or a situation that requires making decisions and using them later to solve a new problem. The deficit is particularly marked in tasks of memory retrieval and especially in their chronological order, which is obviously essential since in an unexpected situation, we have to search our memory to recall a good strategy for solving a motor problem.

These deficits do not prove that these functions are handled by the basal ganglia themselves; rather, they underscore the link between these central nuclei and the frontal cortex via specialized loops whose modular organization has recently been discovered. In short, these disturbances involve analysis, short-term storage and sequential processing of information; development of a plan based on a strategy generated from within the subject himself and not imposed by the environment; the capacity to change plans depending on environmental signals; inhibition of irrelevant responses linked to interfering or distracting stimuli; and continuation of the program until its complete execution [17].

I conclude from all of this that difficulty initiating movements or walking in a patient with Parkinson's is of a piece with the deficits attributable to decision-making disorders that have a cognitive component. In other words, we must seek in these fundamental mechanisms of behavioural decision making the neural basis of the brain's most complex decisions. I do not mean that we should limit ourselves to these mechanisms, but it is impossible to ignore them. They probably constitute a fundamental architecture on which the most complex forms of decision making were worked out over the course of evolution. Moreover, I suggest (without evidence) that each time we make a decision, we live it, we *emulate* its consequences by simulating the actions in the circuits we have just described. All decision making evokes action; all evoked action entails movement.

The postural gamble in immunocompromised patients

Another source of abnormalities in the management of postural strategies is human immune deficiency virus (HIV)–type 1 [18]. Neuroimaging studies have shown that this disease particularly affects the basal ganglia, but also the white matter of the brain, specifically the globus pallidus and the caudal nucleus. The postural reaction to a perturbation is an interesting means of testing neurological effects: the patient is placed on a platform that inclines sharply and tilts the body backwards. These conditions trigger reflexes of different latencies. In particular, a long-latency response is obtained in the leg muscle (anterior tibialis) that bends the body forward. It varies with the amplitude of the stimulation (scaling). Yet in a delayed-task situation this long-latency response can also be modified by a voluntary decision.

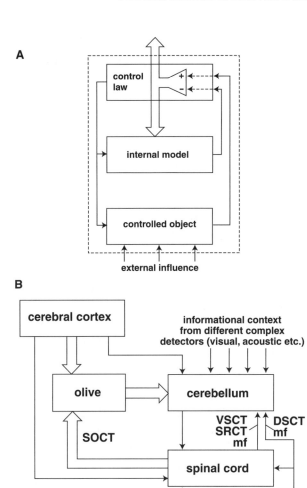

Fig. 5.3 How not to trip over an obstacle: a hierarchical theory of organization of motor decisions. **(A)** The hierarchical theory of K. V. Baev assumes that during locomotion, or any motor activity, the choice of appropriate action is made by stacking modules of different complexities, but whose principle is always the same. For example, this theory postulates such a module in the spinal cord, at the level of the cerebellum, another at the level of the basal ganglia, and so on. Each module is composed of three basic elements 'implemented' differently depending on the level: a regulated object, an internal model, a law of control specific to the level involved. The novelty of this theory consists in assuming that at each level, the controlled object is the internal model of the next level down. This internal model enables genuine dynamic simulation of the activity of lower levels. Thus the brain does not actually 'see' the environment or objects manipulated, but only an internal simulation of the lower level. Think of the head of a company, who doesn't see the machines in his factories but only the simulations of production that are presented by his managers and that are, as it

Usually, when the perturbation is unpredictable, healthy subjects choose a default response, a genuine bet corresponding, in general, to the maximum stimulus they fear they will experience. Patients positive for HIV with no identifiable neurological disturbances exhibit postural disturbances only when the perturbation is unpredictable. The response they choose corresponds to a medium-size perturbation. Their *bet* is thus at the very least different from that of normal subjects. This disturbance could also be related to attentional difficulties in evaluating the predictability of experimental stimuli. The symptoms frequently mentioned are those of bradyphrenia: memory disturbances, decreased concentration, psychomotor slowness—a cluster known as 'subcortical dementia'. These deficits translate into an inability to predict in choosing an appropriate postural strategy. There, the basal ganglia play a major role. Notably, the caudate nucleus, more than the putamen, is responsible for these cognitive-motor disturbances. Indeed, an atrophied caudate nucleus appears to be associated with dementia and with diminished performance in tests of complex motor skill [19]. Actually, these deficits are not due to dysfunction of the basal ganglia alone, but to a problem of functioning of the loops linking the prefrontal cortex to the caudate nucleus, and perhaps even the substantia nigra.

Hierarchical decision making

This analysis of the mechanisms of motor decision making reveals a fundamental property of decision making in biological systems: the nervous system is organized in such a way that the high-level organs that make decisions—changing the mode of locomotion, stopping, avoiding an obstacle, and so on—do not necessarily operate by making direct use of sensory information, which requires analytical processing of a very complex reality and would take a lot of time (see Fig. 5.3). These centres know only the state of lower levels of execution that contain models of the levels that they in turn control and especially that evaluate discrepancies between what is commanded and what gets carried out. These discrepancies are presented in global form without going into the details of execution. In deciding whether to increase or decrease production, or to distribute products differently, the head of a company has no need to know precisely where his trucks are at any given moment, nor every detail of inventory, and so on. He needs more global information that he can use to make simulations and predictions. Remember this principle of hierarchy of

were, a model of how the machines work. **(B)** Organization of the 'object control/internal model/control law' triad for the level of the cerebellum. It is obvious here that the object controlled by the cerebellum is not only the object actually controlled by the subject but also the activity of the spinal cord 'seen' through the ascending spinocerebellar pathways and so on. The cerebellum contains the internal models, and the olive plays a decisive role in the organization of control. After Baev (1997) information and the other principle of replacing empirical reality by internal models for our final analysis of the biological process of decision making, even at its most complex.

Mathematicians maintain that their brain does not only just make analogies. It also constructs analogies of analogies. In other words, each level processes what Sartre would have called an 'analogon', and what today we call an internal model. In Chapter 6, we will see an extraordinary consequence of this sort of operation.

References

1. R Romo and E Salinas, 'Sensing and deciding in the somatosensory system', *Science*, 9 (1999): 487–93; M Flanders, L Daghestani, and A Berthoz, 'Reaching beyond reach', *Experimental Brain Research*, 126 (1999): 19–30.
2. Flanders, Daghestani, and Berthoz, 'Reaching beyond reach'.
3. A Berthoz and T Pozzo, 'Head and body coordination during locomotion and complex movements', in SP Swinnen, H Heuer, J Massion, and P Casaer, eds, *Interlimb coordination: neural, dynamical and cognitive constraints* (London: Academic Press, 1994), pp. 147–65.
4. DM Wolpert and M Kawato, 'Multiple paired forward and inverse models for motor control', *Neural Network*, 11 (1998): 1317–29.
5. S Hanneton, A Berthoz, J Droulez, and JJ Slotine, 'Does the brain use sliding variables for the control of movements?' *Biological Cybernetics*, 77 (1997): 381–93.
6. R Llinás, 'The noncontinuous nature of movement execution', in HJ Freund and D Humphrey, eds, *Motor control: concepts and issues* (Dalhem Conference, Berlin: Wiley, 1990), pp. 223–44.
7. J McIntyre, M Zago, A Berthoz and F Lacquaniti, 'Does the brain model Newton's Laws?' *Nature Neuroscience*, 4 (2000): 693–4.
8. P Crenna, 'Pathophysiology of lengthening contractions in human spasticity: a study of the hamstring muscles during locomotion', *Pathophysiology*, 5 (1999): 283–97; P Crenna and C Frigo, 'Evidence for phase-dependent nociceptive reflexes during locomotion in man', *Experimental Neurology*, 85 (1984): 336–45.
9. Y Prut and EE Fetz, 'Primate spinal interneurons show pre-movement instructed delay activity', *Nature*, 401 (1999): 590–4.
10. R Kawashima, J Okuda, A Umetsu, M Sugiura, K Inoue, K Suzuki, M Tabuchi, T Tsukiura, SL Narayan, T Nagasaka, I Yanagawa, T Fujii, S Takahashi, H Fukuda, and A Yamadori, 'Human cerebellum plays an important role in memory-timed finger movement: an fMRI study', *Journal of Neurophysiology*, 83 (2000): 1079–987.
11. A Pellionisz and R Llinás, 'Tensorial approach to the geometry of brain function: cerebellar coordination via a metric tensor', *Neuroscience*, 5 (1980): 1125–36; M Ito, 'Movement and thought: identical control mechanisms by the cerebellum', *Trends in Neurosciences*, 16 (1993): 447–50.
12. H Poincaré, *The value of science*, trans. GB Halsted (New York: Science Press, 1907), p. 73.
13. Y Takei, R Grasso, and A Berthoz, 'Quantitative analysis of human walking trajectory on a circular path in darkness', *Brain Research Bulletin*, 40 (1996): 491–5; S Glasauer, MA Amorim, I Viaud-Delmon, and A Berthoz, 'Differential effects of labyrinthine dysfunction on distance and direction during blindfolded walking on a triangular path', *Experimental Brain Research*, 145 (2002): 489–97.
14. P Ivanenko, R Grasso, V Macellar, and F Lacquaniti, 'Two-thirds power law in human locomotion: role of ground contact forces', *Neuroreport*, 13 (2002): 1–4.

15 R Grasso, P Prévost, YP Ivanenko, and A Berthoz, 'Eye-head coordination for the steering of locomotion in humans and anticipatory synergy', *Neuroscience*, 253 (1998): 115–18; R Grasso, S Glasauer, Y Takei, and A Berthoz, 'The predictive brain: anticipatory control of head direction for the steering of locomotion', *Neuroreport*, 7 (1996): 1170–4.

16 RD Rafal, AW Inhoff, JH Friedman, and E Bernstein, 'Programming and execution of sequential movements in Parkinson's disease', *Journal of Neurology, Neurosurgery, and Psychiatry*, 50 (1987): 1267–73.

17 S Ammar, S Dubal, and B Dubois, *Humeur et Parkinson* (Paris: PIL, 1998), p. 51.

18 DJ Beckley, BR Bloem, EM Martin, VP Panser, and MP Remler, 'Postural reflexes in patients with HIV-1 infection', *Electroencephalography and Clinical Neurophysiology*, 109 (1998): 402–8.

19 K Kieburtz, L Ketonen, C Cox, H Grossman, R Holloway, H Booth, C Hickey, A Feigin, and ED Caine, 'Cognitive performance and regional brain volume in human immunodeficiency virus type 1 infection', *Archives of Neurology*, 53 (1996): 155–8.

Chapter 6

Deliberating with one's body: me and my second self

> How could we act on things if our mind had no conception of our flesh-and-blood body?
> (*Jean Lhermitte [1]*)

Perception is decision making. Perception of the body, too, follows this rule. Here I would like to propose a new idea: we are basically two beings. Neurologists had this intuition almost a hundred years ago, when they proposed the idea of the body schema [2]. Another example of this intuition is the concept of a mirror image. An infant only recognizes himself in the mirror after a certain age. The image of the double in the mirror becomes possible with the construction of the body schema. The fact that we cohabit with another self is dramatically expressed by the kind of 'out-of-body' illusions exploited by quacks as parapsychological phenomena, but whose neural basis we are beginning to discern (see n. 12). Finally, studying and modelling the neural mechanisms of movement control and of the trajectories of motion led to the idea that our movements are possible only by virtue of internal models of the mechanical and dynamic properties of the limbs and the laws of Newtonian physics [3]. These models enable the brain to internally emulate a phantom body endowed with all the dynamic properties of the physical body to anticipate the consequences of motor commands even before they are produced. Moreover, the need for an anticipatory mechanism was suggested more than 30 years ago in the models proposed by Larry Young [4] at MIT, which assumed that the brain contained a Kalman filter, that is, an optimal estimator, a concept found today in the most sophisticated models.

The double, autoscopy, heautoscopy: to be me or not to be me!

Aristotle was among the first to be interested in the problem of bodily illusions and the 'indecisions' they induce. A nice example of indecision that he describes is the so-called pea illusion.

Take a green pea. Squeeze it between your first two fingers by crossing them (as in Fig. 6.1). You will no longer know which is your index finger and which your middle finger, and you will have the impression that there are two peas. My interpretation of this illusion is that the brain cannot decide which of the two fingers is touching and which is

Fig. 6.1 Illustration by Descartes (1664) of the 'Aristotle's peas' illusion. When you grasp a pea between two fingers, the brain perceives two peas and cannot decide that there is only one. This is perhaps due to the ambiguity between 'touching' and 'being touched' that so fascinated Husserl and Merleau-Ponty.

touched: so it concludes that there are two peas. This is not an illusion but a decision; or rather, the illusion is a solution to a conflict or ambiguity. Indeed, as Merleau-Ponty elegantly stated, the finger has a double status: 'My body, it was said, is recognized by its power to give me "*double* sensations": when I touch my right hand with my left, my right hand, as an object, has the strange property of being able to feel too . . . [But] the two hands are never simultaneously in the relationship of touched and touching to each other. When I press my two hands together, it is not a matter of two sensations felt together as one perceives two objects placed side by side, but of an ambiguous set-up in which both hands can alternate the rôles of "touching" and being "touched" [italics mine]' [5].

Husserl pointed out that if our body is an object just like any other, it has a particular status. Beginning with childhood, our body is the first object of our decisions. The first and most basic of these decisions consists in distinguishing our body from the body of others and of recognizing it in a mirror. It is the problem of the mirror image (see Fig. 6.2). René Zazzo has suggested the following stages of how that decision is made: the baby does not react when placed in front of a mirror; at 3 months, he looks at the reflection of others, but not his own; at 8 months, he turns from the image towards the person, and begins to be interested in his own reflection; 2 months later, at about 1-year-old, the infant is mostly interested in his own image. Yet this stage cannot properly be called one of 'dual personality'. Zazzo writes: 'The 1-year-old infant is introduced, it seems, in the primitive simplicity of this situation where suggestion closes on itself, where the two personages are like an echo of each other' [6]. The infant proceeds to exhibit reactions of confusion, then of explicit recognition (2 years and 1 month). 'The day when the infant blushes before the mirror and turns his head, one may assume that he has lost his initial innocence. Confusion translates an inadaptation. Continuity has been broken. *Now the work of reduplication begins*. Has he been prepared for this carefree game where, nonetheless, the intimate feeling of the movement must shyly find both its translation and negation in the visual image that the infant has of it? . . . Perplexity in front of the mirror seems to us characteristic of a stage of separation, of transition.

Fig. 6.2 The image in the mirror.

Private confusion over the image is lost; recognition is not yet explicit. Finally, towards 2 years 3 months, he recognizes himself and calls himself by name'.

Hallucinations and the double

If illusions reveal certain mental mechanisms, hallucinations take us deeply into mental functioning because they are independent of the external world. They are the brain's creation of a reality that is left to its own devices.

I will borrow from Jean Lhermitte [7] a few nice descriptions of a special hallucination called autoscopy. He underscores the sudden and uncontrollable character of the appearance of a double of oneself. Suddenly, and without warning, a subject apparently wide awake and fully in control of his senses sees his own image appear before his astonished eyes. Actually, it seems to him that the *double* of his physical personality is there before him. And the feeling of reality that foists itself on his mind is so intense that the hallucinator thinks he can touch this illusory image and talk to it. In some subjects, the illusion is so strong that they reach out their hand to the personage that resembles them like a brother, or they try to touch it on the shoulder. In short, the phenomenon manifests in a way analogous to the vision our image offers to us when it is reflected in a mirror. Thus these hallucinations, which seem very strange at first sight, were described as 'mirror hallucinations' or 'autoscopic phenomena' by Charles Féré and Paul Sollier in 1898. In a footnote, Lhermitte remarks that 'the term "autoscopic" can only mean something seen by oneself, whereas "heautoscopic" stresses that the vision is of oneself by oneself, the "seeing oneself" of German authors'. Later, he adds: 'Heautoscopy basically appears as the image of oneself *projected* [italics mine to show the relation to the projective theory of the brain that I have proposed in *The Brain's Sense of Movement*] outside of oneself and giving an illusion of being alive' [8].

Autoscopy was described in both medieval and Renaissance texts. It is found in diverse forms and a first case is that of hallucination of the internal organs of one's own body (internal autoscopy, described by Justinis Kerner in 1824). This rare symptom appears in patients with hysteria and epilepsy.

The second and most frequent case is that of visual hallucinations of one's own body, a 'double', perceived several metres away. In general, the double appears normally sized and moves in ways that often mimic those of the observer (external autoscopy). The clothes, age, and size of the double are not necessarily identical to those of the subject, but the subject is absolutely convinced that it is his double.

A third type (visuovestibular autoscopy) consists—as in the preceding case—of visual hallucinations, but visual hallucinations associated with vestibular sensations connected to the 'I' or 'me' and producing a dissociation of the somatosensory image of the body. This type of autoscopy conveys an impression that I am observing myself from above, at a distance of about a metre or two. A distinction is made between partial autoscopy, in which the parts of the body (especially the limbs) are seen or perceived in positions different from their actual position, and illusional autoscopy, in which the patient meets another person and sees in that person his own double.

Closer to somatic sensations is the illusion of the 'somatic double': the subject perceives half or all of his body reduplicated without seeing it. Reduplication of the body half contralateral to the lesion is usually produced in patients with right-sided lesions. Reduplication of the whole body with a distance between the 'I' and the double, who is not seen, is described as 'somatoparaphrenia'. Temporal lobe epilepsy, exogenous psychoses, and narcolepsy are disorders most often associated with this illusion, which is also seen in mountain climbers hiking at high altitudes.

A final manifestation worth mentioning is 'negative heautoscopy', a bizarre syndrome similar to deficits in face recognition but where the subject no longer recognizes himself in the mirror. Lhermitte writes: 'I had the occasion to follow a subject whose pathological adventure seemed all the more riveting to me because the patient analysed it perfectly. One day, while he was "doing" his beard in front of a mirrored cabinet, his face, he told us, suddenly became deformed. "It was no longer my face. It was no longer me in the mirror; so I threw away my razor and I ran out" ' [9]. Lhermitte continues: 'Patients with schizophrenia are often susceptible to this particular form, and this observation is so common that it is now included as one of the signs of dementia precox: the mirror sign' [10].

For Lhermitte, cenesthesia (also called 'somatognosis'), that is, the awareness we have of our body—our material envelope—assumes a fundamental importance. The feeling of our bodilyness can expand, overflow the limits of our own body, extend to objects that we use, for example, a walking stick, or the scalpel that translates the surgeon's sensitivity into the flesh of the patient. In the same way, our awareness annexes the image of the body in the mirror. There are thus two elements of the illusion of the double: a visual hallucination and the feeling of our physical personality, 'the one conferring its warmth on the other' [11]. The vision of the double is the releasing of the physical image whose construction and maintenance require the cooperation of several different sorts of elements: visual, cenesthesic, and postural.

The double and the image in the mirror

'The specular image seems to be the thershold of the visible world,' wrote Jacques Lacan in 'The Mirror Stage as Formative of the *I* Funtion,' 'if we take into account the mirrored disposition of the *imago of one's body* in hallucinations and dreams, whether it involves one's individual features, or even one's infirmities or object projections; or if we take note of the role of the mirror apparatus in the appearance of *doubles*, in which psychical realities manifest themselves' [12].

An infant's ability to recognize itself in the mirror and the slow development of this ability are related to the appearance of the double.

To see oneself in the mirror is to have the possibility to *project* oneself onto the world. I think that the sudden character of the appearance of the double—its reality—has something to do with the properties that enable us to make perceptual decisions, to reconstruct Kanizsa contours (see Fig. 7.5) that exist not in the physical world but in our head, for example, to perceive a lion in an ordinary rock. The projective power of our brain is such that it does not only put together data from the world but constructs the perceived world according to its *plans*, key to its hallucinations. Moreover, we need to understand whether the double is the projection onto the world of this famous body schema so much talked about for 100 years. In any event, if we do indeed live with a double that we only let out under certain circumstances for fear of losing the ability to contain it, we are going to have to totally reconsider our physiology and models of the brain.

The neural basis of autoscopy

It appears that lesions of the temporal lobe or of the parieto-occipital junction, more often right than left, are associated with these phenomena. Autoscopy of dissociation between 'I' and 'me' (the third type) could be a vestibular syndrome. The fact that the size of the double varies could also be considered as evidence of vestibular origin and

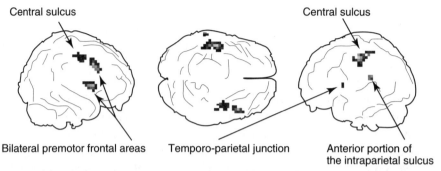

Fig. 6.3 Spatial orientation: a perceptual decision. The vestibular sensors can be activated by a continuous electrical current. Recording the resulting activity with functional magnetic resonance imaging (fMRI) reveals an 'inner vestibular circle': a network of brain regions involved in the processes of spatial orientation. The brain decides how the body is oriented on the basis of integration in the cortical areas of information about head movements coming from the vestibular, visual, and proprioceptive systems. Another neuronal system discovered recently indicates the direction of the head in space (see Chapter 11). After Lobel et al. (1998)

consequently of relations between the vestibular system and the perception of space. For example, epileptic patients presenting temporal focal crises in the region of the cortex (recently called 'vestibular cortex') have the illusion that they are seeing their body from outside it. In fact, they are still lying in their bed, but perceive themselves to be on the ceiling, looking at themselves in bed. We now know the precise location of the area of the brain involved in these illusions. It is situated at the temporo-parietal junction [13] and is part of a circuit that connects vision, proprioception, and the vestibular system and converges on the part of the brain that develops the perception of the physical body and its relation to space (parietal cortex, hippocampus, and so on [14]) (Fig. 6.3) and undoubtedly also involves a perisylvian region [15].

The double is a fleeting impression that may strike people under a variety of circumstances: hysteria, drug induced (marijuana, opium, heroin, mescaline, or LSD), high fever, the delirium tremens of alcoholism, a variety of other intoxicants, hypoxia,[1] epilepsy and especially temporal lobe epilepsy, tumours of the parieto-insular cortex, migraine, and bipolar depression. Hallucination has also been described as resulting from situations of extreme stress. The poet and philosopher Johann Wolfgang von Goethe himself once experienced an episode of autoscopy at Strasbourg. After an extremely emotional parting from a very dear friend, he saw a double of himself on horseback.

The hypothesis of the virtual body

The hallucination of the double is so frequent that I think it must be related to a basic function of the brain. The hypothesis I propose is the following: at the beginning of

---

[1] Mountaineers say that above 6000 metres, it is not uncommon to perceive a companion who does not exist and who accompanies them up the mountain.

this book we saw that the Mauthner cell is able to detect a pattern of stimuli that signifies 'danger' to an animal and takes context into account in triggering an appropriate flight reaction. This neuron is much more than a simple relay in a reflex. It already contains all the ingredients of the self or, as Damasio would say, the 'proto-self'. The Mauthner cell can be thought of as a model of the body of the animal and, in a way—since it is the site where simulation takes place—of the situation of danger, there where appropriate action is decided. One could even say the fish has a double! In this case what we have is a primitive, but effective internal model.

Over the course of evolution a jump in complexity occurred that led primates and especially humans to develop mechanisms enabling them to mentally simulate all the cognitive and motor functions without having to participate in the world. We don't just have a homunculus in the brain; we have a whole other self in there.

The most striking evidence of this biological reality is provided by dreaming. We know that when we dream, all contact with the external world is interrupted. Yet we experience our dreams as perfectly adapted to the reality of the world. Thus the other day I woke up in my bed during a dream in which a person had entered my room and was going to murder me, but I pushed him away with my hand. On waking, I found my hand raised and that my arm was suspended in precisely the same gesture as in my dream. For a few minutes, I was aware both of being in bed in the peacefulness of the morning and feeling my other self trying to escape the murderer. How remarkable that our sensation of ourselves during waking and dreaming differs so little.

So my hypothesis is that we have two bodies: a material one—the one that is 'large as life'—and one that is simulated or, rather, emulated. This is a *virtual body* that has all the properties of the physical body. These two bodies are absolutely identical because they interact constantly during waking.

This line of thinking makes me wonder whether all the theories about spatial neglect—a patient with a lesion of the right parietal cortex consciously perceives only the right half of the visual field; he neglects the left half of the environment, or objects or even his own body—are not wrong. The ease with which remission of symptoms is attained suggests that the brain simply decides to suppress the portion of the field that is out of synch with the internal model of the body and the actual, sentient body. The brain applies the maxim: 'When in doubt, do nothing!'

The object is an extension of my gaze

Merleau-Ponty wrote, 'The thing . . . is an extension of my gaze' [16]. I mentioned Lhermitte's reflections on the extension of the body schema. When we manipulate an object with a tool, for example, when we write with a pencil or a pen, or when we walk on stilts, the tool extends the body. We feel the object not at the end of the tool but at the end of a grouping that consists of the hand and the tool, as if, suddenly, the tool had become part of our body. A person ironing has the same sensation with the iron, the surgeon with his scalpel. A ring on a person's finger has a way of becoming part of the finger itself. We feel the ground under a pair of stilts. Pilots of small planes say that on landing, they feel the wheels as though they were part of their body, and we have the same impression while driving a car: if a tyre smashes into the pavement, we react as though we ourselves had smashed into it. Nonetheless, this very strong impression of extension of the body schema has its limits; for instance, a pilot of a small plane or the driver of a car can feel the wheels of their vehicle as if they were extensions of their body, whereas the pilot of a Boeing 747 told me that the distance between the wheels and the cockpit is too great for him to feel his body 'touch' the runway on landing.

Here, the body does not project onto the world; it extends into it. Perceptual decisions of tactile origin are thus integrated into the body's functioning. If I look for an object hidden under a piece of furniture, I use a stick to do it; the decision that I have indeed found the object is made when I feel the object at the end of the stick as if I were touching it directly. Very little research has been done on this fascinating topic, and we will not linger on it either for want of good data. A patient with a lesion of the right parietal cortex provided an amazing illustration of this extension of the body schema 'at the end of a tool' [17]. Usually, such damage causes a lack of awareness ('neglect') of the portion of space on the side opposite the lesion. According to their symptoms, patients draw only the right half of a scene in front of them, only eat half of what is on their plate, only dress one side of their body, and only read half the numbers on the face of a clock.

This patient presented a special abnormality: if she was given a stick, the lack of awareness extended to the end of the stick; everything to the right of the end of the stick was neglected. In other words, her body was extended into space by the stick, which became part of her physical self.

Sometimes paroxysmal somatognosic disorders are encountered in which the patient feels that a portion of her body is missing or experiences illusions of body displacement that are often of parietal origin. The arm seems twisted, yanked, bent, flung forward. Despite being able to see that the limb has not moved, patients are so thoroughly convinced of the illusion that they beg those around them to hold their arm so it doesn't get ripped off. Usually, the lesions are parietal, although the superior temporal cortex may also be critical in this relationship of the body with space [18].

Moreover, recordings in monkeys of somatosensory neurons [19] showed that when a monkey is given a stick to retrieve fruit, these neurons extend their 'receptive field', that is, the region of space that corresponds to the extremity of the arm, as if it had stretched itself (Fig. 6.4). In the brain, when we simultaneously look at an object and manipulate it with our hand, the information about the object is transmitted via two parallel pathways: the first analyses its visual properties in pathways we will describe in Chapters 7 and 8; the second analyses its tactile properties from skin receptors that project to the postcentral gyrus. These two regions project in turn to a common structure, the posterior intraparietal sulcus. Recording of neurons in the monkey in this

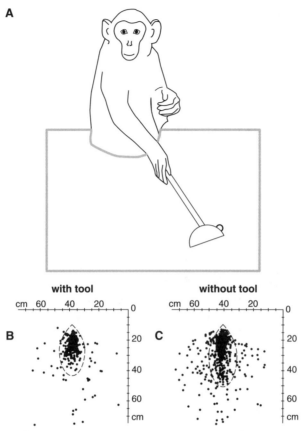

Fig. 6.4 Extension of the body: a perceptual decision. Extension of the representation of the body by manipulating a tool. When we handle a tool, we perceive an extension of our body at the end of the tool. We 'appropriate' the object; it is a basic form of perceptual decision making. **(A)** This image shows a monkey manipulating a small rake. **(B)** Activity in the postcentral area of the monkey's brain is recorded. Without the tool, the neurons are sensitive to stimulation of the hand; their 'receptor field' corresponds to the hand extended in space. **(C)** When the animal uses the tool, the receptor field is observed to extend to the end of the tool. The body schema has been extended. One can also say that the brain has integrated the rake into the body. After Iriki et al. (1996)

part of the cerebral cortex revealed that they are activated both by contact with the hand and by looking at the hand. The reason this result is so interesting is that when the monkey is equipped with a tool that enables it to retrieve objects at a distance, the receptor field, that is, the region of space that activates the neuron, normally limited to the hand, covers all or part of the space occupied by the instrument. The coding of this neuron thus extends to the instrument that is then considered to be part of the body itself. These findings reveal the astonishing plasticity of the body schema.

Phantom limbs

So the brain is capable of creating a double and extending the body schema; it is also capable of replacing a missing part of the body. Indeed, when—accidentally or voluntarily—a limb is amputated, the brain can replace it with a phantom limb [20]. Analysis of the properties of the phantom limb is important for our analysis of the neural basis of decision making. It reinforces our central hypothesis: decision making is a fundamental property of the nervous system based on internal mechanisms for simulating the body and the world that have increased in complexity over the course of evolution. Deliberation is not only a formal and conscious process, it is also an unconscious process of creating scenarios, even apart from all physical reality. The brain can deliberate by emulating action. We need a neurocognitive theory for this capacity [21].

The illusion of amputees who perceive a limb where it doesn't exist was described by Ambroise Paré and René Descartes, who was the first to think that it was the remaining stump that caused the phantom perception. The expression 'phantom limb' was introduced in 1871 [22]. The patient whose limb is amputated still feels it, sometimes in another location. This sensation is not a hallucination; the subject doesn't think it is real: he is aware of the illusory, 'ghostly' character of the phantom limb. What he feels 'is an illusion, not a delusion' [23]. Lord Nelson, the famous British admiral who lost his right arm during a failed attack on Santa Cruz de Tenerife, afterwards considered his phantom arm to be 'direct evidence of the existence of the soul' [24]. Many studies have been devoted to this phenomenon [25]. Around 90 per cent of patients who have lost a limb still feel a phantom limb, and practically all do in cases where pain preceded the operation, or after an extended anticipatory period before the amputation. Phantoms are less frequent among very young children. Only 20 per cent of children under 2, 25 per cent between 2 and 4 years of age, 65 per cent between 4 and 6 years, and 75 per cent between 6 and 8 years experience a phantom, which obviously suggests that a body schema takes shape during the first years of development. That does not exclude the existence of phantom limbs in children born with one or more missing limbs.

Phantoms are surprisingly stable, and to the subject, their reality is without question. Lhermitte emphasized a property—*synkinetic movement*—that later proved important [26]. Based on a memoire by Ludo van Bogaert, who described the phenomenon, Lhermitte described the cases of two patients. The first, whose leg and right arm had been amputated many years previously, reported that it was much easier for him to shake hands with his phantom hand when the movement was executed simultaneously with the healthy hand. Another patient exhibited an *opposite synkinesis*: the phantom hand opened easily if she closed her healthy hand. Indeed, the phantom limb may be 'moved' by the amputee. It is under the control of the will. It is the object of motor decisions, although patients often complain that they cannot prevent it from moving, even if it is congenital. Out of 13 cases, 7 patients were able to move their phantom limb and 4 had telescoped phantoms, that is, phantom limbs whose size was reduced compared with the healthy limb. A 20-year-old woman, born without arms, felt both phantom arms gesticulating when she conversed. They could be activated by corollary motor signals accompanying the desire to move [27].

Phantom limbs reveal themselves in a more striking way during dreaming. They are closer to a hallucination: 'The hallucination of amputees, that is, the prolonged persistence of a phantom limb, as a function of a total image of the body to which the very vivid schema of the mutilated limb attaches, [is that] of an image that is hardly a simple copy of the material our body is made of since this image can undergo very strange distortions, but an image that involves a range of affective complexes, either current or even predating the mutilation' [28]. That is why the vegetative system plays an important role in maintaining the phantom limb.

The time it takes for a phantom limb to appear after an operation is variable. In some cases, it appears immediately after the operation, but this appearance can be delayed several days, even several weeks. In soldiers amputated during the Yom Kippur War in Israel, 33 per cent experienced the phantom immediately after the operation, 32 per cent after 24 hours, and 34 per cent after several weeks.

The duration of the sensation is variable as well. In many cases, the phantom is only perceived for several days or weeks and gradually fades from awareness. Nonetheless, this perception can sometimes persist for several decades. Some patients are able to conjure up the phantom limb through intense concentration or by rubbing their stump.

Phantoms are found all over the body: arms, legs, chest, face, colon, and viscera, and even ulcer pains following a partial gastrectomy. Phantom ejaculations and erections can occur in patients whose penis has been cut off, and Vilayanur Ramachandran has observed menstrual pains in women who have had hysterectomies.

The posture of the phantom limb depends on the circumstances of the accident or the surgery [29]. It can assume a familiar posture, or it can change posture spontaneously. It can also assume an unfamiliar posture experienced as painful. The memory of postures can be conserved: for example, a soldier whose hand exploded while he was holding a grenade retained the memory of a phantom hand holding on to the grenade. The driver of a motorbike whose leg was amputated imagined the phantom leg bent, as in driving a motorbike. Finally, the phantom may provoke voluntary or involuntary movements of the stump, but it can also remain still.

The size of the phantom changes over time. When the phantom fades from consciousness, sometimes it shrinks; for example, the arm gradually becomes smaller until all that remains is the memory of the hand. That may be due to the cortical amplification factor, or the number of neurons per unit of limb surface. The limbs are represented in the neurons of the sensory cortex (areas SI and SII) by a population of neurons that normally receive inputs from the sensory receptors. But the different portions of the body are not represented by equally sized populations. The cortical surface allotted to the penis, for example, is larger than the one devoted to other portions relative to its size. The surface available to the hand is larger than the one available to the forearm.

The gradual disappearance of the stump depends on the following mechanism. Faced with an amputated arm, the brain is subjected to numerous conflicting signals; for example, the frontal regions send motor commands to the phantom that are simultaneously sent to the cerebellum and to the parietal lobe in the form of corollary discharges. In a normal individual, execution of these commands is verified by proprioceptive signals and by visual control of movement. But in an amputee, these signals are lacking, causing a conflict that the brain resolves simply by inhibiting the signals. This inhibition of a mechanism that produces signals inconsistent with the hypotheses formulated by the brain is very widespread. I assume that it also operates during conflicts between vision and the vestibular system following damage [30]. Moreover, since the hand is much more extensively represented than the arm in the somatosensory context, its representation remains active longer.

A little-understood aspect of phantom limbs is the importance of memories associated with the amputated limb just prior to amputation. We saw above the case of the soldier who recalled the grenade he was holding at the instant of explosion; other patients continue to feel a ring or a watch. An unpleasant memory is that of fingernails penetrating the skin, experienced by amputees as being particularly excruciating. It is obvious that to try to understand the nature of internal mechanisms of body representation, the phenomenon of phantom limbs is an exceptional tool.

Several mechanisms may account for the experience of a phantom limb [31], including the 'neuromas' of the stump, that is, the activity of brain centres that correspond to the skin and to the muscles of the stump, and the central reorganization of active mechanisms of the projections of the limbs on the somatosensory cortex. Indeed, in some patients, the awareness of the missing limb is reported in another part of the body. For example, stimulating the skin of a patient's face evoked the sensation of stimulating the finger of her missing limb (Fig. 6.5). The representation of the missing hand was complete on the face of another patient. 'Stimulating' the ears of a woman both of whose breasts had been amputated gave her the feeling that her nipples were being touched. Magnetoencephalographic study of response in the somatosensory cortex of such patients showed that stimulating their face activated regions of the cortex that up to then had been activated by a missing hand. What had occurred was a genuine invasion of the cortical region, normally attributed to the missing hand, by the thalamo-cortical neurons coming from the areas that process the face.

Other mechanisms include the perception of motor activities associated with control of the missing limb (corollary discharge) or even projection of a genetically determined primordial image of the body proper.[2] Finally, somatic memories of unpleasant sensations can be transferred to the phantom. Indeed, emotion felt by the patient at the time of the accident seems to play a very big role.

---

[2] The French is *corps propre*, Merleau-Ponty's term for the living (perceiving) human body. [Trans.]

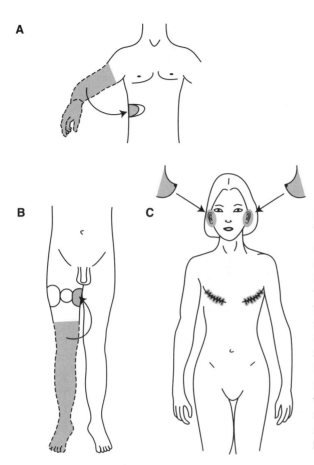

Fig. 6.5 The phantom limb. Three patients underwent amputation. Stimulation of an area of their skin evoked a phantom limb. **(A)** A man whose arm was amputated felt a phantom arm on the skin at the level of his torso. **(B)** A man whose leg was amputated felt a phantom limb at the level of the inside part of his thigh. **(C)** A woman with a bilateral mastectomy felt phantom nipples at the level of her ear lobes. After Aglioti (1999)

Do I have two hands, or three?

How fortunate that we perceive only two arms! The brain could *decide* some other way and have us perceive additional arms, even if we only have two. Cases of multiple phantom limbs do exist. For example, a patient who had had a ruptured aneurysm of the left pericallosal artery that produced a left frontal dorsomedial lesion and a callosal interruption experienced the presence of a third arm several times a day and, sometimes, a third leg (Fig. 6.6). Seeing her actual legs suppressed the phantom. The third limb took a position adopted previously by the actual limbs. If the hand that was on the table moved to another support, the phantom hand remained on the table. The phantom appeared following displacement of the limb and after about a minute. Sometimes, the patient experienced a separation of her body into two: when she stood up, she had the impression that only the right half commenced walking, whereas the left half remained on the seat. This illusion appeared only during repetitive movements.

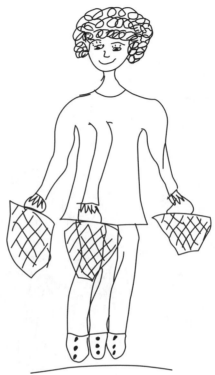

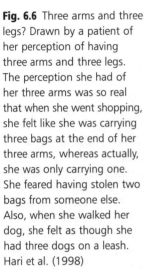

Fig. 6.6 Three arms and three legs? Drawn by a patient of her perception of having three arms and three legs. The perception she had of her three arms was so real that when she went shopping, she felt like she was carrying three bags at the end of her three arms, whereas actually, she was only carrying one. She feared having stolen two bags from someone else. Also, when she walked her dog, she felt as though she had three dogs on a leash. Hari et al. (1998)

The neuromagnetic activity of this patient's scalp was recorded by magnetoencephalography, and the median nerve was stimulated while she provoked her phantom, which enabled recording of the activity in question both in the presence and absence of the phantom [32]. The main finding was that the activity in the left SII (somatosensory) region, for which the maximum was reached 80 ms after right-sided stimulation, was reduced by 50 per cent during evocation of the phantom. The response in area SI was enhanced at 115 ms.

The interpretation of the findings is the following: Area SII is known as a region of multisensory integration and also a region in which damage results in an autotopoagnosia. Alterations in SI could reflect a change in connections between the two hemispheres or also the fact that the neighbouring area 5 was unable to transmit the information that updates the perceived position of the limb when it is moved.

In other words, the appearance of illusory limbs results when some portions of the brain are not informed about what other parts of the brain are doing and do not update their perception, a little like French people in 1945 who had no radio and believed that the Germans were still in the country. Several representations of the limb exist simultaneously in different centres of the brain. Some are, so to speak, 'informed' about changes, others not. This causes an error of 'attribution'. A similar mechanism was proposed by Christopher Frith to explain the impression of being manipulated by someone or the feeling patients with schizophrenia have that they are not the ones moving their own limbs [33].

Because the two hemispheres are no longer communicating and the representations of the limbs—at least one of them—are not properly updated, the patient has access to several images of the body: one part of her body schema that depends on proprioception, and another part that depends on memory of the former position.

Disorders of body recognition: how do you decide whether a limb belongs to you?

We have examined cases where a subject no longer recognizes his own body or, oddly enough, attributes his own sensations to another person outside himself and that he can even perceive. But disturbances of awareness and body recognition are varied and often involve special features that are worth examining.

For example, consider disorders that stem from misrecognition of the body. These are not illusions or delusions, but a false awareness of the body's organization or even which parts are present. A striking example is 'autotopoagnosia', which is the inability to locate parts of the body on verbal command. A patient cannot find his left ear and thinks it is on the table; another believes she has lost her limbs. A patient of the neurologist Josef Gertsmann, who described the syndrome that bears his name, no longer perceived her ears and said they must have been thrown away. She could no longer find her eyes, either, and said she had none. Another patient of Henri Haecan and Juan de Ajuriaguerra presented an abnormality limited to portions of the human body; on the other hand, he recognized the body parts of animals and parts belonging to inanimate objects.

This disturbance may stem from a difficulty in producing and maintaining a schema or spatial representation of the body. It is often observed in connection with other disturbances such as aphasia, apraxia, spatial neglect, and so on. However, patients who cannot designate parts of their own bodies, or those of a doll or an image are able to do it when they are touched by the examiner. These patients give the impression of not being able to deconstruct a whole into its parts, because they also cannot name the parts of inanimate objects. On the other hand, sometimes one encounters patients who cannot name the parts of their own body but are able to reel off those of inanimate objects. So perhaps there are special mechanisms for identifying the parts of one's own body.

It is also possible that the body schema can break down into body parts that, whatever their position, have similar functions. A person's decision to identify one of his limbs is complicated by the fact that the body schema consists in multiple representations [34]. For example, a patient afflicted with Alzheimer's could not point to parts of her body when they were named ('Where is your arm? Where is your nose?') or if they were designated in a nonverbal manner on a doll or a drawing [35]. Yet the patient managed very well to identify objects attached to these parts of her body, for example to her hand. Following this study, four types of representations of the body schema have been proposed:

1. *Semantic or lexical* propositional representations of the parts of the body. Each part is known by its name or by its functional relation to other parts of the body (e.g. the fact that the wrist and the ankle are articulations) or by its particular function, for example the mouth, and so on.

2. *Visuospatial* representations of 'the' body but more generally of 'bodies' that define the position of the parts (the nose sits in the middle of the face), the proximities that exist between the parts (the nose lies to the side of the eyes) and especially the edges of the parts. These representations, which are essential for analysing the relation of the parts to the whole, are linked to the visual nonverbal and somatic memory system.

3. A *body reference system* that supplies information about the relative positions of the parts of the body and the external space. This is a multimodal polymorphic system similar to the concept of body schema defined by Henry Head and Gordon Holmes in 1911.

4. Representations of the *motor programs* themselves, but also, at a more symbolic level, categorical knowledge of the functional and contextual use of the parts of the body.

The variety of cerebral operators that represent the body is suggested by observations regarding a patient who suffered an infarct of the left cerebral artery and an ischemic lesion of the left portion of the corpus callosum. She presented many classic signs of callosal disconnection syndrome, such as lack of awareness of the left side of the body as well as an unusual constellation of symptoms associated with her right hand. Notably, she personified her right hand, speaking to it: 'Grandma, would you please let me go. You are warm, but my hand is sweating and uneasy' [36]. She could name the parts of her body on looking at them but not when blindfolded, nor, still blindfolded, could she point them out with her left hand. Similarly, she could not identify one of her left fingers when someone wiggled it. She could point towards the doctor's body with both hands, but less ably used the left hand to point to objects placed on her own body.

This patient represented a special case of 'tactokinesthetic agnosia' limited to the fingers of the left hand and associated with disorientation of the left hand in naming parts of the body. By analogy with disturbances already described in the literature for the perception of colours, this deficit could be explained by the following mechanism. The tactospatial information of the left hand first must be processed in the right hemisphere. This information must then be transmitted, via the corpus callosum, to the left hemisphere so that the perceived limb can be identified by the language centres. If the corpus callosum is damaged, this transfer cannot take place. Conversely, when the name of the limb was given to the patient, it was processed in her left hemisphere but could not be transferred to the right hemisphere for tactile and proprioceptive identification. The deficit was specific to the hand because of this bottleneck, whereas the information about the more axial parts of the body is partly processed in the left hemisphere. Her difficulty in pointing towards the parts of her body, as indicated by the experimenter, can be explained along the same lines: movement of the right hand, controlled by the right cortex, cannot be guided by the left cortex, which names the parts of the body.

This work suggests the existence of an autonomous and conscious body schema in the left hemisphere that is linked to the semantic aspect of parts of the body. If that really is the case, when the left hand moves and the right hemisphere cannot transmit the left-sided signals, the conscious left hemisphere may find a solution to the paradox that consists in *seeing* the hand move without feeling it: the solution is that the hand belongs to someone else. The illusion of belonging to the grandmother is thus a *solution* to a central conflict.

Aschematia

The idea that we have a body schema inside our brain is supported by many observations. Related deficits fall under the term 'aschematia' proposed by the neurologist Pierre Bonnier and defined in the following way: '*The aschematia* is exactly *an anaesthesia confined to the topographic notion*, to the spatial representation, to the distribution, to the form, to the situation, to the attitude' [37]. One of Bonnier's patients said, 'I keep feeling everything, but nothing is anywhere, and I, too, am nowhere' [38]. What is interesting here is the emphasis on the spatial character of the body schema. It is not only an internal simulation of muscles or parts of the body, but a veritable *body* situated in space.

Patients reported both disturbances of vestibular origin and illusions. One lost the concept of his own existence and could only recover it by shaking his head violently; another had the impression of no longer existing 'as a body'; another, affected with Ménière's disease, felt he was losing his identity; yet another suffered from a labyrinthine vertigo accompanied by agoraphobia. Some patients feel they are becoming enormous. One patient described his head as huge and his entire being reduced to his face. Bonnier lamented the lack of research into these phenomena, which is still the case. He emphasized that these illusions cannot be understood simply by attributing them to labyrinthine dysfunction: what is missing, he said, is intraorganic localization, that is, association to a somatic personality extended and distributed in space, the spatial representation of things whose sensation persists—it is 'the something without some of its parts, the adjective but not the noun, the world without space' [39]. These descriptions fit nicely with the role I have proposed for the vestibular system as a crucial element in calibrating the perception of space and relations between the body and space. Our recent findings on simultaneous motor and perceptual deficits in patients with craniofacial asymmetry also support the idea of a fundamental contribution of the vestibular system to the body schema [40].

Awareness of personal space

The existence of a double in the brain is borne out by many other examples of pathology. For example, 'a patient was utterly incapable of drawing on paper the least geometrical shape nor building with the aid of bits of wood the most elementary structure, although his perception of shapes was intact. He read accurately and could name all the geometrical forms. What he lacked was an image of his own body such that he could not represent positions to himself, the complicated gestures that are required by any motor activity aimed at accomplishing an action whether routine or new, or not yet integrated as part of reflexive activity. We have here yet another example of a disturbance of spatial thought, but . . . it was not the conception of external space that was compromised or weakened, but the awareness of personal space' [41].

This disturbance, termed 'geometric apractognosia', was linked to the so-called closing-in phenomenon described by Wilhelm Mayer-Gross. If you are asked to place wooden blocks in a certain pattern shown on a drawing that serves as a model, you place them on the table beside the drawing. Some patients insist on placing the blocks *on* the model or, if they are given a blank piece of paper to draw the shape on, likewise drawing *on* the model given them. Lhermitte interprets this reliance on the model as an inability to construct an abstract model of a shape. Indeed, for him, our copying activity owes less to the *external* model that our vision supplies us with than the abstract, *internal* model constructed by our imagination and nourished with the perceptual elements we accumulate through sight.

And if—to paraphrase Poincaré—to perceive a shape is to imagine the movements we must make to produce it, we would have to conclude that the deficit is not in the internal model of the shape itself, but is rather caused by a failure of our imagination.

Lhermitte and Mouzon are one with Bonnier when they write that 'the disturbances observed are frequently not gnostic disturbances, or disturbances of spatial representation as such, but especially praxic disturbances of these data, not in a concrete space but in a schematically created and abstract space, the paper one has to write on or the map one is using to orient oneself'. They emphasize the importance of the vestibular system in creating a unity out of perception-action: in pathological cases, the unity of this activity is dissociated, and the activity instead must work with details that are somewhat isolated and added on piecemeal, as if the movement itself were broken and the saccades jerky. It seems to me that this problem of fragmentation of the action of representation must be posed in relation to disturbances of the vestibular apparatus, which are really organizational disturbances.

I think that these deficits result, at least in part, from the patient's difficulty in getting his 'actual body' to communicate with his 'virtual body'. Getting around that requires abandoning an egocentric perception; one has to 'leave one's body'. I once saw a fakir enter a very small (trunk-sized) box and remain there for 20 minutes—amazing! He prefaced his performance by saying: 'I have to leave my body to get it into the box.' It is not necessary to actually leave one's body. One need only mentally manipulate the double that is in us, whether we call it 'internal model', 'body schema', what have you. We have to emulate this virtual body in our brain, as when we imagine ourselves moving to go sit on the other side of the room.

Dear reader, you who have followed me this far, try this very simple exercise. Try to imagine sitting *across from yourself*, and then try to imagine how the room in which you are reading this book would appear to you. You do not have to do what lobsters do in leaving their shell during molting season. All you have to do is to use your virtual body—your second self—to walk you around the room and to watch yourself do it.

The virtual body, the double and decision making

Why is all this so important in a book about decision making? First, because I see in our ability to construct a virtual body, a double, of ourselves the basis for our capacity to deliberate, that is, to create virtual scenarios that involve first us, then, perhaps, others. Second, because this mechanism is probably at the root of our ability to *change our point of view*, to look at the world and especially ourselves from a variety of perspectives. What is primarily a tool for anticipating movement by simulating it internally becomes a tool for emulating diverse strategies, as I explained in *The Brain's Sense of Movement*. As we will see in the chapter on navigation, it is also an immensely useful tool for resolving ambiguities and conflicts. And finally, this mechanism is a cognitive tool for rational deliberation. We recently began to study the neural underpinnings of changes in point of view.

There would be no high-level form of decision making if we could not 'leave our body' and start a dialogue with ourselves. Or rather, in Darwinian terms, perhaps evolution profited from the appearance of internal models to devise this ability to communicate with our double.

Although the goal of this book is not restricted to 'embodying' reason or to 'enacting' thought, to use Francisco Varela's terms, it does support those notions. Varela put it this way: 'We hold with Merleau-Ponty that Western scientific culture requires that we see our bodies both as physical structures and as lived, experiential structures—in short, as both "outer" and "inner", biological and phenomenological. These two sides of embodiment are obviously not opposed. Instead, we continuously circulate back and forth between them. Merleau-Ponty recognized that we cannot understand this circulation without a detailed investigation of its fundamental axis, namely, the embodiment of knowledge, cognition, and experience. For Merleau-Ponty, as for us, *embodiment* has this double sense: it encompasses both the body as a lived, experiential structure and the body as the context or milieu of cognitive mechanisms' [42].

The aim is not to 'rediscover the body', to make it an important, but secondary annex to reason. The aim is really to reverse the proposition and start with the acting body. 'In the beginning was the Act', says Faust. Western theatre is the theatre of the word. Jerzy

Grotowski turned this point of view on its head. Just as Poincaré based his geometry on *movements*, Grotowski's point of departure is the *emotional* body. But be careful! Action and gesture do not reduce to muscles. Indeed, for me, acting is not simply about movements: action and gesture are about plans, intentions, emotions, memories for predicting the future, hopes that extract from past successes and failures a taste for future undertakings. Action and gesture represent both time present and time past, projection of the present moment into the future, deliberations, and decisions.

Reason is always inextricably tied to action. Throughout evolution it became more abstract but remained anchored in space, that is, in the acting body. Yet action proceeds from an exchange between the body and its double; deliberation and decision express this fundamental dialogue. We have two bodies, the physical body and the virtual body. The virtual body consists of all the internal models that comprise the elements of the body schema and allow the brain to simulate and to emulate reality. This body is the one we perceive when we are dreaming. It, too, has a phenomenal reality.

This duality is part of the foundation of consciousness. I think that consciousness appeared in humans at the same time as these two bodies. Consciousness is the activity of the second 'me' watching the first one; it is the product of the dialogue between the 'I' and the 'me'. The body had to be reduplicated for animals to escape from their egocentric experience, from the prison that is the 'acting me'. An *alter ego* had to be created for an internal dialogue between the 'me' and the 'I' to be established, and for consciousnesss to be the focus of my brain on itself. The brain applied to itself the principle of reduplication, not by a process of mitosis—separation into two—but by creating at its core a 'neural double'.

The problem of the double is of course the most difficult among the problems we have to solve. Shame be to him who thinks he has the key! But I am struck by the persistence of doubles since ancient antiquity. The *Epic of Gilgamesh*, one of the most venerable texts of humanity given that it was written 3000 years ago in Sumer, relates how the goddess Arur created a match for Gilgamesh destined to thwart his power. This second self followed Gilgamesh around all his life and, once subdued by him, became his inseparable friend. Together, for instance, they vanquished the monster Khumbaba. The Greek authors Plautus and Sophocles invented the character of Sosia, slave to Amphitryon, and a false Sosia in a game between men and the gods. Taking off from Jean de Rotrou's *Les Sosies* (The Sosias), Molière placed Sosie in a game of four doubles that orbit the loves of the god Jupiter (Fig. 6.7).

Jupiter metamorphoses into a double of Amphitryon to seduce Amphitryon's wife. Mercury, Jupiter's assistant, changes into a double of Sosia to orchestrate the pleasures of his master Jupiter. Completely disoriented on seeing a double of himself, Sosia marvels at his twoness:

> I too was doubtful, and inclined to take
> My doubleness as a sign of mental strain;
> I thought my other self a fraud, a fake;

Fig. 6.7 Amphitryon and his double.

> But at last he made me see that I was twain;
> I saw that he was I, and no mistake;
> From head to foot he's like me—handsome, clever,
> Well-made, with charms no lady could withstand;
>   In short, two drops of milk were never
>   So much alike as we are, and,
> But for a certain heaviness in his hand,
>   I'd see no difference whatever.
>
>   (*Amphitryon,* act 2, scene 1 [43])

Literature abounds in doubles, as is demonstrated in the work of Pierre Jourde and Paolo Tortonese, and of Jean Bessière [44]. The term 'doppelgänger' was the creation of the German poet Jean Paul in 1796. Another example is that of *Peter Schlemihl* [45], who sold his shadow. The idea that we are two fascinated Russian authors, who produced memorable characters: Major Kovaliov in Nicolai Gogol's novel *The Nose*, the odd and disquieting Goliadkine in Fyodor Dostoevsky's *The Double* [46], and Vladimir Nabokov's Hermann Hermann. The idea turns up in Western literature in the personage of Guy de Maupassant's *Horla*, in the work of Jorge Luis Borges and of Julio Cortázar, in Hoffmann, in the double of the chevalier Oluf in Théophile Gauthier, that of Brydon in Henry James and Joseph Conrad's *The Secret Companion*. Alain Robbe-Grillet's Djinn is one of very few female doubles.

Doubles tend to be rather evil [47], like Gustav Meyrink's infamous Golem and especially Robert Louis Stevenson's *Dr Jekyll and Mr Hyde*, or the double in Italo Calvino's *The Cloven Viscount*, who also has a good and a bad side. They may go as far as killing each other, like William Wilson and his double in the work of Edgar Allan Poe. Doubles are frequently rivals in love, like Yousouf in the story of the caliph Hakem by Gérard de Nerval. The double sometimes assumes the face of a twin in *The Devil's Twins* by Marcel Aymé and Michel Tournier's *The Twins*. Nathaniel Hawthorne's *Man in the Mirror* mimics his double down to the last feature.

Alfred de Musset's life was periodically interrupted by 'appearances of "the infant" then "the young man", the "stranger", the "guest", the "orphan", the "unfortunate" always dressed in black' [48]. He wrote:

> A schoolboy, I my vigils kept
> One night, while my companion slept;
> Into the lonely room forlorn
> There came and sat a little lad,
> Poor, and in sombre garments clad,
> As like me as my brother born [49].

Philosophers, psychiatrists, psychoanalysts, and psychologists have all attempted to construct a theory of the double. Otto Rank in *Don Juan and His Double* [50]; Sigmund Freud in 'The Uncanny' [51]; Jacques Lacan, who analyses the question from the angle of the image in the mirror [52]; René Girard, who posits it as the result of the split created by 'mimetic desire' and who explains its violence as the result of this desire [53]; Carl Gustav Jung in his *Two Essays on Analytical Psychology* [54], and, more recently, Clément Rosset in *Le Réel et son double* [55]. Anthropologists have also discovered the 'tona', a constituent of a person that, in some tribes of South America, is externalized in an animal double.

Obviously we have no neurobiological theory of the doppelgänger [56]. But poets should not worry! Even if one day we do manage to construct one, it will not take away any of the double's mystery or its ability to inspire novels.

References

1 J Lhermitte, *Les Hallucinations* (Paris: Doin, 1951), p. 110.
2 Haecan H, 'La notion de schéma corporel et ses applications en psychiatrie', *L'Évolution psychiatrique*, 13 (1948): 75–122.
3 J McIntyre, M Zago, A Berthoz, and F Lacquaniti, 'Does the brain model Newton's laws?', *Nature Neuroscience*, 4 (2001): 693–4.

4 V Henn, B Cohen, and LR Young, *Visual-vestibular interaction in motion perception and the generation of nystagmus* (Cambridge: MIT Press, 1980).
5 M Merleau-Ponty, *Phenomenology of perception*, trans. C Smith (London: Routledge and Kegan Paul, 1962), p. 93.
6 R Zazzo, 'Image du corps et conscience de soi', *Enfance*, 1 (1948): 39. See also 'Image spéculaire, conscience de soi', *Psychologie expérimentale*, special issue in honor of Paul Fraisse (Paris: PUF, 1977), pp. 325–35.
7 Lhermitte, *Les Hallucinations*, p. 124.
8 Ibid., p. 129.
9 Ibid., p. 163.
10 Ibid.
11 Ibid., 200.
12 J Lacan, *Ecrits: a selection*, trans. B Fink, H Fink, and R Grigg (New York: Norton), p. 5.
13 E Lobel, JF Kleine, DL Bihan, A Leroy-Willig, and A Berthoz, 'Functional MRI of galvanic vestibular stimulation', *Journal of Neurophysiology*, 80 (1998): 2699–709.
14 O Blanke, S Ortigue, T Landis, and M Seeck, 'Stimulating illusory own body perception', *Nature*, 419 (2002): 269–70; P Kahane, D Hoffmann, L Minotti, and A Berthoz, 'Reappraisal of the human vestibular cortex by intracerebral electrical stimulation study', *Annals of Neurology*, 54 (2003): 615–24.
15 Kahane, Hoffmann, Minotti, and Berthoz., 'Reappraisal of the human vestibular cortex'.
16 M Merleau-Ponty, *Résumés de cours au Collège de France* (Paris: Gallimard, 1968).
17 A Berti and F Frassinetti, 'When far becomes near: remapping of space by tool use', *Journal of Cognitive Science*, 12 (2000): 415–20.
18 HO Karnath, 'New insights into the functions of the superior temporal cortex', *Nature Neuroscience*, 2 (2001): 568–76.
19 A Iriki, M Tanaka, and Y Iwamura, 'Coding of modified body schema during tool use by macaque post-central neurons', *Neuroreport*, 7 (1996): 2325–30.
20 VS Ramachandran and SW Blakeslee, *Phantoms in the brain* (New York: Morrow, 1998).
21 VS Ramachandran, 'Phantom limbs, neglect syndromes, repressed memories, and Freudian psychology', *International Review of Neurobiology*, 37 (1994): 291–333. See in particular the discussion on pp. 369–72.
22 SW Mitchell, 'Phantom limbs', *Lippincott's magazine*, 8 (1871): 563–69.
23 VS Ramachandran and W Hirstein, 'The perception of phantom limbs', *Brain*, 121 (1998): 1603–30.
24 G Riddoch, 'Phantom limbs and body shape', *Brain*, 64 (1941): 197–222.
25 R Melzach, 'Phantom limbs and the concept of neuromatrix', *Trends in Nurosciences*, 13 (1990): 88–92, and 'Phantom limbs', *Scientific American*, 266 (1992): 120–6.
26 For example, in J Lhermitte and Z Susic, 'Pathologie de l'image de soi. Les hallucinations des amputés', *La Presse médicale*, 23 April 1938, pp. 627–31.
27 Ramachandran, 'Phantom limbs'.
28 Lhermitte, *Les Hallucinations*, p. 33.
29 G Berlucchi and S Aglioti, 'The body in the brain: neural bases of corporeal awareness', *Trends in Neurosciences*, 20 (1997): 560–4.
30 A Berthoz, 'Coopération et substitution entre le système saccadique et les réflexes d'origine vestibulaires: faut-il réviser la notion de réflexe?' *Revue Neurologique*, 145 (1989): 513–26.

31 VS Ramachandran, 'What neurological syndromes can tell us about human nature: some lessons from phantom limbs, Capgras syndrome, anosognosia', *Cold Spring Harbor Symposium on Quantitative Biology*, **61** (1996): 115–34, in particular p. 122.

32 R Hari, R Hänninen, T Måkinen, V Jousmäki, N Forss, M Sepa, and O Salonen, 'Three hands: fragmentation of human bodily awareness', *Neuroscience Letters*, **240** (1998): 131–4.

33 C Frith and E Johnstone, *Schizophrenia: a very short introduction* (Oxford: Oxford University Press, 2003), p. 130.

34 S Gallagher, GE Butterworth, A Lew, and J Cole, 'Hand-mouth coordination, congenital absence of limb and evidence for innate body schemas', *Brain and Cognition*, **38** (1998): 53–65.

35 A Sirigu, J Grafman, K Bressler, and T Sunderland, 'Multiple representations contribute to body knowledge processing: evidence from a case of autotopagnosia', *Brain*, **114** (1991): 629–42.

36 T Nagumo and A Yamadori, 'Callosal disconnection syndrome and knowledge of the body: a case of left hand isolation from the body schema with names', *Journal of Neurology, Neurosurgery, and Psychiatry*, **59** (1995): 548–51.

37 G Vallar and C Papagno, 'Pierre Bonnier's (1905) cases of bodily "aschématie" ', in *Classic cases of neuropsychology*, vol. 2, ed. C Code, C-W Wallesch, Y Joanette, and A Roch Lecours (Hove, East Sussex: Psychology Press, 2003), p. 149.

38 Ibid., p. 151.

39 Ibid., p. 153.

40 D Rousié, J-C Hache, P Pellerin, J-P Deroubaix, and A Berthoz, 'Oculomotor, postural, and perceptual asymmetries associated with a common cause: craniofacial asymmetries and asymmetries in vestibular organ anatomy', *Annals of the New York Academy of Sciences*, **8** (1999): 1–4.

41 J Lhermitte and J Mouzon, 'Sur l'apractognosie géométrique et l'apraxie consécutive aux lésions du lobe occipital', *Revue Neurologique*, 9–10 (1941): 415–31, p. 429.

42 FJ Varela, E Thompson, and E Rosch, *The embodied mind: cognitive science and human experience* (Cambridge: MIT Press, 1993).

43 Molière, *Amphitryon*, trans. R Wilbur (New York: Harcourt Brace, 1995), pp. 54–5.

44 P Jourde and P Tortonese, *Visages du double: un thème littéraire* (Paris: Nathan, 1996); J Bessière, *Le Double* (Paris: Champion, 1995).

45 LCA de Chamisso de Boncourt, *The wonderful history of Peter Schlemihl*, trans. W Howitt (London: Rodale Press, 1954).

46 F Dostoevsky, *Notes from the underground. The double*, trans. J Coulson (Harmondsworth: Penguin, 1972).

47 P Brugger, R Agosti, M Regard, H-G Wieser, and T Landis, 'Heautoscopy, epilepsy, and suicide', *Journal of Neurology, Neurosurgery, and Psychiatry*, **57** (1994): 838–9.

48 Jourde and Tortonese, *Visages du double*, p. 221.

49 A de Musset, 'The poet', in *The complete writings of Alfred de Musset*, vol. 2 of 10, trans. G. Santayana, E. Shaw Forman, and MA Clarke (New York: Edwin C. Hill, 1905), pp. 336–44.

50 O Rank, *The Don Juan legend*, trans. DG Winter (Princeton, NJ: Princeton University Press, 1975).

51 S Freud, 'The uncanny', *The standard edition of the complete psychological works of Sigmund Freud*, trans. J Strachey, with A Freud, A Strachey, and A Tyson, vol. 17 (1917–1919), *An infantile neurosis and other* works (London: Hogarth Press and the Institute for Psycho-Analysis, 1955; reprint, 1986), pp. 217–56.

52 J Lacan, *Écrits: a selection*, trans. A Sheridan (London: Tavistock, 1977).

53 **F Dostoyevsky**, *Resurrection from the underground*, trans. JG Williams (New York: Crossroad, 1997).
54 **CG Jung**, *Two essays on analytical psychology*, trans. RFC Hull (New York: Pantheon, 1953).
55 **C Rosset**, *Le Réel et son double* (Paris: Gallimard, 1984).
56 **P Brugger, M Regard, and T Landis**, 'Unilaterally felt "presences": the neuropsychiatry of one's invisible Doppelgänger', *Neuropsychiatry, Neuropsychology, and Behavioral Neurology*, **9** (1996): 114–22.

Part 3

Perception, preference, and decision making

Chapter 7

To perceive visually is to decide: the physiology of doubt

One might even say, strictly speaking, that almost all our knowledge is tentative at best, and of the few things that we think we know for sure, in the mathematical sciences themselves, the predominant means of approaching the truth, induction and analogy, are based on probabilities.
(Laplace [1]*)*

Perceiving is deciding, and deciding is gambling, to such an extent that we perceive the world not necessarily as an obvious and unique sense datum, but rather a probable one. To say, as Georg Riemann did, that several lines parallel to a given line can all pass through a single point is to make a decision about lived experience. It affirms that perception can give rise to a variety of possible solutions and that reason can arbitrate among them.

In the waiting rooms of the Banque de France, I had occasion to contemplate a painting by Fragonard, dated roughly 1715, that depicts a fair with feasts, couples, children playing, at left a stage set up in the garden, at right a Guignol theatre. Jean-Claude Trichet, the bank's governor, had brought together his former classmates from the École des Mines at Nancy and given them a real-life demonstration of the fact that perceiving is also deciding. Indeed, he told us that the painting can be perceived in two ways. First, it can be taken as an example of the light and joyous French frivolity so characteristic of the art of Fragonard: the colours in the painting have the feel of caresses and the softness of spring; the women look happy and the men charming.

We all found this pleasant air of festivity delightful. But Trichet also drew our attention to a dark man in the centre of the painting gazing devilishly at us, perhaps foreshadowing the social turmoil of the Revolution, a violent wind—harbinger of a storm—that bends and uproots the trees and creates an unsettling confusion. Taken in this way, the painting portrays a disquieting vision of a world in which the established order is about to be turned upside down. But—and this is a critical observation Trichet shared with us—the minute the second vision of the painting is perceived, it is impossible to shake it off and return to the idyllic scene. In so doing, he was, I think, illustrating two principles. First, we can change our perception depending on our point of view; it is limited by the hypotheses that we make about the world. Second, organizing our hypotheses in a certain way creates such a compelling neural organization that it is then difficult to see things any other way. We become prisoners of our assumptions.

To perceive is to remove ambiguities, to choose one interpretation over another; in other words, to decide. In the following chapters I will attempt to show how, more generally, choosing one action over another depends decisively on developing a percept of reality. In Chapter 4 we saw that the toad's decision to approach or flee is made as soon as the neural system tasked with identifying the object in its visual regions has returned its verdict. Perception, as I define it—namely, simulation of action on the targeted object or environment, based on emulation of the relationships between the body and the world in the brain—is thus an integral part of decision making. Decisions do not follow perception; perceiving is tantamount to engaging in the process of deciding one way or the other.

You could argue that this idea of decision making is very primitive; 'real' decision making is a process that requires deliberation, and these low-level mechanisms that handle selection of behaviours necessary for survival have nothing to do with what goes on inside the head of a financier who is about to risk 10 mn Euros on the stock exchange or in the brain of an alpinist who decides to go climbing despite a bad weather forecast. I stand my ground. All decisions require information gathering about the world, and information gathering is always biased, depending on the operator's hypotheses. Moreover, most decisions have to be made quickly. Throughout evolution, the nervous system devised ways of increasing that speed, in particular, making decisions from the earliest stages of sensory data processing in the chain of neural mechanisms that follow one another on the way to a solution.

This point of view is radically different from one, for example, that assigns all decision making to the prefrontal cortex. As we will see, the prefrontal cortex probably comes in at the end of a sequential process of analysing the elements in the world to arrive at a decision, in the same way that a group of jurors considers all the evidence in a case to arrive at a verdict after a process of deliberation. But actually, the brain works the other way around. It formulates hypotheses, chooses from a repertoire of possible decisions, and then seeks in the world elements that confirm or negate the predictions to which these decisions lead.

Fragmentation of the visual world

Truly grasping the neural basis of perceptual decision making requires remembering some ideas that specialists will consider quite elementary and laypeople perhaps quite complex. There are several recent books on the subject [2]. Let us attempt nonetheless to review a few concepts regarding when and where in the brain decisions are made by the visual system.

When we examine a painting, an object, our surroundings, or a person, we assume that our perception is unified. But this impression of unity is an invention of the brain. Indeed, it begins to analyse the visual world by decomposing images into their properties: brightness, colour, shape, movement, orientation, and so on. At first, the visual world appears shattered.

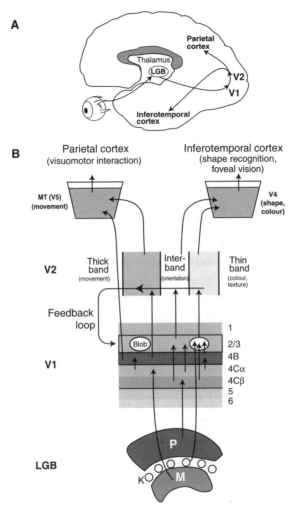

Fig. 7.1 (A) Visual images are transmitted by the lateral geniculate body (LGB) in the thalamus at V1 and V2 of the visual cortex; then the two pathways separate, one going towards the parietal cortex, the other towards the inferotemporal cortex. **(B)** When an object appears in the visual field, its image is transmitted to the first-order relays in the thalamus (LGB). Movement of the object is transmitted by the parvocellular layers (P) and layer K. This information is transmitted to very specialized neuronal layers in the visual cortical area V1, while maintaining its initial segregation. The vague activity caused by the appearance of the image is then transmitted to cortical area V2, and segregated according to three types of activity. But the information contained in the M pathway (which describes the object globally, minus the details) arrives about 20 ms before the other information, because transmission is faster. Owing to two back projections from V2 to V1, integration occurs (at the level of a V1 neuron in layers 2 and 3) between P-type afferent activity and these feedback influences coming from V2. This mechanism enables the brain to improve perception of visual forms through preselection from the earliest relays (see text). J Bullier, *Architecture fonctionnelle du système visuel*. In M Boucart, MA Hennaff, and C Belin (eds), *La vision: aspects perceptifs et cognitifs* (Paris: Solal, 1998), pp.11–41.

Among the central sensory relays of the visual pathways, a nerve centre—the lateral geniculate body in the thalamus—receives all the information incoming from the retina (Fig. 7.1). This first-order relay comprises six layers, of which four dorsal layers are composed of small neurons. The dorsal portion is called 'parvocellular'. It is complemented by a ventral portion, called 'magnocellular', because its neurons are large. Each hemiretina projects to both the parvocellular layers and to the magnocellular layers. The large ganglial retinal cells project to the magnocellular layers and the smaller ones to the parvocellular layers. So beginning with the first-order central relay there is a functional division into two systems. This segregation could be considered as a form of networked decision making. As we will see below, it imposes a provisional judgement on the visual world. But of course, that is an overly simple definition of decision making, open to criticism. One could, for example, suggest that it is an elegant solution found over the course of evolution to simplify neurocomputation. Indeed, at the level of the retina, there are more than a million and a half neurons. The problem of transfer of information towards the first-order relays is very complex, and perhaps the point of segregation is to control the size of the geniculate body by reducing the amount of information. In other words, it is important not to assign the term 'decision making' to all the phenomena of classification that facilitate processing of innumerable sensory data. To better understand why some perceptual decisions are made in the first-order relays of the visual cortex, let us briefly recap a few ideas. For a more in-depth examination, see, for example, Semir Zeki's excellent book [3].

Receptor fields: a window on the world

The neurons of the visual system only 'see' the world in a very restricted area of space called the receptor field. The receptor field is the region of space that activates the neuron when a light appears. For the neurons of the geniculate body, or V1, the first visual area of the cortex, this field is around 1 degree. A very small window! The receptor fields of these neurons are circular and symmetric, but around 90 per cent of them have what is called an 'opposition circumference'. A receptor field can be represented mathematically by the so-called Gabor function, which is nothing other than the product of a Gaussian function and a sinusoid. This function resembles a slightly elongated Mexican hat and behaves like a spatial filter. This particular organization is probably aimed at sensitizing all these neurons to relative discontinuities and variations in the visual field, in particular, contrast. For example, we know that we do not do a very good job of evaluating the absolute value of the light to take a photograph, but we can very easily detect a more luminous point on a less luminous background. The brain readily detects differences.

Centrifugal influences are at work from the earliest stages of visual processing: in humans and animals, the size of the receptor field varies depending on how responsive the brain is. This modulation is probably due to direct action of a thalamic nucleus, the perigeniculate nucleus, which integrates information from the brain stem and the cortex on the brain's state of alertness. The neurons of this nucleus project onto the lateral geniculate body and, accordingly, modify vision at their source [4].

The neural mechanisms that organize the receptor fields are not fixed [5], as was long thought. The receptor fields are not only spatial windows onto the world. Their diameter and their dynamic properties of encoding can change depending on the visual context of the surroundings and perhaps also on attention.

The general shape of these receptor fields is also modified as signals advance through the visual areas. The window through which the neurons 'see' the world expands, allowing them to integrate larger and larger elements from the environment and the visual context. What is amazing is that the visual system begins by breaking the world down into attributes and looking at them in a very fragmented way. Later, it puts everything back together again by connecting up the parts; at least, that is the impression one has when considering only centripetal pathways. We will see later on that centrifugal (top-down) messages can organize the properties of the first-order visual relays and their functioning, depending on what the brain predicts or what it is looking to analyse. It may be this double movement—afferent and efferent, centrifugal and centripetal—that holds the secret of our ability to scope out the world with such speed, precision, and richness by making decisions about it very early on.

Magno and parvo, two of the visual world's analytical pathways

In general, neurons of the magnocellular pathway have receptor fields two to three times larger than those of the neurons of the parvocellular pathway. A difference is also found in the detection of contrast. The neurons of the magnocellular pathway are more sensitive to weak contrast than are those of the parvocellular pathway. Another property that distinguishes these two pathways is detection of colour.[1] Moreover, these neurons do not respond equally to speed: indeed, the cells of the parvocellular layers have little sensitivity to movement, in contrast to those of the 'magno' layer.

Thus, at the level of the cortical primary visual area V1, signals coming from the retina via the lateral geniculate body of the thalamus are separated into two pathways that terminate in the separated layers of the cortex.[2]

---

[1] Around 90 per cent of the parvocellular cells are sensitive to differences in wavelength. The cells of the retina are sensitive to three colours, red, green, and blue, which are combined in the neurons of the parvocellular region. In contrast, the magnocellular system is insensitive to colour. Thus, 'parvo' is equivalent to 'sensitive to colour', with a remarkable ability to differentiate colours, and 'magno' is equivalent to 'colour blind'

[2] The cells of the magnocellular pathway project to layer C4α, which itself projects onto 4B, which in turn projects towards V2 and the structures that concern the perception of movement (MT, for example). In contrast, the neurons of the parvocellular layer project towards 4Cβ, which projects

V1 also contains neurons that are important for perceiving the geometric shape of objects. For example, some are activated by short lines or edges, but are not activated by longer versions of the same lines. They appear to detect a limit of length and are called 'end-stop'. They are always activated by the end of a line, or even by corners or by irregular areas that have a very sharp curve. So they can detect the direction of a line as well as its end. These properties allow the brain to carry out a geometrical analysis of natural forms in the visual field beginning with the primary visual relays.

Taken together, these data reveal a division between a system that is sensitive to movement, but with large receptor fields and neurons that are insensitive to colour; this is the pathway that projects mainly to the parietal cortex and guides action in space. A second system is sensitive to colour but not very sensitive to movement, with receptor fields that are small and respond to shape. This is the pathway that projects to the temporal cortex and plays a decisive role in identifying objects, faces, and so on. However, the magnocellular pathway also projects onto the ventral pathway.

The brain analyses the visual world by segregating information according to categories (colour, contrast, and so on), geometry (shape, length, edges, and so on) and dynamics (movement). The brain deconstructs the world based on a repertoire of predetermined properties and then reconstitutes, links, combines, and synthesizes the pieces depending on its objectives. This mode of processing simplifies neuro-computation and enables very rapid extraction of information that is indispensable for guiding action and making decisions from the complexity of the visual world (Fig. 7.2).

When I assert that decision making is a basic property of the nervous system, I mean that this chopping up of the world into relevant features is a perceptual decision. The brain chops up the world, so to speak, according to categories of properties. The more we know about the physiology of animals carrying out complex cognitive tasks, the more we will find that this process of classification is influenced by hierarchical (top-down) mechanisms that alter perception at its source.

Is visual perception a gamble?

Visual perception is not an absolute process; it is a mechanism for probability testing. For example, we ask a subject to pick out a bright point shown very briefly to him on a screen in the dark, and we measure the relation between the number of points seen and

---

mainly and very strongly towards the upper layers of V1, whose neurons are, once again, differentiated: there are oval regions in V1 called 'blobs' that resemble pillars about 0.2 mm in diameter. The blobs are separated by regions called 'interblobs'. Blob neurons are particularly sensitive to colour. They are activated by the parvocellular pathway. They are excited by one colour, inhibited by another, and sensitive to orientation. There is a dissociation between geometry and colour. Interblob neurons are selective for orientation, but not for colour. They respond to a moving line based on its orientation, but independent of its colour. They thus tend to subserve the perception of high-resolution shapes.

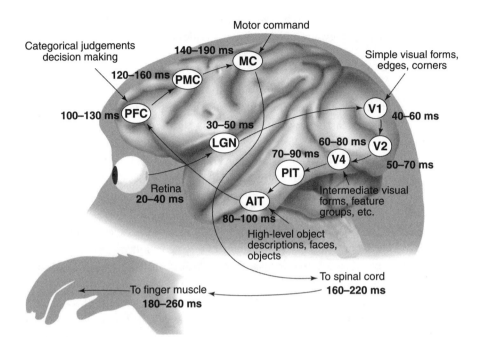

Fig. 7.2 Identifying natural shapes. The brain identifies natural shapes very rapidly. This scheme shows the steps of this recognition in the different areas of the cerebral cortex that are involved. For example, the image of an animal or a face is transmitted to the lateral geniculate nucleus in the thalamus beginning at the retina. Around 40 to 60 ms later, the V1 neurons process the basic features of the image (simple shapes, contours, edges); after progressing to V2, colour and some details of the shape are processed in V4, and movement in V5 (MT) (not shown here). Eighty to 100 ms after the arrival on the retina, the global properties of the shape (face, and so on) are processed in the areas of the temporal lobe. This activity influences the prefrontal cortex around 130 ms, and around 190 ms, the motor cortex, after a relay in the premotor cortex, activates the motor reactions. As can be seen here, identification is fairly rapid, but the motor response and decision making, which involve the premotor cortex, are fairly slow. This rapid reaction has to be contrasted with the even greater speed with which the 'short' pathway, in the rat at least, goes directly through the amygdala to trigger the fear response (see LeDoux's diagram, Fig. 2.3). After Thorpe and Fabre-Thorpe (2000)

the brightness of each point. We find that for a certain threshold of brightness, subjects sometimes see the light, but mostly they do not. Similarly, Ernst Weber invented the concept of 'just noticeable difference' to evaluate the smallest variation between two stimuli. But when one is at the threshold, the perception is quite indistinct.

Genuine perceptual decisions are made in the primary visual centres. Perhaps these decisions employ Bayesian processes, in which variability and noise play a decisive role. They rely on encoding of visual space that involves populations of neurons [6], and today it is possible to model these decisions using highly sophisticated methods of computational neuroscience [7]. But we do not intend to review these methods here. Let us simply consider a few examples without complex calculations.

Shorter or longer? In front or behind?

You are busy reading this book: look around you! How does your brain decide whether the chair is in front of or behind the table? How does it decide whether an object is closer (in front of) or farther away from (behind) another object in space? You can answer because it is very simple. Either one object hides another, or the brain calculates the distance between the two objects and makes a decision. But does the brain really calculate distances? The question is not a trivial one. For James Gibson's 'ecological' school of psychology, the brain does not always need to measure distances in a Euclidean framework. In a game of pétanque,[3] all you have to know is which of two balls is closer to the wooden jack; you do not need to know the exact number of centimetres between the two balls except to be able to distinguish them.

For us, the question is to determine whether the brain constructs this categorization very early in visual processing or whether it requires more elaborate working out. The surprise contributed by modern neurobiology is that even the neurons of V1 have the ability to signal the distance between two objects in space. It is possible to directly compare the discharge of neurons in area V1 based on the subjective perception of distance.

Experiments carried out with animals show that in visual areas V1 and V2, some neurons are active during a visual evaluation of the relative position of two objects. One might then conclude that decisions are made at this level because neurons can tell whether one target is in front of or behind another.

Obtaining experimental proof of this relation between neural activity and perception means developing very complex and subtle paradigms [8]. For example, a monkey is required to fixate a small spot, and a recording is made of the response of the monkey's neurons to a stimulus whose disparity—or distance—is varied[4] (Fig. 7.3). The monkey thus has to make a decision about what it has perceived, for which the objective evidence is the monkey's shift in gaze after presentation of the stimulus. It is possible to record what happens while the monkey makes its decision and to construct a 'psychometric'

---

[3] Pétanque is a French lawn game. You throw a small wooden ball called a 'cochonnet' a few metres away in front of you and the participants in the game have to throw heavy metal balls as close as possible to the cochonnet. Deciding which ball is closer is often a cause of vivid discussions for lack of precise measurement.

[4] The stimulus is surrounded by a ring of points having the same disparity, that is, the animal is shown a constant basic disparity and variably disparate points. Finally, using an optical technique, the monkey is shown two points that appear to be located on both sides of the point of fixation on the screen. If the animal has perceived the background stimuli, it must make a shift in gaze—a saccade—towards one of the points; the saccade will be made towards the other point if it has perceived the stimulus in the foreground.

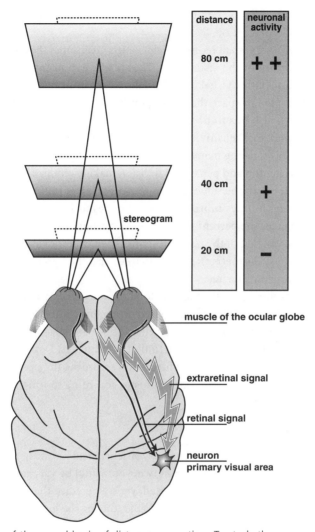

Fig. 7.3 Study of the neural basis of distance perception. To study the neuronal mechanisms of decision making in the perception of distance, a monkey is placed in front of a screen on which a 'stereogramme' is projected, that is, an image that gives the impression of depth. This stereogramme creates two slightly shifted images that are perceived by each eye owing to a special image separator device. This screen can be placed more or less far from the animal. Here, three positions of the screen are shown. Recordings are made of the activity in the primary visual areas caused by the representation of points that are perceived by the monkey at various depths. The neuron is shown to be activated at the furthest distance. It is not active when there is a slight discrepancy between the two stereogramme images. In addition to the visual information, evaluation of the distance involves the angle of convergence of the two ocular globes. Thus the brain makes a decision about depth based on both visual signals and signals linked to the movements that it controls. After Trotter (1992)

response curve that reflects the probability of the response, that is, the monkey's decision linked to its perception.

V1 neurons have a very small receptor field, which means that they are sensitive only to a very small portion of the visual field. That is why exploring the properties of a neuron requires distributing visual stimuli on the screen in such a way that they are situated in the tiny region of space that corresponds to the receptor field. The preferred distance of the cell also has to be measured. Carrying out the test, then, requires determining the maximal sensitivity of the cell very precisely. But defining the preferred distance involves taking many measurements. And figuring out the relation that exists between this neuron and perceptual decision making requires knowing the range of its responses. However, the same neuron sometimes fires 10 times in response to a disparity, sometimes 5 times more than that. The result is a so-called neurometric function that indicates the percentage of responses produced by the neuron to a disparity. The two curves, that of the neural response (neurometric) and that of the monkey's perception (psychometric), are compared, and the two curves are found to correspond to a surprising degree.

The experiments show that it is possible to know the perceptual decision (target in front or behind) of the monkey's brain based on the psychometric curve of a single neuron. That is, a global response can be predicted from a single neuron. The performance of humans in this test is very close to that of monkeys. Data from this kind of experiment, like those we will describe further on regarding the perception of movement, support the idea of 'analytic isomorphism', according to which it is possible to find a neural basis for any percept.

The bisected dog

If, each time your dog passed behind a chair, obscuring all but its head and tail, your brain were to conclude it had been cut in two, you would be very upset. In everyday life, we see objects that are partly concealed, yet we perceive them in their entirety (Fig. 7.4). This cognitive ability develops in babies during the first year. Piaget called it 'object permanence'. I think of it as perceptual decision making.

When a moving object passes behind an obstacle, how do we decide that it is a single object when the V1 neurons each only perceive a single piece of it? What mechanism governs this perceptual connection? The visual system very often performs this grouping of fragments of the world that it perceives. Is this tying together a complex cognitive property of the brain?

Recordings of V1 neurons in the monkey [9] have shown that a neuron can be activated by the perception of a moving rod and that it continues to be activated when a patch is introduced whose visual properties are such that the animal perceives it as being farther away than the rod and thus behind it. In contrast, this neuron stops firing when the patch has disparity such that it is perceived by the monkey to be in front of the rod. This experiment demonstrates that the property of depth perception enables V1 neurons to decide whether two fragments belong to the same object.

Fig. 7.4 The bisected dog. This dog is cut into two by the tree, but you continue to perceive his body as if it were indeed behind the tree. This perception without sensory information is a decision made by the brain, a form of 'filling in' that ensures that the world we see is coherent.

The mechanisms for these decisions are thus present from the first-order visual relays. We will consider another example using the phenomenon called 'filling in'.

Construction of illusory contours

Let us proceed in our analysis along the visual pathways and take a look at what is going on in level V2, Brodmann area 18. As we saw earlier, here organization occurs in layers or stripes defined by their appearance as thin, pale, or thick. Each contains neurons sensitive to different aspects of the visual world—shape, colour, and so on. Other neurons in the thick stripes of V2 encode distance and depth owing to *binocular disparity*, which causes two images of the same object not to project to the same location on each of the two retinas.

Studying these neurons in the monkey involves putting glasses on the animal and placing filters—red and green, respectively—before each eye. Two images, one red, one green (seen separately by the two eyes, which restores depth by shifting the images slightly sideways), are projected onto a screen, inducing a disparity on the two retinas. The monkey looks at the object in the images. Because of the glasses, the monkey has the impression that the object is either in the plane of the screen, in front of it, or behind it. The neuronal activity in the visual areas is recorded. Varying the disparity changes the apparent depth of the object. It is observed that some neurons discharge when the image is nearer ('close' neurons) and others discharge when the visual object or the image is at a greater depth ('far' neurons). I use the term 'proto-decision' for this activity which assigns a relative position to an object. Positioning is already deciding; it introduces a discontinuity—a choice—into the world.

The ability of these neurons to detect depth is not uniquely the result of visual information; it is also due to motor convergence of internal signals about the position of the eye and information about the visual disparity. So there is an additional mechanism that allows the brain to make a perceptual decision regarding the distance of an object. But this time, it is action that is used to characterize the depth.

A famous illusion, Kanisza figures (Fig. 7.5), demonstrates the brain's extraordinary ability to fill in the physical world. Here, the brain completes the actual contours of triangles by contours that are perceived but that do not exist in the physical world—using illusory contours. Is this ability a high-level cognitive property or is it inscribed in the properties of neurons from the primary visual centres?

Indeed, even in the absence of physical contrasts, the brain constructs illusory contours. Cortical area V2 contains neurons that select for orientation of the edges of

Fig. 7.5 Kanizsa's illusory contours. It is very easy to perceive a triangle in the top shape, a curved line in the left lower shape, and a rectangle in the shape on the right. This perception is accompanied by a very particular kind of oscillatory activity in the visual cortex that has been demonstrated using magnetoencephalography, suggesting that this perception is a decision made in the earliest visual areas.

contrast as well as *illusory contours*.[5] This perceptual illusion thus derives from a mechanism of visual processing at a very early stage. The conscious percept of illusory contours is associated with gamma band oscillation in the occipital cortex and seems to involve mainly the right striate cortex [10].

Gestalt in V2

Another aspect of the *detection* of visual shapes is the ability of our brain to *recognize* shapes. If you draw on paper a cloud of arbitrary points whose diameter increases going down the page, you will get a perception of depth like that between a ceiling and a floor. If the points are illuminated in an intermittent way (using 'strobe' light), there is absolutely no percept of depth; on the contrary, making the points move induces an illusion of acceleration and possibly three-dimensional movement.

Take a piece of paper and draw on it five or six points in a row. If you move the sheet of paper on the table, you perceive the points move as if they were on a virtual line that you perceive as actual. This illusory line is created by your brain, probably owing to the coherence of the movement of the points. I say that the brain 'decides' that the points are in a line. They form a 'gestalt' that is perceived as actual [11]. V2 contains neurons that respond to a shape made by points in movement as if it existed. Another interesting property of V2 neurons is that in 42 per cent of them, contours can be modified by vestibular inputs. This vestibular influence on the detection of contours explains how, when we bend our head, we continue to perceive the door or the window in front of us as vertical, whereas its image on the retina has changed. In this case the invariance of the perception is a decision: the decision that the external world is stable, despite rotation of the image.

Decision making by comparing patterns

The description I have just given of visual centres could lead one to think that vision is a one-way process that goes from the retina to the brain. That is not the case, and moreover, we have already described above the example of mechanisms that enable perception of a unified object when a rod passes behind an obstacle. At each level of the visual system—the lateral geniculate body, V1, V2—this processing is subjected to central centrifugal influences, that is, influences that instead of going from the retina towards the brain, go from the brain to the primary visual relays (V1, V2, V3).

An especially interesting hypothesis was recently formulated that is linked to the fact that the speed at which information is transmitted is not the same in all the visual

---

[5] A third of these cells prefer longs bars with oblique orientations like the ones that occur in Kanisza figures. V1 neurons, however, only respond to true contours, not to illusory contours, which V2 neurons do respond to. This means that if an illusory contour is introduced into the V2 neuron receptor field, it will respond to it as if the contour were actually present.

pathways. In the system formed by V1 and V2, information is rapidly transferred by the magnocellular system [12]. It reaches V2 before information of the parvocellular type (see Fig. 7.1).

The key to the speed of the visual system may well be that information coming from the retina does not propagate like a single wave that makes its way up through the hierarchical levels, but rather like three successive waves. A first wave is formed by the neurons in the magnocellular system, which are independent of colour and form but encode movement and have, as we specified above, large receptor fields. A second wave originates in the parvocellular system, which encodes colour and fine detail. Finally, a third wave is formed by K neurons, which were only discovered recently and are still not well understood.

This mechanism enables predefinition of the shape of the perceived object in the V2 neurons, which receive fine details from the parvocellular pathway.[6] It also chooses among several interpretations of the retinal image of the object.

What is interesting about the theory is that decisions are made not by decision-making neurons but via interaction with re-entrant loops and by a pattern-comparison mechanism (since the patterns proposed by the descending wave are compared with those of the ascending wave) that gradually evolve what I consider to be a decision.

No doubt there exist many mechanisms of dynamic preparation such as these. Recall that the receptor fields of the parietal cortex shift before an ocular saccade. This is a huge effect, because it is found in 50 per cent of V3a neurons, 30 per cent of V3 neurons, and 10 per cent of V2 neurons. This anticipation is constructed in the parietal cortex and returned in a centrifugal way towards V2.

In humans, the visual areas of the cerebral cortex are activated just by summoning objects mentally, even when the eyes are closed. Consequently, the brain appears to be able to activate the visual areas by memory and in this way to prepare visual analysis. What we know biases what we see. The structures of the cortex such as the precuneus nucleus, for example, are activated when the brain imagines a visual scene in the absence of the actual object. Paul Fletcher calls this structure 'the mind's eye'. Perceiving is deciding what one wants to see.

Filling in

We saw earlier, with the example of Kanizsa figures, that the brain fills in holes. The visual system detests emptiness! It performs so-called filling-in operations, that is, it has visual sensations even when the retina sends no message. The most well-known

---

[6] A loop mechanism that first generates a global representation prepares the detailed analysis. But this mechanism is also *selective*. When the wave arrives in V1, there are several solutions: either the information coming from the parvo pathway (P) corresponds to what was predicted by the magno pathway (M) and the signals emerge reinforced, or there is an incongruity and the visual information is consequently eliminated. This introduces a genuine choice of important and, especially, coherent information.

example is that of the blind spot, a tiny region located around the middle of the retina. It is impossible to stimulate this spot by light since it contains no neurons, being the origin of the optic nerve fibres. So we should *perceive* a black spot in our visual field. But that is not the case, because our brain 'fills in' this zone. The precise mechanism is unknown, but it does seem to be located in the primary relays in the visual cortex.

The percept linked to the absence of a hole in the visual field can be interpreted in two ways: one is concerned with neural underpinnings, the other considers that the brain does not 'see' the hole because there is nothing to see. The proponents of this last interpretation write that 'one of the most surprising characteristics of consciousness is its discontinuity. Another is its apparent continuity. It would be a serious mistake to try to explain apparent continuity by assuming that the brain "fills in" empty spaces' [13]. Similarly, the philosopher Daniel Dennett has criticized analytic isomorphism, which, as I mentioned earlier, seeks a neural explanation for every perception at all costs [14].

Experiments also show that although no visual activity is produced by the blind spot, subjects perceive the same characteristics in this region of the visual field as in its immediate surroundings. This *filling-in* activity has been studied in detail by psychologists who classify it into two types, modal and amodal. Amodal filling in enables mental reconstruction of parts of objects that are hidden; but the objects completed in this way are not perceived as having visual reality (contours, and so on). An important function of filling in is to complete the visible world when the world gives only partial information. Modal filling in is even more powerful: it gives completed visual objects a visual structure and attributes.

This filling in is not only the result of the blind spot. Experiments carried out with monkeys have shown that very fine lesions of the retina made by a laser also give rise to perceptual filling in very quickly after the injury [15]. That speaks for a remarkable compensatory mechanism. But it is still mysterious.

Human patients whose visual field includes scotomas caused by lesions of the visual pathways frequently report that the patterns that occupy the rest of the visual field fill in the scotoma. Moreover, this filling in can occur independently for colour and texture. You can observe the phenomenon yourself by staring at the sailboat at the centre of Claude Monet's painting *Impression, Rising Sun*. After a few seconds, the sun, perceptible at the periphery of the visual field, seems to be replaced by the colours and luminosity of the surrounding sky. Here, filling in erases an element of the scene. The question of filling in is thus only a part of the more general question: How does our brain construct a unified perception of the world, when what reaches it is so fragmented? One answer is that perception really is a *projection* of a preperception, that perceiving is merely *verifying*, and so perceiving is deciding whether a preperception is correct or not, or even useful. The brain chooses from the world what it deems suitable.

It could be argued that filling in is not strictly speaking a process of decision making. Yet it is a process that creates a percept based on available information. It is already interpretation of the visual world and perhaps the reflection of a more widespread activity of selection, of *inference* and a certain independence of perception with respect to physical reality, like the Kanizsa illusions, or others that have been shown to involve very early mechanisms in visual processing.

Gibson drew attention to the fact that the switch over the course of evolution from lateral and panoramic vision to frontal, fovea-based vision profoundly changed the way we see the world. Animals that have no fovea see the world less precisely but more globally. The fovea brought precision, but at the cost of panoramic vision, which gave way to saccades that offer a library of successive frames that must be stitched together. So here again we chop up the visual world into little scenes that have to be rejoined. This process must involve some decision making since clearly there are different ways of assembling the scenes.

Change blindness: to see or not to see?

Researchers have proposed a scenario along the following lines: Suppose you are given a task of viewing a series of photographs on a computer screen and are allowed a certain amount of time to look at each of them. You are told that now and again details of the photos will change as you are examining them, and that you should give a sign when you notice a change. Carefully conducted experiments have shown a surprising blindness to change. For example, during a filmed scene, many observers failed to notice that two people had switched hats or, even more startlingly, that they have switched heads [16]. The stratagem was to present subjects with two different images separated by a blank interval of about 150 ms. The researchers conclude that the change goes unnoticed because the blank interval prevents the brain from perceiving it.

This blindness is also produced when stimuli are introduced that briefly disrupt attention, for example, splashes of mud on a screen [17] or a flicker, and prevent attention being focused on the region of change. It is well known that filmmakers use such mechanisms in putting scenes together; the cuts are imperceptible to most viewers.

For a long time it was thought that these processes were carried out by high-level cognitive modules, and it is clear that they depend on attention [18]. But in fact, the attentional filtering of information is probably done from the earliest levels of processing based on the intentions and predictions of the perceiving subject. Underscoring the change in perspective brought about by contemporary research, Michael Posner, one of the most eminent specialists of attentional mechanisms, writes: 'The classical notion concerning attention effects in visual cortex has held that the strongest effects are seen at the highest levels of the visual pathway and that in primary visual cortex there is no effect of attention. The report . . . of a functional MRI study indicating primary visual cortex (V1) modulation by instructions to attend illustrates the change that has taken place in recent years' [19].

Indeed, it is now known that the activity of V1 neurons depends on the task, on the amount of attention paid to it, on the presence of competing stimuli in the visual scene, and on the need to integrate the context. In other words, the V1 level—the first visual relay of the cerebral cortex—contains many elements that could be the basis for *perceptual decision making*.

Preselection of neural properties in V1 is probably partly responsible for the phenomenon of change blindness. It is hard to understand in the context of a transformational and representational theory of the brain because from that perspective, the brain does nothing more than passively process sensory data. It is more obvious if, as I suggested in *The Brain's Sense of Movement*, perception is simulated action (an idea to which psychologists have recently warmed [20]), and if, as I am doing here, the theory is extended to include the idea that perception is always decision making—that any perception is already a choice, and that this choice translates into a process of selection at the level of the primary visual relays. Perceiving is choosing. And choosing is deciding.

References

1 **PS de Laplace**, *Théorie analytique des probabilités* (Paris, 1812).
2 **SE Palmer**, *Vision science from photons to phenomenology* (Cambridge: MIT Press, 1999); ET Rolls and G Deco, *Computational neuroscience of vision* (New York: Oxford University Press, 2002).
3 **S Zeki**, *Vision of the brain* (Cambridge: Blackwell, 1993).
4 **F Wörgötter, K Suder, Y Zhao, N Kersher, U Eysel, and K Funke**, 'State-dependent receptive-field restructuring in the visual cortex', *Nature*, **396** (1998): 165–8.
5 **Y Frégnac**, 'Dynamics of functional connectivity in visual cortical networks: an overview', *Journal of Physiology*, **90** (1996): 113–39; Y Frégnac and V Bringuier, 'Spatio-temporal dynamics of synaptic integration in cat visual cortical receptive fields', in A Aertsen and V Braitenberg, eds, *Brain theory: biological basis and computational principles* (Amsterdam: Elsevier, 1996), pp. 143–99; V Bringuier, F Chavane, L Glaeser, and Y Frégnac, 'Horizontal propagation of visual activity in the synaptic integration field of area 17 neurons', *Science*, **283** (1999): 695–9.
6 **A Pouget, P Dayan, and R Zemel**, 'Information processing with population codes', *Nature Reviews Neuroscience*, **1** (2000): 125–32.
7 **P Dayan and LF Abbott**, *Theoretical neuroscience: computational and mathematical modeling of neural systems* (Cambridge: MIT Press, 2001); Y Trotter, 'Bases neuronales de la perception visuelle tridimensionnelle chez le primate', *Journal français d'orthoptique*, **27** (1995): 9–20; Y Trotter and S Celebrini, 'Gaze direction controls response gain in primary visual-cortex neurons', *Nature*, **398** (1999): 239–42.
8 **BG Cumming and AJ Parker**, 'Binocular neurons in V1 of awake monkeys are selective for absolute, not relative, disparity', *Journal of Neuroscience*, **19** (1999): 5602–18.
9 **Y Sugita**, 'Grouping of image fragments in primary visual cortex', *Nature*, **401** (1999): 269–72.
10 **F Brighina, R Ricci, A Piazza, S Scalia, G Giglia, and B Fierro**, 'Illusory contours and specific regions of human extrastriate cortex: evidence from TMS', *European Journal of Neuroscience*, **17** (2003): 2469–74.
11 **J Bullier**, 'Architecture fonctionnelle du système visuel', in M Boucart, M-A Hénaff, and C Belin, *Vision: aspects perceptifs et cognitifs* (Paris: Solal, 1998), pp. 11–41.

12 JM Hupe, AC James, BR Baynes, SG Lomber, P Girard, and J Bullier, 'Cortical feedback improves discrimination between figure and background by V1, V2 and V3 neurons', *Nature*, **394** (1998): 784–7; J Bullier, 'An integrated model of visual processing', *Brain Research: Brain Research Reviews*, **36** (2001): 96–107.
13 I Murakami, H Komatsu, and M Kinoshita, 'Perceptual filling in at the scotoma following a monocular retinal lesion in the monkey', *Visual Neuroscience*, **14** (1997): 89–101.
14 DC Dennett, 'Filling in versus finding out: an ubiquitous confusion in cognitive science', in HL Pick Jr, P van den Broek, and DC Knill, eds, *Cognition: conceptual and methodological issues* (Washington: American Psychological Association, 1992), pp. 33–49.
15 L Pessoa, E Thompson, and A Noë, 'Finding out about filling in', *Behavioural and Brain Sciences*, **21** (1998): 723–48.
16 D Levin and D Simons, 'Failure to detect changes to attend objects in motion pictures', *Psychonomic Bulletin and Review*, **4** (1997): 501–6; D Simons and D Levin, 'Change blindness', *Trends in Cognitive Sciences*, **1** (1997): 261–7.
17 K O'Regan, R Rensink, and J Clark, 'Change blindness as a result of mudsplashes', *Nature*, **398** (1999): 34.
18 R Rensink, K O'Regan, and J Clark, 'To see or not to see: the need for attention to perceive changes in scenes', *Psychological Science*, **8** (1997): 368–73.
19 M I Posner and C Gilbert, 'Attention and primary visual cortex', *Proceedings of the National Academy of Sciences of the USA*, **96** (1999): 2585–7.
20 JK O'Regan and A Noë, 'What it is like to see: a sensorimotor theory of perceptual experience', *Synthèse*, **129** (2001): 79–103.

Chapter 8

Decision making and shape recognition: ambiguity and rivalry

There is indeed one human act which at one stroke cuts through all possible doubts to stand in the full light of the truth: this act is perception, in the wide sense of knowledge of existences.
(*M. Merleau-Ponty [1]*)

An old French expression cautions against 'mistaking bladders for lanterns'. I imagine that in villages, in the past, animal bladders were hung out to dry in front of houses and then used as containers. Their shape and texture must have been similar to lanterns. Today, the expression refers to letting oneself be fooled. Deciding whether an ambiguous visual form is a bladder or a lantern is a fundamental task of perception.

Here, decision making settles a contest between two possible interpretations of the same shape. The ambiguous forms of Gestalt theory are classic examples of decisions made by the brain to interpret the same visual form. It really is a matter of perceptual decision making because the percept can be changed although sometimes only with great difficulty. Once a percept has been acquired, the brain persists in its interpretation, in the same way that some patients with frontal lesions have a hard time changing the rules of a game (see Chapter 3).

Henri Poincaré observed that we must constantly be making decisions when we observe things. For example, a change in the shape of an object's image on the retina can be interpreted in two ways: the shape may have altered, or our perspective may have changed. How does the brain decide between these two solutions [2]? What is the role of attention [3]?

Visual illusions furnish an inexhaustible source of examples demonstrating that the brain makes perceptual decisions. So many books have been devoted to these illusions [4] that I will not add to them. The impossible drawings of Maurits Cornelis Escher also supply us with situations in which the brain can voluntarily alter its percepts. One can either mentally manipulate these views to grasp their absurd character, or let oneself be drawn into the ambiguity of the scene and enjoy the resulting vertigo.

In *The Brain's Sense of Movement* I maintained that illusions are solutions. Here I would like to say that these solutions, which remove ambiguities, are actually decisions—judgements the brain makes whose neural basis is worth trying to understand. For instance, the famous cylinder illusion, illustrated in Fig. 8.1, is probably a

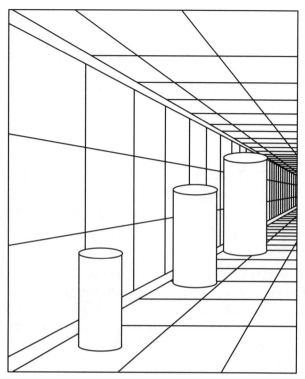

Fig. 8.1 The height of objects: a perceptual decision. The three cylinders arranged in the corridor are of equal height, which can be verified by measuring them on the paper. Yet the brain decides that they are of different heights owing to an illusion of perspective. This illusion disappears when the image is lit under certain conditions (equiluminance). Illusions are not only the brain's way of solving conflicts and ambiguities. They are real decisions. After Gibson (1950)

decision the brain makes to resolve the conflict introduced by the perspective. Similarly, the interpretation of moving figures [5], shown in Fig. 8.2, where the point moves following an ellipse in a plane but is perceived as a circular movement in a different plane, is produced by competition between perspective and indicators of movement. When the velocity of a cloud of points on a screen is set correctly, the subject perceives a sphere.

Are these decisions made in the primary visual areas V1 and V2? Are they developed in the frontal or temporo-parietal cortex? Are they the result of feedback from the frontal or temporal areas to the primary areas? Or of a competitive mechanism between many populations of neurons distributed throughout the brain, with winners and losers [6]? Of Bayesian processes [7]? Recordings in the parietal cortex of monkeys suggest that neurons in the caudolateral part of the intraparietal sulcus play an important role both in perceiving three-dimensional visual features and in cognitive functions related to processing them [8].

The examples above concern an ambiguity in interpreting visual forms. But another category of perceptual decision making results from a rivalry between two perceptions: when each eye—and thus each cerebral hemisphere—is simultaneously presented with two different images, the brain chooses one or the other. This choice also reveals a genuine process of decision making.

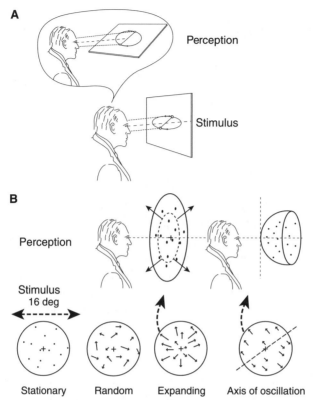

Fig. 8.2 The trajectory of a moving point: a perceptual decision. **(A)** When a bright point moves in an elliptical trajectory in a vertical plane, the brain interprets the movement as being a circular movement in an inclined plane. This surprising perceptual decision has to do with the fact that the brain is used to seeing objects moving in space. **(B)** Perception of a three-dimensional volume based on an image of moving points in a plane (two-dimensional). The stimuli shown to the subject are indicated in the four circles at the bottom of the drawing. First the subject sees the points not moving. Then the points are animated in a random way (as on a television screen with no image). After that, the points are made to move in different ways, which causes a perception of expansion in the plane. Finally, the velocity of each point is regulated in such a way that it causes a perception of volume, here, a sphere. Brain imaging helps to identify some of the regions involved in perceiving three-dimensional objects based on two-dimensional cues. After Paradis et al. (1999)

In this chapter, I will review the empirical evidence for the perception of visual objects and related theories. The question is not so much to summarize them as to show that decision making is always at work.

Geon theory

The idea that the brain deconstructs visual forms according to a repertoire of canonical shapes suggests that it does not possess a general system for recognizing objects of variable geometry. Indeed, objects have multiple representations, formed in various parts of the brain, each specific to the transformations required either for perception or for action. Recognition of the prototypical members of a category of objects (a pot, a hammer, a pair of scissors, a face, and so on), encoding their size and orientation, identifying the individual members of a class of homogeneous objects and planning movements or actions that are generally associated with other familiar objects depend on representations formed in many nerve centres or in their connections.

The brain is subject to strict constraints in recognizing shape. For example, it is hard to identify an upside-down face. This limitation does not exist in the monkey, which is used to seeing faces in a number of orientations.

These constraints do not prevent the brain from constructing constants and generalizing perception based on particular points of view. Except for the disorders described in Chapter 3, object identification is independent of perspective; it seems to be centred on the object itself, whose properties have been in some way 'generalized', as psychologists say. But it does also sometimes depend on point of view. Cognitive psychology has not yet settled this question. David Marr is one of the first to have suggested that the visual system progressively reconstructs objects beginning with their analysis in the primary relays. I would like to examine a recent version of this idea proposed by Irving Biederman [9], for whom perception of objects is founded on the decomposition of images into elements that he calls 'geons' (Fig. 8.3).

The first deconstruction is carried out in V1, V2, V3, and so on. This step comprises extraction of the contours of the object based on information such as brightness, texture, and colour. It delivers information about contour based on edges (see Chapter 7). Another variant suggests that the 'skeleton' of objects is extracted, that is, a series of lines that make up the architecture of the form and that are used especially in analysing moving, nonrigid objects such as animals. Next comes detection of 'nonaccidental' properties such as colinearity, symmetry, continuous curves, parallelism, and so on, and the division of objects into regions.

The next step is comparison of these constituents with geons, stored representations that are essential in this process of analysis. A geon is a volume defined by the displacement of a closed curve in a plane along an axis assumed to be at a right angle to the plane of the surface. Examples are cylinders, cones, cubes, parallelepids, and so on that correspond to the Gestalt criteria for 'good shape'. A library of around 36 geons enables identification of quite a number of objects, considerably more than those we encounter or that we could hope to store and recognize.[1] The problem is whether contemporary neurophysiology can corroborate this decomposition into elementary shapes. At present, two distinct major visual neuronal pathways—'dorsal' and 'ventral'—are known to be involved in the development of visual information and its relation to action. There

---

[1] An important consequence of geon theory is that the Gestalt principle is applied to each geon and not to the entire figure: it is the constituents that are stable in the case of perceptual noise.

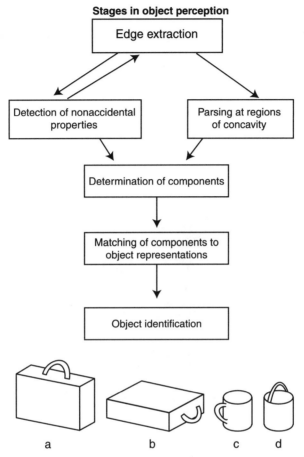

Fig. 8.3 Assumed steps in recognizing an object. This diagram proposes several steps for recognizing an object from the breakdown of its visual properties (shape, movement, colour, and so on) in the earliest relays of the visual pathways of the brain. The edges of the objects activate neurons whose activity is then analysed to determine whether the properties are 'nonaccidental', on the one hand, and, on the other, which curvature fits with which edges. The resulting 'components' of objects (cylinders, spheres, cones, doughnut shapes) are the standard basic forms whose assembly defines the object. Each arrangement of these basic forms is then compared with the memories of different objects, enabling identification. The lower part of the picture shows how (a) a suitcase, (b) a drawer, (c) a cup, and (d) a bucket can be characterized very simply by two components for each object. After Biederman (1987)

is now consensus regarding the importance of the ventral pathway in identifying objects. Moreover, the two hemispheres, right and left, analyse different properties of objects. It is claimed that the right hemisphere is more concerned with global properties of the shapes of objects and processes in a spatial frame of reference, whereas the left hemisphere is only concerned with the local attributes of each object (Fig. 8.4).

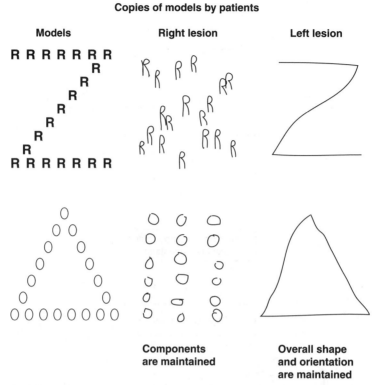

Fig. 8.4 The left brain processes details, and the right brain processes global form. This drawing is a reproduction of drawings of shapes (the letter 'Z' and a triangle) by patients with brain lesions. The novelty of this test is that the two shapes are themselves composed of elements (the 'Rs' for the Z, and circles for the triangle). If patients are asked to copy the two shapes, patients with damage to the right side of the brain (and thus an intact left side) can reproduce parts of the forms, but not the entire shape. In contrast, patients with lesions of the left side of the brain (and thus an intact right side) can reproduce the global form but not the details. The two hemispheres do not play the same role in perceptual decision making during identification of visual forms.

Another interpretation of this difference is that the right brain analyses the spatial properties of 'coordinates', and the left brain is specialized for language, 'category' properties [10].

Perceptual decision making

A little anatomy

Anatomy reveals the choices made by the organism in processing information. In *The Brain's Sense of Movement* I showed that the connectivity of neural axons handles the geometry of an ocular saccade and that the distribution of projections of a pyramidal neuron on the motor centres of the muscles determines motor synergies. I now think that anatomy also subserves the process of perceptual decision making because it classifies, separates attributes, and assigns specific categories. We will see an example that shows the distinction between the dorsal and ventral pathways. Visual objects are processed in the areas of the cerebral cortex that are located roughly along these two pathways, or 'streams', to suggest the idea of flow.

The *dorsal pathway*, fed mainly by the magnocellular system (see Fig. 7.1), links the visual cortex to the parietal lobe, then to an area of the mediotemporal lobe called MT (V5) and several other areas that are important in perceiving movement (see Chapter 9). The neurons of this dorsal pathway participate in visually guided action.

The *ventral pathway*, whose inputs derive mainly from the parvocellular system, links the primary visual cortex to the centres of the temporal lobe. It participates in identifying objects and shapes such as faces and plays an essential role in conscious perception. It converges on the inferotemporal (IT) cortex (Brodmann areas 20 and 21; Fig. 7.1), and is hierarchically the top-most level in the chain of visual processing of objects because at its core are found the neurons that respond, for example, to an entire face, whereas in the lowest levels the neurons respond only to parts of a face (eyes, moustache, mouth, and so on). These temporal areas are divided into many regions each with its own function. The brain is not an immense public space where everyone talks to everyone else; it is a busy urban centre where each workshop and boutique has its speciality and function, even if each exchanges substantial quantities of materials, knowledge, and ideas with the others! Initially, the temporal areas were divided into two: TEO and TE.[2]

Areas TEO and TE project to area 12 of the prefrontal cortex and area 45, and towards the limbic cortex, the striatum, and the amygdala. The signals prepared by the temporal areas are then sent to the prefrontal cortex and the limbic system, where the most cognitive—even emotional—aspects of decisions are worked out. These two areas also project to many subcortical structures, TE towards the mediodorsal nucleus of the thalamus, and TEO towards the colliculus, where it may influence the direction of gaze and thus of attention, but also approach and flee responses.

Once reconstitution has been achieved along the ventral pathway, the signals regarding identification of objects and natural shapes are sent to all the centres that influence

---

[2] Today they are divided into five parallel areas: TE1, TE2, TE3, Tem (which abuts STS, the superior temporal sulcus, so-called polysensory area) and Tea. TEO receives information from the primary areas, especially V4, the colour region, and less from V2, V3, and MT or V5, the mediotemporal cortex region that contains neurons sensitive to movement. V5 is also in contact with other temporal regions: TE, TH, TG, the parahippocampus, the perirhinal cortex (Brodmann areas 35 and 36). Finally, TEO fibres feed back ('centrifugal projection') towards the primary visual areas (V1, V2, V3, V4), which permits anticipated control of visual perception in these same areas. Information from TEO is also sent to the dorsal pathway (MT, FST, IP, PP, LIP), probably to guide action.

reasoning and plan action, combine perception and emotion, and even feed back towards the primary sensory visual relays where they prepare and modulate perception at its source. Perceptual decisions made during this development of constancy and of identifying shapes and objects in the environment are then subject to decisions at a higher level. The hierarchy I presented at the beginning of this book is clearly detectable. But it is equally clear that each level of this hierarchy influences the level below and no doubt competes with other decisions made in parallel.

But enough anatomy! Now let us see how signals relating to a face or an object are analysed. To simplify things, we will refer to the group of areas described below as IT (inferotemporal).

Face recognition

To decide whether the face of a person we encounter is familiar to us, the face first has to be apprehended as a whole, that is, as a face and not simply as a constellation of shapes: eyes, lips, and so on. This unity of the object 'face' is constructed along the temporal pathway which, if damaged, as we saw in Chapter 3, causes prosopagnosia, or the inability to recognize a face.

How is the percept of a face reconstituted following the splintering of the visual world into colour, shape, brightness, and so on? These areas are arranged functionally in a sequential manner.

A basic property of neurons along the temporal pathway is that their receptor field becomes bigger and bigger. In Chapter 7 we saw that V1 neurons have a receptor field of around 1 degree. So the window through which they see the world is very small. When processing reaches the temporal areas, the size of the receptor field increases substantially. The brain can integrate in a single neuron information continued in a larger region of the visual world. The size of the receptor fields along this pathway increases from 2 to 3° in V2 up to 30 to 50° in TE and TEO, suggesting that local properties of objects are gradually integrated in the more global properties of the environment. For example, a person's face is integrated with her posture and a vase of flowers with the room it is in.

Although V4 already contains neurons that are sensitive to shape, the largest proportion of neurons responding to variations in complex forms are found in the IT cortex. The IT cortex thus contains all the elements needed to recognize objects. The question is whether these are represented by a few 'gnostic' neurons, or by the joint activation of small groups of neurons each representing different attributes or geons of the object or even by combinations of discharges in large populations of neurons.

Moreover, optical imaging of portions of the IT cortex in monkeys recently showed that in a region of around 1 square millimeter neurons encode successive aspects of, for example, a shape or face from several angles [11]. Inferotemporal neurons react to animate objects such as faces, hands, or other parts of the body. They are sensitive to identity, expression, direction of gaze, and body shape. So faces and certain parts of the body are encoded by individual neurons that activate when an animal is presented with a specific point of view.[3] How does an IT neuron manage to recognize a pattern of indicators characteristic of the complex properties of an inanimate object? To answer this question, monkeys were trained to recognize artificial objects [12]. Recordings of IT neurons showed that they are sensitive to objects seen in a particular way. Some neurons are sensitive to parts of faces, but others to the entire form, because they discharge differently when the pattern changes. Inferotemporal neurons can acquire the same properties for objects as for faces.

Some neurons respond simultaneously to the face of a monkey and to that of Albert Einstein. Consequently, these neurons respond to complex groupings of characteristics of stimuli. Some also respond gradually to the face or the body of a person in profile. The neuron is said to be 'tuned' to a profile of around 45 degrees. When a person turns his head, the neuron activates less and less. Some cells are so attuned to particular shapes that they can be shown 600 objects, but respond only to a specific face. The system is very selective.

This neural network must also make decisions. Suppose that we present each eye with a different face, or one eye with a face and the other eye with a flower. A conflict situation, to be sure, yet we are only aware of one of the two images at a time. Let us look a little more closely at the mechanisms underlying this perceptual problem solving.

Conscious perception

Generally, along the temporal pathways, neurons respond to larger and larger objects and environmental context (the receptor field gets bigger); they become increasingly invariant to size (they respond to the same object in different sizes); and, finally, they are more and more taken with the complexity of shapes. But especially, they are all the more active because there is a conscious perception of the stimulus. Currently, many researchers think that the ventral pathway is one of the fundamental constituents of conscious perception.

Consider a few examples of what we might call 'conscious perception'. If you look at the image of an infant holding a hoop drawn as an ellipse, that is, seen at an angle, you 'consciously' perceive the hoop as 'circular'. The brain decides that in fact the hoop is circular, although the picture on paper, like that on the retina, is elliptical. Here, consciousness is indeed perceptual decision making.

---

[3] It is the group of these neurons, each sensitive to one profile, that generate the representation of the face. Neurons have not yet been found that respond abstractly to 'Paul's face'. According to geon theory, responses that can be evoked by objects can also be suggested by simplified versions of these objects or even by basic shapes (circles, spheres, T-shaped figures, and so on).

Binocular rivalry: the Cheshire cat

Ambiguous shapes are useful for analysing decisions. Removing an ambiguity amounts to making a decision. By 'ambiguity' I mean, for example, the two possible interpretations of a Necker cube; the different Gestalt images, depending on what one chooses as shape or background; and so on. Yet neurophysiologists have recently been inclining towards another paradigm: binocular rivalry [13].

Here is an example of the new paradigm: If a subject is shown two different pictures, one to each eye, his brain will choose to perceive one or the other. In humans, this rivalry involves alternating between two perceptions, that is, the subject perceives the images one after the other; the brain hesitates. 'How happy my brain could be with either', to paraphrase a French proverb. Binocular rivalry is a good paradigm because one can train an animal to communicate what it perceives by pressing on levers. Moreover, this rivalry has a specific dynamic: the temporal sequence follows a distribution that obeys a simple law; the time course plot of the switches is very stable and especially quite comparable between monkeys and humans. The maximum alternation is about 1 s per stimulus.

Movement can also cause a part of the visual field to disappear. For example, suppose that owing to a device like that in Fig. 8.5, the left eye is shown a constant image of a face and the right eye a hand that sweeps the visual field. These conditions give rise to the 'Cheshire cat phenomenon' [14], by analogy with the story of *Alice in Wonderland*: if the subject concentrates her attention on the face's mouth, for example, the appearance of the hand will wipe out the entire face except for the mouth. In other words, attention conserves a part of the visual field for which the two eyes are competing. If now the subject looks at the eyes, the mouth and the rest of the face will disappear when the hand crosses the field. The binocular rivalry paradigm can be used in studying monkeys.

For example, first a monkey's right eye is shown a picture of a lion and his left eye is shown a picture of an inanimate object or an abstract shape. The monkey presses a button with its left hand if it perceives the picture on the left and another button with the right hand if it perceives the picture on the right. It is also trained not to respond if it sees a mixture of the two. As mentioned earlier, generally what is observed is an alternation roughly every few seconds. But a very small number of V1 neurons modify their activity with the alternating perception thus revealed. So the primary visual cortex doesn't seesaw; it remains indifferent to what the animal perceives in 85 per cent of cases. At V4 or MT (V5) along the visual pathway, a slightly greater proportion (30 to 40 per cent) are increasingly activated in direct correlation to the perception. The decision that leads to perceptual alternation between the two images seems to be made at the level of the temporal lobe. Indeed, 90 per cent of IT neurons change their activity just as the animal chooses one or the other image. Fig. 8.6 illustrates the way the IT contribution is revealed in this process of decision making.

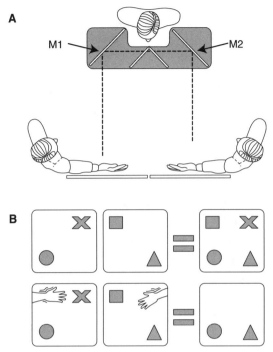

Fig. 8.5 The Cheshire cat. **(A)** Mechanism for demonstrating the disappearance of objects. The subject sits before two screens, S1 and S2, that he sees by reflection in two mirrors, M1 and M2, set at 45 degrees. Two visual objects are displayed on each screen (a triangle and a square on S1, and a circle and a cross on S2). The merging of the two images gives the subject the perception of four objects at the four corners of the screen. **(B)** If two experimenters situated at each side of the screen simultaneously introduce their hand as shown (without hiding the objects), the two upper objects (the square and the cross) seem to disappear, whereas they are perfectly visible. This disappearance is linked to competition between the perception of the hands and perception of each object. The brain decides to 'perceive' only one or the other, in this case, the hand. After Duensing and Miller (1979)

Even if V1 is not the seat of decision making, the precise mechanism still remains to be discovered. Detection of a congruence at this level might yet play an essential role in working out decisions in successive stages. It has been suggested that the activity of binocular neurons might be synchronized in V1 when stimuli are congruent. In contrast, if there is a discongruity, these neurons would activate asynchronously. This is the asynchrony that, arising in V1, produces the rivalry in successive areas of processing [15]. Centrifugal neural projections, feeding back from the temporal lobe to V1, could block at V1 the transmission of noncongruent signals. Yet there may be a 'collective vote', that is, the brain may decide in accordance with the response of a majority of the neurons, whatever the choice of the minority [16]. Erik Lumer and colleagues [17] have used brain imaging to show that activity of the prefrontal cortex in humans is also

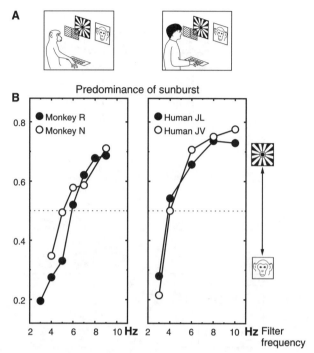

Fig. 8.6 Binocular rivalry and the perception of visual forms. **(A)** Each eye of a monkey is shown a different image (thanks to polarizing glasses) to create a situation of binocular rivalry. The left eye sees a sunburst, and the right eye sees a monkey's face. The image of the sun is manipulated to make it increasingly fuzzy by filtering its visual properties (low-pass filtering). At 2 Hz the image is very vague; at 10 Hz it is very clear. The monkey must press on a button to the right if it perceives the sunburst, and a button on the left if it perceives the face. During any session, the time it takes a monkey to record its perception of the sunburst gives an estimate of the 'predominance' of this percept. **(B)** It is also possible to construct a curve that shows the evolution of the predominance of the perception of one of the two images over the other according to the clarity of the sunburst image. The same experience is carried out with humans to show that recording of the neurons of the temporal lobe in the monkey corresponds well with a perceptual mechanism that exists in humans, and that perceptual decision making is essentially similar in monkeys and in humans. After Sheinberg and Logothetis (1997)

associated with alternation of percepts during presentation of visual gratings that could give rise to one perception or another.

This description is obviously a little technical, but it is important because it suggests that decisions are produced in the brain by a complex play of reciprocal relations between the structures in contact with the material world and those where decisions are made based on a global analysis of context, and probably also of memory and emotion.

We discussed above the fact that rivalry leads to a choice between two images. The experiment can be refined by using another property of rivalry: the influence of one image on another. Try it yourself. Make a tube with a sheet of paper and hold the tube in front of your left eye with your left hand. Now, move your right hand beside the tube about 20 centimetres in front of your right eye. Instead of seeing your entire hand with the right eye, you will often have the impression that there is a hole in your right hand! This occlusion is a transfer of interpretation from one eye to the other.

You can do another experiment: On the left side of a sheet of paper, draw an ellipse with a line going through its centre from left to right, perpendicular to the plane of the ellipse. The drawing is in perspective to suggest the geometry of the axis of the line. On the right side of the paper, draw an ellipse without the line going through it. If you look at the ellipse on the left for 15 s, you will have the clear impression that to the right, the other ellipse is also transsected by a line going right to left. The left-sided view has caused a very strong percept on the right. This experiment was even done with monkeys and showed that they respond when they perceive the line crossing the ellipse: the activity of the IT neurons correlates perfectly with this illusion.

How are these experiments related to the more general problem of perceptual awareness? When we have lost our keys and they are in reality in front of our eyes on the table, we are doubtless encountering a problem similar to that of suppression during binocular rivalry. We *see* the keys, but do not *perceive* them. Our brain has not decided, 'Those are your keys.' The processing in these areas is probably one of the components of perceptual consciousness.

So reconstitution in this temporal lobe is sequential and hierarchical, and three major processes for developing perceptual decision making take place there. First, the whole of an object, face, or scene is gradually put back together and identified. Second, changes are made to frames of reference: neural activity in the primary relays (V1 and CC) represents the object in a frame of reference linked to the retina, but further forward in the pathway the neurons adopt a frame of reference linked to the observer or even to the object itself. Finally, when several objects are in competition, a choice is made among them. In addition, IT neurons interact reciprocally with the structures of the rhinal system involved in memory and probably in conscious processes.

Perceptual decision making does indeed appear to be a distributed and hierarchical process.

Rapid recognition of natural shapes: categorization

The average duration of a television image is scarcely more than 1 or 2 s, especially for very brief commercial clips. Indeed, we know that we have only to be exposed to images for a few hundreds or even tens of milliseconds to be influenced even if we have not consciously perceived them. These so-called subliminal messages would be used for commercials if they were not prohibited by law. Still, it is surprising how quickly the brain can identify the content of an image, especially when it includes objects, animals, or real people.

That the brain more rapidly recognizes natural objects and people than artificial ones has been shown experimentally [18]. For example, if a subject is shown an animal or a face, recordings of the subject's brain activity using the technique of evoked potentials shows that natural images give rise to more rapid potentials. In a monkey, it takes around 40 to 60 ms for an image to arrive at the primary visual areas; the neurons of the posterior temporal lobe, which are sensitive, for example, to the eyes and mouth in a face, are activated 70 to 80 ms after presentation of the image. It takes another 10 to 20 ms for the information to be transferred to the anterior portions of the temporal lobe, where activities linked to the global properties of faces, and so on, are encoded.

Following this reconstitution, the brain still has to categorize things, to decide, for example, whether the animal it perceives is a dog or a cat. Categorizing requires independently identifying the physical properties of the whole object, animal, or person. Thus a banana and an apple are both foods, though they are very different. The question is where in the brain that categorization is performed.

The areas of the temporal cortex communicate with the regions of the limbic system (such as the enthorinal and perirhinal cortex), with which they exchange information essential for comparing perceived shapes with memory and trying out associations to determine whether an image has already been seen [19]. They also exchange information with the prefrontal cortex, which probably plays a major role in identifying the general properties of shapes.

For example, a monkey is shown pictures of a cat and a dog on a computer that gradually transforms the image of the cat into a dog and vice versa [20]. After a delay, the monkey is shown two images in succession and is required to judge whether they belong to the same category. For some in-between images, the monkey will have to make a real decision, namely, whether it is a cat or a dog. The neurons of the prefrontal cortex respond in a clearly dichotomous manner to these two categories and choose one or the other despite the amorphous character of the modifications that are introduced by the computer. I should not say 'choose', because we have no evidence that a choice is actually made at that level.

We need to understand more about the differences between the neurons of the temporal cortex and those of the prefrontal cortex. It is possible that a real dissociation exists between neurons that encode categories and others that encode more specific things. One example of that has already been given in another context: neurons have been found in the hypothalamus that respond to all stimuli associated with 'food'; but in the temporal lobe, neurons that fire in response to particular foods do not relate these stimuli to the more general category of food.

What is most striking about the activity of neurons in the prefrontal cortex is how quickly they respond. Inferotemporal neurons respond in about 100 ms; those of the prefrontal cortex take only slightly longer (100 to 130 ms). This speed allows the animal a reaction time of around 180 to 260 ms. It takes some time for the neural outputs of the prefrontal cortex to be processed by the premotor and then the motor cortex, and finally for the motor commands to be sent to the spinal cord to guide the monkey's hand. Generally speaking, monkeys respond more quickly than humans (humans take about 50 ms longer), although processing in humans is still very fast.

What happens when we see impossible shapes like those drawn by Escher? Is there a mechanism that allows us to decide that the shape is absurd? Brain imaging allows us to ask the following question: What areas of the brain are involved in concluding whether a shape is possible or not? Or more modestly: Does a subject's brain react differently when presented with a possible geometric shape as opposed to an impossible one? The data on this subject are still too tentative to extract a theory from them [21].

It has now been established that in the monkey, the major areas for identifying objects, faces, and so on are found in the IT cortex. What area of the brain in humans fulfils the same functions as the IT area in the monkey? Lesions of the temporal cortex lead to distortions in the appearance of perceived objects. But the data cited above concerning the speed of recognition of natural forms (in particular animals) suggest that perhaps the brain does not process faces and buildings in the same structures. Indeed, it is known that in humans, the 'fusiform gyrus' [22] is one of the centres that is critical to this recognition. It is activated in a specific way by faces and, especially, shows only very weak activity when an individual is presented with familiar objects or scenes from his environment, such as places or houses. On the other hand, the parahippocampus is activated by houses and only slightly by faces [23]. Showing each eye separately a face and a house creates a perceptual rivalry, and the brain has to make a decision. If a subject is asked to respond only when she perceives a face or a house, the fusiform gyrus or the parahippocampus is activated according to the percept chosen by the brain. That suggests that these regions contain correlates of conscious perceptual decision making and of object classification. But the mechanisms underlying their contribution to perceptual awareness are still mysterious.

Recently, magnetoencephalography has revealed the great variability of areas activated among subjects and sometimes in the same subject during tasks that alternate percepts [24], and moreover that several areas are involved in the process of decision making that resolves ambiguity linked to rivalry. A so-called neurodynamic approach was developed by Varela and his collaborators [25]. It consists in recording brain activity using electroencephalography. Mathematical analysis of the signals identifies areas of synchronized activity. This synchronization is itself evidence of a temporary network of relations among areas of the brain that participate in a perception or elaboration of a particular mental process. It reveals the dynamic reconfiguration that is produced in the brain at the instant it resolves a perceptual ambiguity, for example, in the case of observing a shape that hides a profile. It is thus possible to trace a succession of changes in global states of these networks. Surprise caused by a face seen out of context that stupefies and suspends decision making is amenable to analysis. It is beyond the scope of this book to discuss these theories, which assume that the unity of perception of an object or a face is due to temporal synchronization of the neurons that encode the attributes of the object. Literature cited in several chapters reviews these theories and the new concepts they bring.

References

1. M Merleau-Ponty, *Phenomenology of perception*, trans. C Smith (London: Routledge and Kegan Paul, 1962), p. 40.
2. HB Barlow, 'The absolute efficiency of perceptual decisions', *Philosophical Transactions of the Royal Society B*, **290** (1980): 1–82.
3. WT Maddox and SV Bogdanov, 'On the relation between decision rules and perceptual representation in multidimensional perceptual categorization', *Perception and Psychophysics*, **62** (2000): 984–97.
4. J Ninio, *The science of illusions*, trans. F Philip (New York: Cornell University Press, 2001).
5. G Johansson, 'Visual motion perception', *Scientific American*, **232** (1975): 76–88.
6. JH Reynolds, L Chelazzi, and R Desimone, 'Competitive mechanisms subserve attention in macaque areas V2 and V4', *Journal of Neuroscience*, **19** (1999): 1736–53.
7. D Ascher and NM Grzywacz, 'A Bayesian model of temporal frequency masking', *Vision Research*, **40** (2000): 2219–32.
8. K Tsutsui, M Jiang, H Sakata, and M Taira, 'Short-term memory and perceptual decision for three-dimensional visual features in the caudal intraparietal sulcus (Area CIP)', *Journal of Neuroscience*, **23** (2003): 5486–95.
9. I Biederman and P Kaloczat, 'Neurocomputational bases of object and face recognition', *Philosophical Transactions of the Royal Society B*, **352** (1997): 1203–19. About decision making in object recognition, see, for example, TJ Lloyd-Jones and GW Humphreys, 'Perceptual differentiation as a source of category effects object processing: evidence from naming and object decision', *Memory and Cognition*, **25** (1997): 18–35.
10. JB Hellege and C Michimata, 'Categorization versus distance: hemispheric differences for processing spatial information', *Memory and Cognition*, **17** (1989): 770–6; SM Kosslyn, O Koenig, A Barrett, CB Cave, J Tang, and JDE Gabrieli, 'Evidence for two types of spatial representations: hemispheric specialization for categorical and coordinate relations', *Journal of Experimental Psychology: Human Perception and Performance*, **15** (1989): 723–35; M Okubo and C Michimata, 'Hemispheric processing of categorical and coordinate relations in the absence of low spatial frequencies', *Journal of Cognitive Neuroscience*, **14** (2002): 291–7; SD Slotnick, LR Moo, MA Tesoro, and J Hart, 'Hemispheric asymmetry in categorical versus coordinate spatial processing revealed by temporary cortical deactivation', *Journal of Cognitive Neuroscience*, **13** (2001): 1088–96.
11. G Wang, K Tanaka, and M Tanifuji, 'Optical imaging of functional organization in the monkey inferotemporal cortex', *Science*, **272** (1996): 1665–8.
12. NK Logothetis, J Pauls, HH Bûlthoff, and T Poggio, 'View-dependent object recognition by monkeys', *Current Biology*, **4** (1994): 401–14; NK Logothetis, 'Objection vision and visual awareness', *Current Opinion in Neurobiology*, **8** (1998): 536–44.
13. F Tong, K Nakayama, JT Vaughan, and N Kanwisher, 'Binocular rivalry and visual awareness in human extrastriate cortex', *Neuron*, **21** (1998): 755–59; ED Lumer, KJ Friston, and G Rees, 'Neural correlates of perceptual rivalry in the human brain', *Science*, **280** (1998): 1930–4.
14. GC Grindley and V Townsend, 'Binocular masking induced by a moving object', *Quarterly Journal of Experimental Psychology*, **17** (1965): 97–109; S Duensing and B Miller, 'The Cheshire cat effect', *Perception*, **8** (1979): 269–73.

15 DL Sheinberg and NK Logothetis, 'The role of temporal cortical areas in perceptual organization', *Proceedings of the National Academy of Sciences of the USA*, **94** (1997): 3408–13, fig. 3.

16 Lumer, Friston, and Rees, 'Neural correlates of perceptual rivalry'; ED Lumer and G Rees, 'Covariation of activity in visual and prefrontal cortex associated with subjective visual perception', *Proceedings of the National Academy of Sciences of the USA*, **96** (1999): 1669–73; F Tong and SA Engel, 'Interocular rivalry revealed in the human cortical blind-spot representation', *Nature*, **411** (2001): 195–9.

17 Lumer, Friston, and Rees, 'Neural correlates of perceptual rivalry'.

18 S Thorpe, D Fize, and C Marlot, 'Speed of processing in the human visual system', *Nature*, **381** (1996): 520–2; S Thorpe and M Fabre-Thorpe, 'Seeking categories in the brain', *Science*, **291** (1998): 260–3.

19 Y Miyashita, 'Neural correlates of visual associative long term memory in the primate temporal cortex', *Nature*, **335** (1988): 817–20; K Sakai and Y Miyashita, 'Neural organisation for the long term memory of paired associates', *Nature*, **354** (1991): 152–5.

20 DJ Freedman and M Riesenhuber, 'Categorical representation of visual stimuli in the primate prefrontal cortex', *Science*, **291** (2001): 312–16.

21 A Uecker, EM Reiman, DL Schacter, MR Polster, LA Cooper, LS Yun, and K Chen, 'Neuroanatomical correlates of implicit and explicit memory for structurally possible and impossible visual objects', *Learning and Memory*, **4** (1997): 337–55; DL Schacter, E Reiman, A Uecker, MR Polster, LS Yun, and LA Cooper, 'Brain regions associated with retrieval of structurally coherent visual information', *Nature*, **376** (1995): 587–90.

22 N Kanwhisher, J McDermott, and MM Chun, 'The fusiform face area: a module in human extrastriate cortex specialized for face perception', *Journal of Neuroscience*, **17** (1997): 4302–11; N George, J Driver, and RJ Dolan, 'Seen gaze-direction modulates fusiform activity and its coupling with other brain areas during face processing', *Neuroimage*, **13** (2001): 1102–12.

23 E Maguire, N Burgess, JG Donnett, RSJ Frackowiak, CD Frith, and J O'Keefe, 'Knowing where and getting there: a human navigation network', *Science*, **280** (1998): 921–4; E Maguire, C Frith, N Burgess, JG Donnett, and J O'Keefe, 'Knowing where things are: parahippocampal involvement in encoding objects located in virtual large-scale space', *Journal of Cognitive Neuroscience*, **10** (1998): 61–76; N George, RJ Dolan, GR Fink, GC Baylis, C Russell, and J Driver, 'Contrast polarity and face recognition in the human fusiform gyrus', *Nature Neuroscience*, **2** (1999): 574–80; B Jemel, N George, E Olivares, N Fiori, and B Renault, 'Event-related potentials to structural familiar face incongruity processing', *Psychophysiology*, **36** (1999): 437–52.

24 GM Edelman and G Tonini, *Consciousness: how matter becomes imagination* (London: Allan Lane, 2000).

25 F Varela, J-P Lachaux, E Rodriguez, and J Martinerie, 'The brainweb: phase synchronization and large-scale integration', *Nature Review Neuroscience*, **2** (2001): 229–39; E Rodriguez, N George, J-P Lachaux, J Martinerie, B Renault, and F Varela, 'Perception's shadow: long-distance synchronization of human brain activity', *Nature*, **397** (1999): 430–3.

Chapter 9

Sensory conflict: perception of movement

In fact, the image of a constituted world where, with my body, I should be only one object among others, and the idea of an absolute constituting consciousness are only apparently antithetical; they are a dual expression of a universe perfectly explicit in itself. Authentic reflection, instead of turning from one to the other as both true, in the manner of a philosophy of understanding, rejects them as both false.
(M. Merleau-Ponty [1])

Imagine you are sitting in a train. The train on the track beside you departs. Watching it move out your window, you have the illusion that your train is the one leaving. This illusion can also be felt on a bridge, looking at a river. The movement of a visual scene may be due either to movement of the body, or to movement of the world, in that case, with the body remaining still. How do we decide which interpretation is correct? We are in the same situation as Solomon, except that we cannot cut the solution in two.

Sometimes, a conflict is the result of several simultaneous movements in the visual world. For example, in the train, if you look out the window and a blizzard is raging, the clouds race by. The visual field consists of several contradictory movements: that of the environment in a sense contrary to the train, of the snowflakes, of the clouds—how does the brain decide that some elements of the environment are moving and others not? This decision is complex [2]. Analysis of its neural underpinnings can put us on the path of general mechanisms by which the brain solves complex problems, since these relate to the fundamental mechanisms of perception and of action.

Deciding implies choosing among several solutions. The dilemma of Buridan's ass was that it had to choose between two identical solutions—two equally inviting piles of hay. We have considered cases where there is an ambiguity between two or several solutions. But often solutions come into conflict. One clashes with the other, and it is not always easy to decide. Is it possible to find in the basic mechanisms of perception the roots of the decisions that we make in complex cases? The brain does not remain passive in the face of conflict, nor even when it is presented with simultaneous stimuli. It chooses. It can even 'turn off' stimuli, as attested by the phenomenon of 'extinction', which arises when two bright points are presented at the same time to each of the hemispheres [3]. How does the brain decide between contrasting perceptions? In this chapter, I would like to probe the idea of ambiguity and sensory conflict. I think that the mechanisms developed over the course of evolution to resolve these problems are also used by the brain to make cognitive decisions.

Do not get me wrong. I do not claim that formal logic is reducible to perceptual decision making. What I am saying is that a neurocognitive theory of decision making must be anchored in neurobiological processes, and that the processes that subserve perception are fundamental. And perhaps all logical thought is anchored in these processes to a more or less greater degree according to the person or even their gender. In *The Brain's Sense of Movement*, I suggested that mathematical concepts are an integral part of the brain's functioning and are not located outside of us. Here, I suggest that we can also find perceptual mechanisms useful for building a theory of decision making. We must escape from the tyranny of language and reintegrate thought among the other forms of brain activity, in particular perception and action.

The cortical pathway of visual movement

Seen from the train, the landscape detected by the retina is processed via several parallel pathways. One pathway goes through the colliculus (Fig. 4.5); the 'dorsal' visual pathway is linked to the magnocellular system (see Chapter 7), which reaches several cortical areas [4], in particular V5, a mediotemporal cortical area that corresponds to the medial temporal (MT) cortex in the monkey, and is specialized in processing visual movement. The neurons of this MT region have preferential orientations and a receptor field (the small region of space that is 'seen' by the neuron) that is slightly larger than the movement neurons of V1 and V2. This area projects to another mediotemporal area that in the monkey is called medial superior temporal cortex (MST). There is a big difference between the neurons of MT and those of MST. In MT, neurons stop their activity when the visual stimulation disappears or the movement stops; they are tightly coupled to visual movement and to the external world. In MST, on the other hand, the neurons continue to be active, even if the visual stimulus is briefly disrupted. At the level of MST, information about the external world and signals coming from the brain itself (extraretinal) converge.

Moreover, the receptor field of the MST neurons—their visual span—is much greater than that of MT or V1. As visual processing moves along the dorsal or ventral pathway, more neurons gain in properties of invariance and contextualization. Local gives way to global. This idea is important because it gives us a theoretical framework for developing one aspect of a theory of decision making.

How can one study perceptual decision making in these structures? Imagine that you are looking at a computer screen. On the screen is a cylindrical surface covered with two series of moving points. Some of the points are moving in a way that suggests a concave shape, and the others in a way that suggests a convex shape. Under these conditions, experiments show that the brain perceives the surface as alternately convex and concave. The strength of these percepts can be altered, for example, by changing the percentage of points or their luminosity. Ambiguity can be removed by a genuine perceptual decision.

The experiment has been tried with monkeys. The monkey was first put in a situation where the two groups of points were balanced. There was an equal probability that the monkey would decide the surface was concave or convex. Then the relationship between the number and contrast of the two groups of points was changed to force the monkey to make a decision. The monkey indicated its perceptual decision by pressing one of two buttons in front of it [5]. To try and distinguish the effects of sensory processing of information and the brain's preparations for responding, a delay was introduced between the stimulus and the response to clearly isolate a moment of deliberation [6]. Once the monkey had decided, recordings were made of the neuronal activity of various structures of its brain.

The areas of the parietal cortex that process movement contain neurons that are activated in a very different way depending on how an animal is responding. These neurons are thus involved in the process of decision making. One can also stimulate, that is, deliver a very brief electric discharge that causes no pain but activates the neurons located around the electrode. The electrical stimulation alters the perception of the monkey, which presses a different button during the stimulation. Its response indicates that it has perceived a change of direction in the visual pattern, confirming the existence in this region of a mechanism that contributes to development of the percept.

The mechanism of decision making

To make a decision, the monkey's brain can weight the effects of all the points by vectorial addition. In other words, each point on the screen produces a little activity in the neuron, or causes it to deactivate depending on which way the point moves; the neuron adds up the activities it receives. There is a threshold to this addition beyond which the neuron is activated, thus resulting in a percept in one direction or the other.

The decision could also be made by an 'all-or-nothing' process; when a population of neurons becomes statistically dominant, it leads all the others to respond in the same way. The 'catastrophic'[1] and nonlinear feature of this mechanism is appealing, because it takes into account the sudden and quasi-discrete way in which a percept appears. It is a sort of statistical decision, by majority vote, to which the brain can easily adapt, according to the same principle by which our democracies operate.

Yet these neurons are probably not the ones that make decisions, since they do not respond when the stimulus is absent: a genuine decision must be maintained even in the absence of stimuli. These neurons are always linked to the presence of the stimulus and lack the abstract character of decision making.

One might then ask whether a hierarchy exists between the successive visual areas from the primary visual relays (V3a) going through areas MT and MST all the way to the areas of the temporal cortex (superior temporal sulcus, STS) that process movement [7].

---

[1] The name given to René Thom's theory of the ways in which abrupt changes occur.

The activity of the neurons in these four areas of the brain have been recorded in the monkey during presentation of a stimulus whose movement is ambiguous, that is, containing two populations of points moving in opposite directions in the receptor field of the neuron. The monkey had to decide about the direction of movement. The correlation between the activity of the neurons and the number of judgemental errors made by the monkey was calculated: there seems to be no real hierarchy among these areas in perceptual decision making. They do not 'make' decisions.

Still, because the region of space (the receptor field) whose movement the neurons signal gets larger and larger at each step of processing, context is increasingly taken into account, which sets the stage for more global decision making. To express what happens in the visual pathways in everyday language, one might say that the brain 'expands its horizons' or 'gets some distance'.

Since no structures have been found that arbitrate, so to speak, in this portion of the brain, the next likely place to look for mechanisms specific to perceptual decision making in ambiguous situations of detecting global movement in a visual scene is the frontal and prefrontal cortex. Indeed, pathology suggests precisely that (see Chapter 3).

The prefrontal cortex and perceptual decision making

The pathology of decision making involves the frontal and prefrontal cortex in the process of decision making. Only the analysis of neuronal activity in monkeys during tasks that require cerebral decision making can reveal the mechanisms at work. All other approaches, cerebral imaging, neuropsychology, and so on, are only approximations.

The hypothesis for a role of the neurons of the prefrontal cortex in the processes of decision making is supported by their behaviour in tasks where response is delayed to observe brain activity while the decision is being processed. This paradigm makes it possible to separate activity due to the presentation of a stimulus that requires a decision from the period of reflection during which the decision is made (the delay) and finally the period of response [8]. The critical observation is that the neurons remain active during the *delay* between presentation of the stimulus and the decision. Several interpretations have been proposed to explain this cognitive activity during the delay. In the dorsolateral frontal cortex, it is believed to play a role in working memory [9]. It could also reflect motor intention or suppression of undesired movements or even attention to prominent features of the visual scene. The commonality in these interpretations is *selection* of perception or action. To decide is to select—to choose.

However, the existence of activity in the prefrontal cortex during the delay is not sufficient to attribute to these structures a decisive role in decision making. Indeed, the neurons of the parietal cortex [10], the superior colliculus, and so on are also active during delayed-response tasks. This reinforces the idea of a cognitive network among several areas as opposed to isolated activity [11]. Once again, it comes down to activity that mobilizes a network of interconnected brain areas.

Let us consider several experimental examples that suggest a role for the prefrontal cortex in mechanisms of perceptual decision making.

Intention-to-choose neurons

Clever experiments [12] that it would take too long to detail here reveal the activity of a neuron involved in choosing among a group of moving points by requiring a monkey to shift its gaze towards one target or another based on its perception (Fig. 9.1). In addition, a delay is introduced. The activity is seen to increase with the choice the monkey makes.

A mathematical model based on the theory of decision making has been proposed to explain the experimental findings. Information about the moving points is assumed to be relayed to the prefrontal cortex through the MT and MST centres of the parietotemporal cortex. A neuron of the prefrontal cortex receives information from *two* separate MT or MST neurons that signal the two opposite directions of movement of the points of stimulus. The model evaluates the probability that the neurons of the frontal cortex 'connect' the information given by the MT and MST. The weighting of these neurons of the frontal cortex by an all-or-nothing type mechanism causes the neurons to make a decision at a given moment. The processes involved here are thus not logical but constitute a cascade of probabilistic neuronal events that lead to a decision.

The prefrontal cortex is well placed to handle associations and make decisions because it is located at the intersection of cerebral pathways that process sensory data. For example, the 'dorsal' pathway, which concerns spatial aspects of the sensory environment, and the 'ventral' pathway, where objects, faces, and so on, are identified, converge towards the prefrontal cortex. Thus dorsolateral areas 9 and 46 receive information from the dorsal pathway (posterior parietal lobe); ventrolateral areas 12 and 45 receive information from areas TEO and TE. In these regions, neurons sensitive to shape have been located; some maintain their activity even after removal of the stimulus. Brain-imaging studies have confirmed the existence in the frontal lobe of distinct regions that process information linked to memory of places and of objects and faces. So it does seem that, for example, combining object and place is carried out at the level of the prefrontal cortex [13].

It is in the prefrontal cortex where the different aspects of the world meet up. And as we will see in Chapter 11, it is also at this level (particularly the orbitofrontal cortex) that information supplied by the senses based on the value of reward or punishment, or the value of emotion, is modified. Here is really where the processes for developing a decision that takes into account the many facets of relations between the intention to act, the goal a person or animal is pursuing, and the data regarding the acting body all come together. Again, remember that the prefrontal cortex is not an isolated operator but is networked into the loops that link it with the centres that control movement and perception: the basal ganglia, the thalamus, and the cerebellum.

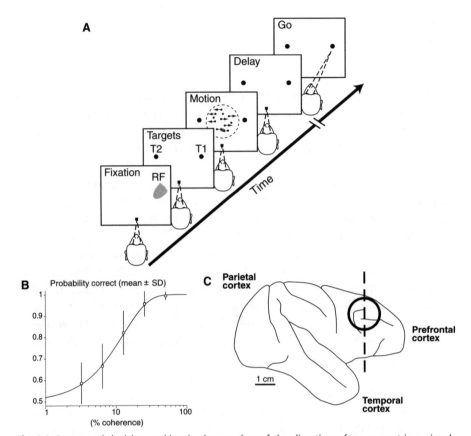

Fig. 9.1 Perceptual decision making in the monkey of the direction of movement in a visual scene. **(A)** Examples of stimuli used to explore the choices a monkey makes. In a first stage, it stares at a point on the screen (point of fixation). The neuron recorded in its prefrontal cortex is sensitive to a small part of the visual field called the 'receptor field' (RF), shown in grey on the image. Two targets, T1 and T2, are then introduced on each side of the screen. Next a visual scene appears that contains points moving left or right. Depending on the density of the points, the monkey perceives global movement in one direction or the other. The monkey is given time to make a decision through a delay that enables registration of the neuron linked to decision making or at least to the fact that the monkey has chosen. Finally, the animal is asked to shift its gaze (Go) to either target T1 or T2 to indicate its choice.
(B) The curve indicates the probability of correct responses made by the monkey for the movement in one direction. This probability is expressed as a function of the coherence of the movement of the scene, that is, of the percentage of points that are moving in one direction. One hundred per cent means that all of the points go in the same direction; in this case, it is obvious that the monkey always responds correctly. **(C)** Localization in the prefrontal cortex of the region where recordings were made of the neurons that discharge in relation to the decision making and not only with the stimulus or the motor gaze response.
After Kim and Shadlen (1999)

To respond or not to respond: choosy neurons

In everyday life, we not only have to choose between two contrary movements. Perceptual choices often involve a combination of decisions, for example, signals that change colour, like stoplights at city intersections. We thus must adapt our decision based on the importance we assign to one or another aspect of the world around us and give other instructions. Other models have been proposed to explain decision processes in the brain. For instance, Shadlen and his group suggest that the brain accumulates sensory information over time and makes a decision once the information reaches a certain threshold [14]. Building on prior work done with monkeys, Hauke Heekeren and colleagues have proposed that higher-level regions in the human brain use a relatively simple subtraction mechanism to compute perceptual decisions [15].

In the monkey, the dorsolateral prefrontal cortex plays an essential role in tasks of complex discrimination that require a decision to respond either to a visual movement or to a colour. In particular, the 'oddball' paradigm consists in displaying a series of targets arranged in a circle and changing something about them, for example, the colour of one of the targets, consequently rendering it atypical. The prefrontal neurons are activated during the period where the monkey tries to register the presence of this unfamiliar target—the oddball [16]. But they are also involved in giving different orders.

In another experiment, a monkey is placed before a screen on which bright spots, all the same colour, are moving around. The experimenter changes either the colour or the direction of movement of the spots. The monkey must make a decision about one or the other. This *change of orders* is indicated by a coloured marker in a corner of the screen: yellow if the monkey must respond to the colour and ignore the movement, purple if the monkey is supposed to respond to movement without considering the colour. In the case where the colour is what matters, the animal must depress a lever if the illuminated marker on the screen is purple and not respond if it is yellow. In the case of movement, it must signal if the direction is oriented to the right and not signal if it is directed to the left. It is a difficult cognitive task for a monkey; it requires activating, choosing, and inhibiting responses, all at the same time. Moreover, it takes six months of training to obtain a 90 per cent success rate. Recordings have been made in Brodmann areas 6, 8, 9, 12, 45, and 46. Two types of neurons have been identified.

The first type are neurons activated by the instruction 'colour' (C cells) and the choice of corresponding behaviours. C neurons are located in the inferior prefrontal cortex, that is, the portion of area 46 that is ventral with respect to the principal sulcus and the upper portion of area 12. These two regions receive projections from the IT area, which carries information about colour to the inferior part of the dorsolateral prefrontal cortex.

The second type are neurons sensitive to the instruction 'movement' (M cells) in the regions of the superior prefrontal cortex, that is, the superior portion of area 46 and the superior arcuate area (8, 45, 6). These regions receive information from the parietal cortex in the dorsal pathway, which transmits via the MT region information about the movement of visual targets to the superior part of the dorsolateral frontal cortex.

A final type are neurons sensitive only to the decision to respond or not to respond (go/no-go) independent of the instruction 'colour' (C) or 'movement' (M). These neurons (CM) respond to both colour and movement. In contrast, changes of activity in either of the two conditions reveal a preference of some CM neurons for the stimulus that involves a response (go) and others for the stimulus that involves no response (no-go) [17].

Colour and movement cells have shorter latencies than CM cells. The former have a slight sensitivity to contralateral space, whereas the latter are indifferent to the presentation of such stimuli. Colour and movement cells are generally found in the regions of the prefrontal cortex that receive sensory information, whereas CM cells are located in the areas that appear to integrate this information.

These data suggest the model shown in Fig. 9.2. A cell that processes information about colour is designated as a C cell, M is a cell that processes movement, white circles designate sensory neurons, and grey circles designate go/no-go neurons, that is, neurons that actually code instructions regarding the behaviour the animal must adopt in a given situation. The CM cell receives information from the C and M cells, and decides based on its state and, probably, on context. This model assumes that modular processing of the sensory information is maintained right up to the prefrontal cortex. This level is where the behavioural significance of the sensory data is developed. These neurons thus play an essential role in associating sensory and behavioural information. Nonetheless, a complex process of decision making may depend on the context. So we have to suppose that carrying out this work of discriminating and adapting to the context requires that the neurons of the prefrontal cortex receive information from multiple sources and simultaneously organize them in a specific and hierarchical way.

It would be premature to conclude from this analysis of the prefrontal cortex that contextualizing the patterning of signals about the world and the body is only done in this region of the brain. For example, it is possible that some of the sensory neurons of the thalamus, the major centre for acquiring sensory information, are actually concerned with the content of information given by the senses, whereas those of another part—the mamillary bodies—work in cooperation with the sensory thalamus to do other kinds of processing.

In summary, it seems that decisions are made at all levels of processing. But decisions made in the final stages concern *global* aspects of the perceived world, whereas in the initial stages, it is the *local* aspects that are the object of decision making. Moreover, decisions made in the lower-level sensory relays involve categorizing and segregating into *attributes*, whereas the higher-level relays are concerned with situating perception in the *context* of an action or a reward. Decision making is initially about *deconstructing* and *categorizing*; afterwards, it is about *preference*. Indeed, it is at the level of frontal

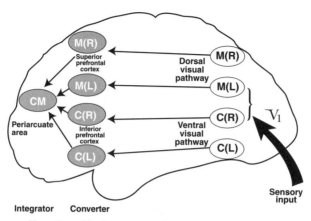

Fig. 9.2 Combining colour and movement in perceptual decision making: a theory. A monkey is presented with stimuli that tell it to respond (go) or not to respond (no-go) depending on the colour and movement of a target, or the combination of these two properties. Neurons in the prefrontal cortex are recorded to detect the ones involved in the decision (go or no-go). This scheme summarizes the results obtained and the proposed theory. Different neurons are activated by colour (C) and movement (M) in the primary visual areas (V1) of the left (L) and right (R) cerebral cortex. Information about colour is transmitted by the ventral pathway towards the inferior part of the dorsolateral prefrontal cortex, and the dorsal pathway transmits information about movement to the superior portion of the dorsolateral prefrontal cortex. In a first stage, this sensory information is transformed into activity having behavioural meaning in specialized regions of the dorsolateral prefrontal cortex in a parallel fashion: the meaning linked to colour in the inferior prefrontal cortex and that linked to movement in the superior prefrontal cortex. Each hemisphere processes the information of the images detected in the controlateral visual field. All this information is finally integrated in the periarcuate prefrontal cortex in a unique representation that contains the behavioural meaning of the combination of information about colour and movement. At this stage, a decision can be considered to have been made. The cells of the prefrontal cortex that are indicated in grey can intervene in the go/no-go process. Sagakami and Tsutsui (1999)

structures that a link is forged between information regarding the tangible world, intended action, and assignment of a value, as we will see in Chapter 11.

Decision making and conflict solving: spatial anxiety

The need to resolve conflicts is important not only in the functioning of the visual system. We saw in Chapter 5 that to maintain balance, a hierarchical cascade of mechanisms aids in resolving many problems and in making decisions. What happens when sensory data conflict? Or when the brain has trouble constructing a coherent representation of the relationships between the body and space? Many people suffer from disorders of orientation such as agoraphobia, panic attacks, and vertigo. More simply, imagine that an elderly person wants to keep her balance while the visual world is moving. This situation often results in fear of falling. We know that vision plays an essential role in maintaining balance and that visual perturbations can throw people off and make them fall. This vision-induced perturbation of equilibrium stems from the brain's difficulty in integrating contradictory sensory information when a conflict arises. In certain cases, postural reactions that normally prevent falling may be completely blocked. For example, if, during a fall, the visual world is moving along with the head, giving the brain the illusion that the head is not moving, the reflex activity of the muscles of the legs that stabilize posture fails [18]. Everything happens as if the brain 'believes' vision to the detriment of balance. To my mind, it is a mechanism of genuine perceptual decision making that arbitrates the conflict. I have suggested that the trouble elderly people have making the decisions needed to quickly solve this type of conflict is the cause of many falls, and no doubt it was the reason for the premature death of my beloved uncle, the architect André Bruyère, after a fall down the stairs. The brain, otherwise, is amazingly quick at developing strategies for resolving conflicts using its ability to predict [19].

We also know that anxious people are particularly vulnerable to sensory conflicts [20] and that differences exist between men and women [21]. Women reportedly depend more on the visual environment for orientation. Moreover, responses vary greatly from person to person. Some, as Jacques Paillard has suggested, are more 'visual' and others more 'proprioceptive', that is, they use or have more trust in so-called somatosensory information, data provided by the sensors of the muscles and joints.

We have also shown that anxious mice fall more than nonanxious mice in a situation of imbalance caused by sensory conflict [22], and that the precursors of serotonin enhance the capacity of mice to solve such a conflict.

Most of the studies carried out until recently have tried to define mechanisms that resolve these conflicts as though they were low level, that is, involving very little spatial cognition. On the contrary, I think that the alterations in perception of the vertical that often accompany bending of the body when, for example, a person is placed before a rotating disk [23] can be manipulated by the imagination.

Indeed, we have placed subjects before a disk rotating in the frontal plane and measured changes in the vertical subjective using the classical method of having an indicator in the centre of the disk that the subject must manipulate. A small region of the disk is bare of patterns such that the subject imagines movement in that region. Simply imagining movement in this neutral region changed the vertical subjective.

In other words, we showed, for the first time, that imagination could significantly change the way the brain resolves conflicts. We are talking about real mental decision making. I am convinced, for example, that astronauts who manage to overcome space sickness also often use cognitive strategies to resolve sensory conflicts caused by the absence of gravity. Although this experimental work is very preliminary, it suggests a new line of research into cognitive strategies for resolving conflicts and opens the possibility to behavioural therapies for relearning balance and walking [24] after vestibular lesions or dysfunction. There may be an essential role for the mechanisms that involve the hippocampus in compensating deficits of spatial disorientation [25].

My theory is that these mechanisms for adapting to and resolving sensory conflicts bring into play not only automatic processes of interaction and balance between the sensorimotor elements involved but also, and perhaps especially, involve *cognitive decisions* that spur sensorimotor systems towards solutions that, possibly, will substitute another system for an impaired system and choose the most appropriate strategy. Very little research has been devoted to this problem. Of course, the precise neural basis of these mechanisms would have to be determined before taking it on, at the risk of getting lost in vagaries. But the way is open for a neurocognitive theory of resolving sensorimotor conflicts that includes decision making.

References

1 M Merleau-Ponty, *Phenomenology of perception*, trans. C Smith (London: Routledge and Kegan Paul, 1962), p. 41.
2 A Berthoz, 'Neural basis of decision in perception and in the control of movement', in AR Damasio, H Damasio, and Y Christen, *Neurobiology of decision-making* (Berlin: Springer, 1996), pp. 83–100.
3 GR Fink, J Driver, D Phil, C Rorden, T Baldeweg, and RJ Dolan, 'Neural consequences of competing stimuli in both visual hemifields: a physiological basis for visual extinction', *Annals of Neurology*, **47** (2000): 440–6.
4 P Dupont, GA Orban, B De Bruyn, A Verbruggen, and L Mortelmans, 'Many areas in the human brain respond to visual motion', *Journal of Neurophysiology*, **72** (1994): 1420–4.
5 CD Salzman, KH Britten, and WT Newsome, 'Cortical microstimulation influences perceptual judgements of motion direction', *Nature*, **346** (1990): 174–7. For an analysis of the possibility of removing uncertainty in the detection of conflict between two movements by recording populations of neurons, see S Treuer, K Hol, and HJ Rauber, 'Seeing multiple directions of motion: physiology and biophysics', *Nature Neuroscience*, **3** (2000): 270–6.
6 CD Salzman and WT Newsome, 'Neural mechanisms for forming a perceptual decision', *Science*, **264** (1994): 231–7. The possibility that Bayesian processes are involved in decisions regarding perception of movement in MT was suggested by E Koechlin, JL Anton, and Y Burnod, 'Dynamical computational properties of local cortical networks for visual and motor processing: a Bayesian framework', *Journal of Physiology*, **90** (1996): 257–62.
7 A Thiel, C Distler, and K-P Hoffman, 'Decision-related activity in the macaque dorsal visual pathway', *European Journal of Neuroscience*, **11** (1999): 2044–58.
8 JM Fuster, RH Bauer, and JP Jervey, 'Cellular discharge in the dorsolateral prefrontal cortex of the monkey in cognitive tasks', *Experimental Neurology*, **77** (1982): 579–694; JM Fuster, *The prefrontal cortex: anatomy, physiology, and neuropsychology of the frontal lobe* (New York: Raven Press, 1997); JM Fuster, 'Network memory', *Trends in Neuroscience*, **20** (1997): 451–9; S Funahashi, MV Chafee, and PS Goldman-Rakic, 'Prefrontal neuronal activity in rhesus monkeys performing a delayed anti-saccade task', *Nature*, **365** (1993): 753–6.

9 S Funahashi, C Bruce, and PS Goldman-Rakic, 'Mnemonic coding of visual space in the monkey's dorsolateral prefrontal cortex', *Journal of Neurophysiology*, **61** (1989): 1–19.

10 JW Gnadt and RA Anderson, 'Memory related motor planning activity in posterior parietal cortex of the macaque', *Experimental Brain Research*, **70** (1988): 216–20.

11 PS Goldman-Rakic, 'Topography of cognition: parallel distributed networks in primate association cortex', *Annual Review of Neuroscience*, **11** (1988): 137–56.

12 J Kim and MN Shadlen, 'Neural correlates of a decision in the dorsolateral prefrontal cortex of the macaque', *Nature Neuroscience*, **2** (1999): 176–85. See also JI Gold and MN Shadlen, 'Representation of a perceptual decision in developing oculomotor commands', *Nature*, **404** (2000): 390–4.

13 SC Rao, G Rainer, and EK Miller, 'Integration of what and where in the primate prefrontal cortex', *Science*, **276** (1997): 821–4.

14 ME Mazurek, JD Roitman, J Ditterich, and MN Shadlen, 'A role for neural integrators in perceptual decision making', *Cerebral Cortex*, **13** (2003): 1257–69. See also J Ditterich, ME Mazurek, and MN Shadlen, 'Microstimulation of visual cortex affects the speed of perceptual decisions', *Nature Neuroscience*, **6** (2003): 891–8.

15 HR Heekeren, S Marrett, PA Bandettini, and LG Ungerleider, 'A general mechanism for perceptual decision-making in the human brain', *Nature*, **14** (2004): 859–62.

16 M Sagakami and K Tsutsui, 'The hierarchical organization of decision making in the primate prefrontal cortex', *Neuroscience Research*, **34** (1999): 79–89.

17 Ibid., p. 83.

18 L Nasher and A Berthoz, 'Visual contribution to rapid motor responses during postural control', *Brain Research*, **150** (1978): 403–7; PP Vidal, M Lacour, and A Berthoz, 'Control of vision to muscle responses during free fall: visual stabilization decreases vestibular dependent responses', *Experimental Brain Research*, **37** (1979): 241–52.

19 J Droulez, A Berthoz, and PP Vidal, 'Servo-controlled (conservative) versus topological (projective) mode of sensory motor control', in W Bles and T Brandt, eds, *Disorders of posture and gait* (Amsterdam: Elsevier, 1986).

20 I Viaud-Delmon, Y Ivanenko, A Berthoz, and R Jouvent, 'Anxiety and integration of visual vestibular information studied with virtual reality', *Biological Psychiatry*, **47** (1999): 112–18.

21 I Viaud-Delmon, Y Ivanenko, A Berthoz, and R Jouvent, 'Sex, lies, and virtual reality', *Nature Neuroscience*, **1** (1997): 15–16.

22 EM Lepicard, P Venault, F Perez-Diaz, C Joubert, A Berthoz, and G Chapouthier, 'Balance control and posture differences in the anxious BALB/cByJ mice compared to the non-anxious C57BL/6J mice', *Behavioural Brain Research*, **117** (2000): 185–95.

23 F Mast, SM Kosslyn, and A Berthoz, 'Visual mental imagery interferes with allocentric orientation judgements', *Neuroreport*, **10** (1999): 3549–53.

24 E Vitte, A Semont, and A Berthoz, 'Repeated optokinetic stimulation in conditions of active standing facilitates recovery from vestibular deficits', *Experimental Brain Research*, **102** (1994): 141–8.

25 A Berthoz, 'Parietal and hippocampal contribution to topokinetic and topographic memory', *Philosophical Transactions of the Royal Physiological Society, Biological Sciences*, **352** (1997): 1437–48; A Berthoz, 'Hippocampal and parietal contribution to topokinetic and topographic memory', in N Burgess, KJ Jeffery, and J O'Keefe, eds, *The hippocampal and parietal foundations of spatial cognition* (Oxford: Oxford University Press, 1999), pp. 381–99.

Chapter 10

Fountains

The elegant gastronomic soirees of the Taste-Vin confrérie at the renowned Clos de Vougeot vineyard in Burgundy offer a feast of delectable dishes and fine wines. But palate fatigue quickly sets in, and the pleasure runs the risk of diminishing if the palate is not given a break. Tradition demands that the meal be interrupted by a lemon sorbet whose pallor and simple roundness contrast with the thousand subtleties of the dishes that preceded it. Its acidity provides a reference for better appreciating both the foregoing flavours and those to come. Floating in alcohol, the sorbet takes on the milky colour of a light fog suspended in the silver goblet. The combination of simplicity and sharpness renews the desire of the tongue to venture further. This chapter, devoted to fountains, has a similar role.

Written on a whim, it is intended to remind the reader that the brain enjoys the multiple play of water and stone, of light and moss or leaves, of rigid shapes and liquid flows. Indeed, the brain is not just a tool for making decisions in dealing rooms where fortunes are gained and lost, in operating rooms where life and death are in the balance, or even in a court of law where the stakes are liberty or imprisonment.

Fountains will be our momentary distraction, a little digression on water, agent of diversity if there is one, whose fluidity refuses the bounds and restraints imposed on it by the fountain's solid architecture. Water reminds us that decision making is not a rational process that can be tidily summed up in an equation, even when introducing probabilities or principles of uncertainty. $\pi\alpha\nu\tau\alpha\ \rho\epsilon\iota$—'Everything flows'—wrote Heraclitus. The early models of the brain were hydraulic models.

Thinking about fountains helps us to understand the mechanisms of decision making. The surge of water belongs to the history of intercourse between humans and nature. It contrasts with the continuous and boring drift of everyday events.

The impetuous character of decision making is akin to the coursing rivers that feed glaciers and cascade within their depths through chasms called 'ovens'. The final gush of the torrent through the gaping mouth of the ice monster reveals the contained force and the patient impatience of the water that, like Snow White, only pretends to die and go cold all the while conserving its strength. This force is like William James's third type of decision making (see Chapter 2), passion which 'combines with ebullient activity, when by any chance the passion's outlet has been dammed by scruples or apprehensions' [1].

Decision making is like travelling on a river of uncertainty. The river Styx, say, which serves as a backdrop for Orpheus to regain his Eurydice, and to take her from hell and the abyss back into the light.

I have systematically studied the fountains of Italy. I am certain that eminent specialists of architecture have things to say about their structure. But I write ignorant of the fountains' ingenious designs. I have, I believe, discerned a few basic principles on which they appear to be built, and especially the effect that their presence and thinking about them has on me. Because a fountain is not just something to be looked at; it is to be approached, listened to, breathed in, and touched. One must yield to its dizziness and clamber over its edges, follow and experience its movements and guess their meaning. A fountain shares itself with others. It is the essence of encounter; it changes with the hours of the day and with the moon, which gives it a character that it forgets under the midday sun. Especially, one must drink its water. The equivalent of the French proverb for 'Never say never' means literally 'Never say, "Fountain, I will not drink your water."'

Water

The primary spirit of a fountain is one with the water it emits. Consider the water that descends from the mountain, still loaded with minerals, now cool and clear from the winter snow, now turbid with the mud of storms. At the tap it is all stirred up with oxygen and bubbles, eager to be caught in mid-stream by a mouth or two outstretched hands. Or it remains cupped in the basin, where it rests—quiet, heavy, serene, but ready to disappear through the outlet, its return awaited by the earth that offered it after receiving it from the sky. It caresses the hands that soak in it, amazed by its coolness. This water is a generous gift of gravitational acceleration; it flows rather than falls.

Fountain water captures the flavour of the ground it moves over. To taste the water from a fountain is to know the world that surrounds it as a wine reveals the soil that fed the grape. Heading back down after an ascent, as the sun rises and melts the snow, mountaineers know to watch out for the pernicious cold of glacier water. But they love the cool and sparkling water of the high forest springs, imbued with minerals from granite or shale, and the softness of its lather, refreshing as it enters, sliding gently into the mouth where its velvet touch contrasts with the dryness brought on by exertion. They love the aroma of the mist rising from the lazy river that crosses the meadows as they regain the warmth of the valley. Finally, they know they are home from the acrid taste of the village fountain, which tells of its journey through metal pipes.

Geometry

A fountain is architecture, geometry too, play of forms and forces, image of hierarchy or democracy. Three tiers of circular basins for some, elliptical or square bowls for others, combinations of pure geometrical shapes, or suggesting bathtubs, sighs, leaves, whatever. The simple cylinder of a hollowed-out tree trunk, a liquid staircase, spitting mouths, frog heads, horse heads: human imagination has found all possible combinations of shapes suited to the rhythm of water that shoots, rests, or slides. Some structures evoke a cupped hand for quenching thirst—a scallop or cockle, for example, whose container mimics the open-palm gesture of offering one makes to others.

Sometimes a fountain is a grotto, a sanctuary, sometimes a simple slab, and sometimes phallic, a jet of water pulsing towards the sky until it remembers the direction of gravity. It is a subtle play of the circle, the triangle, the square, the basic forms beloved of the adherents of Bauhaus who failed to see that such shapes must be buried within a living subtlety and that geometry can never be separated from murmurs and hesitations. The fountain reassures by obeying the horizontal and vertical and, when it bends, acknowledges the desires of the brain, which tolerates only a few departures from the oblique. It knows how pleasing the convergence of a circle and an ellipse can be.

Gushing

Fountains surge. Water wells out from a thousand places. Take the modest pipe, whose unpretentious cylindrical geometry imparts to all the wonders that follow an air of simplicity and rationality. Hiding a fountain's ingenuity in its innermost workings bespeaks humility: it is only water in a pipe. But it can happen that the water spouts from the mouth of a dragon, a horse, or a turtle. It never fails to astound me; this limpid regurgitation is not the usual picture we have of the fluids coming out of our mouth, except among lamas, as Hergé recalled in *Tintin au Tibet*. Animals generally spit out poisonous foods or foods they don't like. Why are animals so often the vehicles of the fountain's display? I don't know. Fountains are a veritable bestiary: rearing horses, fish, lions, turtles, mermaids and reptiles, not so many birds. Some of these animals share the spectacle with us; others are the source of the flow; still others take part in celebrating the water. They receive the stream and give it the shape of their body, which appears to come alive under the magic of moving water.

Sometimes, the stream emerges from a seashell, symbol of the sea and of fortune. Other times, as for the Belgians' Manneken Pis, humour reminds us—perhaps too coarsely—that our body also has fountains.

I like fountains whose hidden stream emerges from the basin without rising very far. It only creates a bit of bubbling that suggests invisible forces yet to emerge from illusory depths. This guarded flow of warbling water is calming, yet powerful.

The vertical thrust of such a stream represents a battle between the hope of life and the gravity that pushes Ariel towards Caliban. The leap heavenward is quickly reined in by the planet, which refuses to allow the precious liquid to escape. This struggle between two opposite forces, the yen to jump into space and the mechanics of terrestrial

188 | FOUNTAINS

Fig. 10.1 The fountain at Lake Geneva.

attraction, ends in the victory of the latter, and the conquered water falls back. But for one miraculous moment—which the Japanese call 'ma'—the water appears to hang suspended between these forces. At such times, it frequently assumes extravagant shapes that the wind disperses and the night light makes into a myriad shimmering stars. For that one moment, the stream could become the Amazon.

The vertical jet of Lake Geneva (Fig. 10.1) is also beautiful in its return to Earth. On this path, again governed by the laws of mechanics, it happily encounters the madcap whims of the air, which blows it about, bends its trajectory, crushes the fluid pipe, shapes it or transforms it into a gossamer veil streaked with light that suggests the charming evanescence of a wedding gown in the wind. Complex, nonlinear phenomena, turbulence and vortices trifle with the stream and make of its geometry the irregularities so pleasing to the brain.

When the jet is horizontal, the battle with gravity is less violent; it is heavy rather than combative; it suggests acquiescence. But above all, it is about movement. If the vertical stream is isolated in its struggle, the horizontal stream is often constructed to be drunk, for lips to approach it, for a face to splash in it, for the cupped palm to receive it, or bruised and swollen feet to be soothed in it. In this instance, the water is even experienced by our gaze as an integral part of our body, our movements, our sorrows, our desires. This water does not challenge the laws of Newton; it is renewal.

Plunging and splashing

A stream is not always carried away by the wind. Sometimes it stays strong and falls back into the plane of still water. It enters forcefully, like a diver, with a violence that smashes and splatters into a thousand drops and the rhythm of the waves. The shock of entry turns into a creative act. If you carefully observe this remarkable event, you will see the water reverberate there: scattered drops that make other drops, circles on the surface of the water making a ballet of concentric ripples that meet, intersect, greet each other, are absorbed, and harmonize.

If the fall occurs on stone, the dance of the water is more animated. It spreads in flying drops or is flattened into rivulets. It may bounce off naked stone and illuminate it with its damp sparkle. Sometimes it falls onto stone coated with moss, and the water both gets lost in the moss and transforms it into a tinted tumescence.

The fountains of the region of Forcalquier, in Provence, are constructed on the same model: a circular pedestal situates the fountain slightly above the square, the step low enough for a child to manage; then a large basin, often square or octagonal, whose edges rise just to the beltline, in such a way that you can gently lean your pelvis against it; it joins four streams of water from four horizontal pipes protruding from the four sides of a column. The flow of water from the pipes is not particularly forceful, but it is strong enough to fill a bucket while leaving time for a smile and a few words without anyone getting impatient. Indeed, it is possible to place a bucket under at least two of these streams thanks to metal bars that connect the column to the edges of the basin. Very simple, very beautiful. Frequently, two of the streams emerge from the heads of lion-men, the others just from stone.

It is the fall of the water created in this way that holds our attention: contact with the water of the basin has the same effect as that of a high diver entering the water of a swimming pool. The water falling from high above enters the water of the basin and creates turbulence, churning up water with air, which resolves into bubbles that vanish, as in a witch's cauldron. The cool water seems to boil, and this agitation spreads a ripple in the basin that suggests the quivering effect of pebbles on the smooth flow of a creek, which ravishes the eye and the brain. This extremely erotic entry of male water into female water, and the bubbling stream of the cascade, emit a quaver that fills the square with a tender rumbling.

Trickling

A fountain also trickles. On the rounded edge of an overflowing basin, the water coils and no longer resembles a lake but light sheets drying in the wind, a flowing drapery that caresses the stone while enveloping it. Sometimes, the trickling is produced on the surface of stone that it gently moistens, making it glisten with humidity tinted the same soft green and yellow that in spring, in the space of a few weeks, illuminate the leaves and the fields. Trickling is essential to the fountain; it gives it the languor and variety of delicate encounters. It brings the stone to life, lights it with reflections

Fig. 10.2 The fountain in the Mouffetard section of Paris. Courtesy J. P. Martin.

that change with every hour, awaits any hand that happens to be placed there, which in an instant becomes itself moss and stone, sloshing gently. Trickling amounts to a mental caress.

Froth is the ultimate ornament for a fountain. It is the product of trickling, sometimes light and almost black; it carpets the stone discreetly and allows one to see through the fine layer of gliding water. Sometimes it is voluminous, swollen, turgid, shining with monochrome coatings of greens and browns, capturing the sun, puffed with water that it seems to guard jealously.

Froth is the opposite of reason. It is delicate proliferation: it contrasts its humid smoothness with the rigidity of marble. The finest example of this contrast is to be found at Digne. Within the city limits, by the side of the road to Barcelonnette, a fountain has been built into the walls of the hills. Its architecture is an almost life-size columned temple that emerges from the mountain as if to extend it, in the manner of the troglodyte churches of Cappadoce or Petra, in the land of Solomon, in Jordan. But a surprise awaits us there. Between the columns of this temple, huge, frothy blisters claim the space and spill out of the frame—great big tears of foam and stone on which the water drips its concretions, contrasting with the classical order of the columns and the wall of the temple. You'd think you were looking at those paintings where the artists have blended the ruins of ancient buildings with the natural environment that has reclaimed them. Is this incongruity the result of neglect? And do these chubby foam ogres conceal statues? Or are they a symbol of the struggle between the mind's twin tendencies of finesse and geometric precision?

The basin of serenity

After having gushed, then trickled, water remains a while in the basin or bowl. Now calm, like the ocean when the tide is slack, the water waits to begin flowing gently. It symbolizes peace, calm reflection, and those moments of deliberation when all thoughts are at rest. It is then that, Narcissus for an instant, I see my face—my double is floating there, mirror image of myself and the sky. I see, too, the reflection of the fountain, deformed by the laws of mechanics, blown by the wind, shattered by the hand that dips into it.

The water of the fountain is a well into which one tosses a few coins to appease the gods. It tempts bathers. I have often contemplated the red, white, and black fish that bask or swim there, never ceasing to search for what they will not find, just as, often, our own mind hesitates in a goalless quest and turns over its thoughts. The water is a fluid mirror that reflects the light of the sky on the stone from top to bottom. It animates the shadow of the trees with the rhythm of its movements, which follow the wind, the fall of the streams, and so on. The water creates a universe of depth. This impression breaks the rigid ground of the city and gives the illusion of a possible void, from the other side of the mirror.

At the crossroads

The realm of the fountain invites encounters. Those who draw near to it are united in the enchanted bosom of the water. It is a common attractor that unifies like the thread of a game of pass the slipper. Around the fountain, intrigues are concocted, love knots tied, gossip passed, and even indifference shared. People wait for each other at a fountain, and find each other and separate there. The paths that lead to the spot find their intersection. The centrifugal motion of the water that goes from the centre of the fountain to its perimeter contrasts with the convergence of paths that connect at the fountain. The fountain encourages both rest and departure, and its streams are as much suggestions of possible routes to take as compass points. The fountain inspires decisions; it is the centre of a crossroads. It marks out their possibilities without imposing a choice.

There is no better example of the role of the fountain in a city than the fountain of the four dolphins, at Aix-en-Provence (Fig. 10.3). The old town of Aix is based on a plan like that of Miletus designed by Thales. It is a precise grid of streets arranged like Manhattan: several broad parallel boulevards, of which the cours Mirabeau is the widest, and others perpendicular to it. The entire thing is perfectly Euclidean and rational, and one's sense of subtlety would founder if something did not break the implacable logic. One of the intersections was widened just so to give the impression that within the narrowness of the streets, all of a sudden some space set apart in this way could offer vistas, trees and their shade to the eye, the rustling of wind in the ear, and room for children to play.

Four streets converge here, oriented more or less along the cardinal points. As a result, the sun enters the biggest street in one direction in the morning, and in the other in the evening. A square base of stone was dropped into the centre of the site, duplicating its geometry. On this pediment sits the fountain, a hip-high basin, almost large enough to accommodate a little sailboat and filled with water blown into it by four dolphins, each facing one of the four streets. Their puffy faces, popping eyes, and arched bodies, terminating in a tail that seems to whip the air, contrast with the calm, exact streams of water, both strong and gentle, that escape their mouths, which are equipped with a plain pipe that looks like a cigarette. The rhythm is set—square pedestal, circular basin, dolphins facing the four streets—an alternation of square and circle. The dolphins are arranged around the square stone, which is topped with an obelisk, itself also square.

But the builder of the fountain turned the obelisk 90° such that each angle is directed towards each street like an arrow. The square has thus become a diamond shape for the eye; the geometric monotony of the site is interrupted to gratify the mind.

Finally, the alternation between the rigidity of the right angles and the pleasure of the curve is supplied by a shape that dominates the obelisk: balanced at the pointed summit, an object that looks like an egg or a flower seems to offer to the sky its voluptuous roundness and to negate the self-assuredness of the Euclidean grid of the city.

Fig. 10.3 The fountain of the four dolphins in Aix-en-Provence.

How wonderful! And the pleasure only increases over the course of the day and night. Indeed, at dawn, the dolphin to the east is lighted in front, as is its stream, which takes on a different quality seen from this or that angle. At midday, the shadow of the trees projects onto the stone the body of the fish, the water, the basin, and the changing frivolities of the shapes of the leaves. If, by luck, the mistral, the wind that brings madness, kicks up, and its breeze embraces the city, then a ballet and visual music cast a spell over the spot.

Of course, I have seen passers-by drink at this fountain, others dip an arm into it, but I must also tell you the best story about the fountain of the four dolphins. Late one evening I saw a man walk up to the fountain, at the outlet where the water flows towards Caliban's cave. The man carefully rolled up a sleeve, then, instead of simply dipping in his arm to cool himself, thrust it to the bottom, looking for something—I thought perhaps a coin. Wrong! It was the key to his house, whose magnificent wooden, door, facing the square, he was going to open. What better testament could there be to the fountain's intimacy?

That is how, at a crossroads, an architect managed to introduce into the humdrum layout of the city a monument rich in fantasy that still remembers its original purpose—namely, to choose among four streets—but that provides the brain innumerable opportunities to pause for a moment, to gaze or to linger, to let thoughts wander, or to sigh, and that offers the right angle an occasion to fertilize the imagination and to encourage meetings.

Subtlety

A fountain is all about nonlogical intuition. How could we think without fountains? What cloister—place of reflection and walks, of isolation and meditation—could exist without one? In such a setting, a fountain is the sign and inspiration of fluid and fertile reasoning. Only a fountain allows the mind to abandon fixed, fossilized ideas, and stubborn preconceptions. The fountain alone encourages the leaps of logic that spark innovation, that test theories without shattering them.

The awfulness of our urban architecture, proof of the mediocre, unrefined thinking that I decried in *The Brain's Sense of Movement*, goes hand in hand with the disappearance of fountains. Architects have abandoned roofs; they have also given up on meeting places. Water has no place in a world of filing cabinets, those endless rows of buildings aligned with military precision. With such architecture, everything is predetermined; one is on the first floor or the fifth; one is in building C or F. In contrast to these cities, which we used to love, in the village space is not assigned. The fountain symbolizes a confluence where imagination is stirred by the water that flows and surges, settles down and burbles forth, a model of the pliancy that, in a petrified universe, is our reason for living.

Persuasion

Making a decision also involves persuasion. The theme of flowing water turns up in the efforts of Louis Jouvet to teach the art of persuasion to his student Claudia as she rehearses Doña Elvira's famous scene in *Don Juan* (act 4, scene 6). Doña Elvira is attempting one last time to persuade Don Juan to give up his dissolute life or be damned. In arguing her case, she affects detachment from the mixture of love and hatred that she feels for him:

> Do not be surprised, Don Juan, to see me again so soon, and in these clothes; but I have urgent reasons for coming, and what I have to say to you will admit of no delay. My anger of this morning is all gone, and I come to you now in a very different frame of mind. I am no longer the Doña Elvira who prayed for your punishment, and whose outraged feelings found an outlet in

threats of vengeance. Heaven has banished from my heart all that was unworthy in my love for you; the heady violence of a criminal attachment, the shameful transports of a gross and earthly love. All that remains for you in my heart is a flame purged of sensuality, a holy affection, a pure and disinterested love which, with no thought of self, thinks only of your good [2].

Jouvet counsels Claudia to deliver this monologue in the following way: '*Feeling* must motivate your reading of the text. That's the art of acting. For a piece like this one, if the words evoke the feeling, that's all well and good for the actor and the audience. But to my mind, it is the actress's emotional state that compels her utterance . . . *Feeling is what will make your actions meaningful* . . . You arrive at a gallop. You run towards your lover, but the instant you enter the house, you compose yourself. Your walk takes on a measured, almost somnambulant quality. This woman who could not move fast enough, who was in such a hurry, suddenly assumes a preternaturally calm and quiet state. That is the feeling you must capture, or you won't say your lines convincingly. You argue, you reason; but the text doesn't have to be logical. What it needs is to be delivered with the ease that comes from the heart. I think it is the most extraordinary monologue in classical theatre' [3].

Jouvet's grasp of emotion suggests none of the cold, inflexible logic that makes the structure of a thought predictable, but rather the agility of running water that the fountain builder guides by leaving it free. That is what Jouvet tells Claudia in another lesson on the same text: what matters is '*the ability to goad one's own emotions to find new starting points*. This text is a good example. It is like a flood of water that arrives and courses. Water doesn't always move with the sort of slow majesty you find in Bossuet;[1] water surges, it splashes off a rock, it flows less quickly depending on the incline of the terrain, or suddenly it cascades down. This inrush of feelings does not have the consistency you are trying to give it. In addition to Don Juan's own gestures, there should be a churning of emotion that I don't see in your acting' [4].

The sorbet has been eaten. For those who have been kind enough to stay with me, we must now return to the thread of our discussion. Decision making is like a fountain. It is often described as a logical ordering of accumulated facts dictated by geometry and architecture. But actually, the world of decision making is immersed in the liquidness of a sentient body. We talk of 'cascades' or 'streams' of arguments, we comment on the 'rippling effect' or 'trickling down' of political and economic decisions, and a simple solution is said to 'flow'.

Decision making is the result of a subtle game that, behind the apparent simplicity of the conclusion, hides many hesitations, false starts, and choices. In the deceptively solid framework of a hierarchical and formal geometry, these choices themselves conceal a host of impulses and confusions, the effects of the wind, the reflection of the world and of oneself. Deciding entails a compromise among the laws of gravity, the art of the

---

[1] Jacques-Bénigne Bossuet, a French bishop of the seventeenth century renowned for the eloquence of his oratory. [Trans.]

fountain maker, the power of the water, and the caprices of the moment, the mating season, or the anger of Manon of the spring.²

References

1. W James, *Psychology: the briefer course*, vol. 14, *The works of William James* (Cambridge, MA: Harvard University Press, 1983), p. 371.
2. Molière, 'Don Juan', in *Don Juan and other plays*, trans. George Graveley and Ian Maclean, ed. Ian Maclean (New York: Oxford University Press, 1968), pp. 79–80.
3. L Jouvet, *Molière et la comédie classique* (Paris: Gallimard, 1992), 5th lesson.
4. Ibid., 1st lesson.

---

² From the title of a revenge tragedy (*Manon des sources*) by Marcel Pagnol involving control of the water supply in a rural French village.

Part 4

Magical thinking

Chapter 11

The physiology of preference

> It is a misunderstanding of the phenomena of pure expression when a certain psychological theory makes them originate in a secondary act of interpretation and declares them to be products of 'empathy'. The main weakness in this theory . . . is that it reverses the order of the phenomenal data. It must first kill perception by making it into a complex of mere sensory contents, before it can reanimate this dead matter of sensation by the act of empathy.
> *(Ernst Cassirer [1])*

Unable to decide between two bales of hay, Buridan's ass[1] died a victim of its inability to choose. This happens to us when, faced with a sumptuous buffet, dazzled by so many dishes, we do not know where to start. Feeling gluttonous, but also desiring not to gain weight or to appear greedy to the other guests, we must decide what will satisfy us for an evening. Deliberation is difficult; it means choosing between the possibility of returning for a second serving, the fear that the dishes will be removed or consumed, and taking our 'preference' into account.

Preference and emotion play an essential role in the process of decision making [2]. What happens in the case of the buffet also happens in everyday life. Most of the time, we have the choice of acting one way or another based on our preference. Is it possible to develop a physiology of preference? In what structures of the brain do these neuronal deliberations take place? Where do we get the idea that another person is the man or woman of our dreams? What produces the inhibition that keeps us from eating oysters if we have once gotten sick from them? Where is the decision made not to flee from a reptile that we have identified as a harmless garden snake? Where and how does judicial or penal action keep a delinquent from relapsing? 'The heart has its reasons of which reason knows nothing' wrote Blaise Pascal. Where does the encounter between cognition and emotion, between reason and feelings occur? Where does the choice of action—based on expected reward, failure, or past success—take place?

---

[1] Contrary to what one might think, John Buridan (c.1300–c.1358) was not a farmer. This philosopher intuited, even before Galileo, that Aristotle was wrong, and that to move, bodies once launched have no need of force to constantly propel them.

A major problem is to know whether the information given by the senses is imprinted with emotion from the outset. When we are very angry, is it that the sensory analysis of the world is profoundly altered by emotion to the point of 'seeing red', or is it only at the level of the frontal-most areas of the cerebral cortex where perception and emotion meet [3]? In other words, is decision making influenced by emotion at its source, since we have shown that basic decisions are already made at the level of the primary centres of perception? Or is emotion a late arrival in the complex processes that shape our perception of the world?

The physiology of fear

You are walking in the forest when, suddenly, a long, supple form glides across your path. Your companions shriek in horror: 'A snake!' It takes only a fraction of a second for an uncontrollable impulse to flee to register. But a more careful examination allows you to see that what you are looking at is perhaps not a venomous snake but a harmless one. Fear gives way to curiosity. The brain switches strategy. A structure of the brain called the amygdala plays an essential role in the initial reaction, automatically identifying danger and triggering flight. Yet if our first reaction to a snake is flight, the second may be to understand that it poses no danger and to want to look at it more closely. What is the neural basis of this play of hierarchical behaviours that opposes reactions of curiosity and survival, approach and flight, where the decision to flee is replaced with the desire to understand? Today the links between cognition and emotion are better understood [4]. Let us start with some data concerning a critical brain structure: the amygdala [5] (Figs. 2.3 and 11.1).

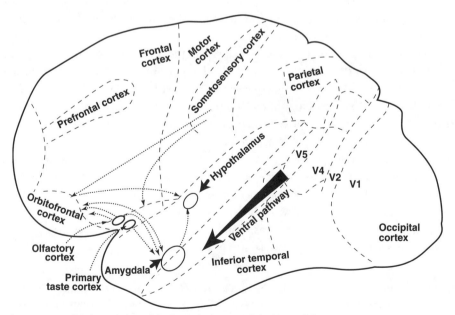

Fig. 11.1 Localization of the orbitofrontal cortex and the amygdala.

The first clinical observations of the function of the amygdala date from the end of the nineteenth century, when it was noticed that following damage to the temporal lobe, monkeys were unusually docile. Fifty years later, it was confirmed that bitemporal lobectomy, including the two amygdalas, caused a hypersexual and hyperoral syndrome in monkeys that led them to eat foods they would normally reject. The monkeys put everything in their mouth. Notably, they showed no fear: they remained calm in the most frightening situations, such as the appearance of a snake in their cage. This so-called Klüver-Bucy syndrome has also been described in humans [6].

In fact, the amygdala is involved [7] in two mechanisms that involve at least two different neural pathways, one short and one long (see Fig. 2.3). The rapid pathway enables an instantaneous avoidance reaction to danger stimuli. For instance, the sight of a snake or a spider or a threatening person activates the neurons of the amygdala. In 75 ms, we can trigger a ready-made reaction to flee or to stay put. Some of these reactions are innate; but others are learned through conditioning, that is, by association with a disagreeable experience at the appearance of a stimulus. This learning of fear is important in formulating a biological theory of decision making. For to decide is also to remember past successes and failures, rewards, and punishments.

Once is enough to retain the memory of a stressful stimulus, for example, a burn. So this learning is powerful and long-lasting. A rat that has learned to associate a sound and a slight electrical shock remembers it beginning with the second time, and has only to hear the same sound to show fear. This test was widely used to understand how the amygdala works. The auditory signals project onto the lateral nucleus of the amygdala, the principal nucleus for incoming sensory information. These signals are then projected to the central nucleus of the amygdala, directly and indirectly by way of the basolateral nucleus. Learning in the nuclei of the amygdala may well be produced by a synaptic mechanism such that, if two signals are transmitted simultaneously to the same neuron, the molecular mechanisms that associated the two signals are reinforced. Association is thus formed at the level of the neuron itself. It involves so-called long-term potentiation: the response of the neuron to the same stimulus steadily increases. The response is potentialized.[2] The acquisition of fear conditioned by the amygdala depends on other very subtle synaptic mechanisms that we will not consider.

The activity of the neurons of the amygdala does indeed, then, reflect preference and especially learning of a connection between a stimulus (an image) and a punishment (a disagreeable stimulus). For example, the neurons' response is altered both when the monkey is given a piece of melon and when the melon is salted, which monkeys do not like at all. However, the range of values of the stimuli recognized by the neurons is relatively rigid. For instance, they respond to reinforcers such as food, but do not change their response when the relation between stimulus and reward is modified. If a

---

[2] Injection of AP5, an agent that blocks NMDA glutamate receptors. (NMDA is a basic modulator in learning and memory). Thus, NMDA-induced long-term potentiation suppresses conditioned fear.

sound and a piece of melon are presented at the same time, the neurons of the amygdala associate them. But if the same sound is associated with salted melon, the neurons continue to discharge for a bit, which shows that they have only a limited capacity to adapt and to reverse their response when the context suddenly changes, in contrast to the orbitofrontal cortex (see later).

Nonetheless, the bulimia of Klüver and Bucy monkeys is not only due to their loss of preference; it is also linked to the intervention of the amygdala in the regulation of satiety. Monkeys with ibotenic acid damage to the amygdala exhibit a very specific abnormality: following damage, they continue to eat indefinitely. The same disturbances have been observed in humans.

The amygdala is also involved in identifying certain emotional characteristics of faces. Patients who have suffered bilateral destruction or disconnection of the amygdala can identify faces but cannot recognize their emotional expression or realize that a person is looking at them. The properties of faces are processed in several different areas of the brain [8]. But brain imaging shows very strong activity of the amygdala during presentation of faces showing emotion [9]. More generally, the amygdala is involved in recognizing visual scenes with strong emotive content, for example, erotic scenes or past memories of an emotional nature.[3] Preference, and thus decision making, are intimately linked to memory.

Direct perception of a face is important. When we have a major decision to make concerning a person in our private or professional life, we prefer to do it 'face to face'. How often do we say, 'I would like to see you for a minute?' Perhaps you have noticed that when someone looks at you directly, you experience a particular emotion, a curious sensation of being caught or seized by the gaze of the other person. In certain cultures, people do not look at each other directly, or at least, women are not supposed to look men in the eyes. Everyone knows that it is better to look away in many circumstances, especially to avoid being influenced by someone else's arguments in making a decision. To reinforce our powers of persuasion, we often stare at a person. This 'capturing' the gaze of the other person is accompanied by very strong activity in the amygdala [10].

The amygdala and pessimism

When the amygdala detects danger, it sends a message to the hypothalamus, which activates the pituitary gland, which itself secretes the hormone adrenocorticotrophin. In the cortex, this hormone induces secretion of steroid hormones that in turn modulate the activity of the neurons of the hippocampus and the prefrontal cortex. It is one of the mechanisms of stress induction. It may link stress and decision making. However, one must guard against extrapolating to humans mechanisms at work in the

---

[3] A possible mechanism for this involvement of memory is the following: the basolateral nucleus facilitates *consolidation* of the cortical processes of memorization via the basal anterior brain and through interaction with peripheral adrenergic changes.

rat. In this animal, a sound can trigger an affective and behavioural reaction along LeDoux's short pathway. But these mechanisms are much more subtle in primates and humans. This subtlety was amply revealed in the recent work of Edmund Rolls, to which we will refer for data relevant to this chapter, as well as to the books of Jaak Panksepp, and of Daniel Tranel, and John Aggleton, cited earlier.

Nonetheless, the workings of the amygdala remain largely a mystery—an enigma. An increase in its activity is observed in subjects inclined to pessimism [11]. But whereas in the monkey destruction of the amygdala produces loss of affective behaviour and the catastrophic impairment of social interaction, such changes have never been observed in humans [12]. Probably in humans many parallel mechanisms make it possible to rapidly compensate for these deficits.

The orbitofrontal cortex: flexibility and contextualization

Suppose that one of the merchants in your neighbourhood prepares superb baked goods and you have taken to buying your favourite pastry in his shop. You have created an association between the reward you expect and the act of going into the bakery, which involves several brain structures: the amygdala, the nucleus accumbens, the hippocampus, and so on. But now suppose that one day you find out that the baker holds political positions that are racist, or that he insulted your child when your dog, drawn by the smell of the cakes, entered the store. So you decide never to go into his shop again. For that, you have to modify the association you created, even though it has been reinforced by several batches of cakes. This flexibility of choice and its contextualization are partly a function of the orbitofrontal cortex, which, as we will see further on, also allows us to work out *relative preferences*.

Making a decision requires associating events, sensations, memories, and so on. This work of association is carried out in part in the amygdala. But making a decision also requires deliberating over it, changing one's point of view, mentally altering the relations between associated elements, and simulating different possible realities. So it has to be somewhat flexible. For example, if the image of an apple is associated with sugar, then with vinegar, the value of the association created in our brain has to be reversed—between the apple and the sugar (agreeable) and the apple and vinegar (disagreeable). This flexibility is limited to the level of the amygdala.

If the hand of a subject is stroked with a piece of velvet, very soft and pleasant, and a piece of rough wood, the orbitofrontal cortex is activated more strongly by the velvet than by the wood [13]. This activation is linked to the degree of subjective pleasure evoked by the velvet. It really is a judgement of pleasure and not a sensation developed in the cortex that is at work here in evaluating the consciously emotional perception of a sensory stimulus. It is probably also related to the disagreeable character of pain, for patients with lesions in this region report that they perceive a painful stimulus but that they do not experience the pain as disagreeable!

Cognitive functions of the orbitofrontal cortex

Damage to the orbitofrontal cortex causes euphoria, social irresponsibility, and sometimes a loss of affect. Patients experience difficulty changing strategies owing to consecutive reinforcement of environmental contingencies or to changes of rules imposed on them. The orbitofrontal cortex is not only involved in evaluating pleasant and unpleasant aspects of foods or other sensory experiences, but also in the most complex cognitive functions linked to predicting the consequences of actions, and to more abstract evaluations such as winning or losing money [14], or violations of social norms [15].

Patients presenting with lesions of the ventral prefrontal and orbitofrontal cortex show a marked deficit in this ability to switch strategy.[4] This characteristic involvement of the orbitofrontal cortex in the reversal of values of judgements in decision making—called reversal learning—is clear in the monkey; but the ventromedial cortex in humans appears to possess the same function [16]. Recent reports and reviews have now established the variety of rules the orbitofrontal cortex uses in decision making and cognitive processes related to evaluating our actions [17]. The trouble patients have performing reversal-learning tasks is associated with difficulty inferring a rule based on successive observations. For example, during a card game, patients are asked to infer the rule that determines the order in which they show the cards. The examiner says only 'right' or 'wrong' each time a card is laid down and can change the rule at will. Patients struggle to infer the rule and to think of what strategy to adopt.

These deficits are clearly linked to symptoms exhibited by these patients in their social life: uninhibited or maladaptive social behaviour, violence, lack of initiative, errors in interpreting the behaviour of others and, sometimes, disregard for their own situation.

The impact of these disturbances on daily life is significant. If one of these patients has a car accident and receives financial compensation, he tends to spend his money lavishly, without thinking of the future. Such patients can perfectly well estimate the positive or negative value of a stimulus; what they cannot estimate is implications for future actions or changes in rules or context, which are characteristic of functions of the prefrontal cortex. They also have difficulty doing spatial tasks. A subject might be shown a maze drawn on paper and be asked to follow a path with a pencil. At each intersection, a sound signals whether the subject has made the correct choice [18]. The right decision is known only to the experimenter. The patient must infer it. But he has

---

[4] For example, they are shown visual stimuli on a computer screen that is sensitive to contact (touch screen). They must touch certain stimuli with their finger and not touch others. They win a point when they react to the appearance of a good stimulus. So they must create an association (stimulus-contact). They lose a point when they make a mistake. Once they have learned the sequence of stimuli, the association is reversed. They must restrain their hand at the appearance of the stimulus that, previously, resulted in a reward when they touched the screen.

difficulty understanding the rule. Sometimes such a patient can even correctly verbalize the strategy, but he is incapable of executing it.

What cerebral activity is linked to decisions concerning financial gain or loss [19]? To find out, based on a protocol derived from that of Antoine Bechara and Antonio Damasio, a subject is shown one red card and one black one and must choose one or the other. In deciding, the subject knows only that half of the time, the red card is the right choice, and half of the time, the black one is, and must thus guess which card is the right one based on no rational information—which is what we do when we play cards or the stock market, but also, generally, in everyday life. To the right of the computer screen is displayed a bar indicating a reward score. Each wrong choice diminishes the bar by the same amount. After the experiment, subjects are asked to describe their degree of pleasure or disappointment.

The interest of this experiment is in separating two variables: the first and simplest is the level of reward; the second is the tendancy of the rewards to increase or decrease. The situation is similar to what we see on television when the stock analyst displays each figure with an arrow that shows the upward or downward direction of a stock. Everyone knows that the direction of this arrow determines our feelings about our financial security and guides our investments.

For a subject, winning activates most of the areas that belong to the limbic system for evaluating rewards.[5] When the subject has won a lot and has just given a correct answer (analogous to the stock market at the height of the boom at the end of the 1990s), the pallidum, the ventral thalamus, and the subgenual cingulate are activated. Damage to the subgenual cingulate is involved in depression [20], which dampens optimism and makes it hard for patients to see the good side of things. A basic optimism is fundamental in decision making. In contrast, losing, and thus accumulating penalties (which is the case today with the stock market), activates the hippocampus.

Psychologists have shown that decision making is influenced by mood [21]. The orbitofrontal cortex is one of the areas involved in positive and negative emotions. It has been suggested that the *left cortex is particularly involved in positively evaluating* events in life, and that is why it seems that patients with impaired activity in this region are especially vulnerable to depression [22]. A lesion of the orbital cortex in the monkey produces emotional changes somewhat similar to those caused by lesions of the amygdala, such as a decrease or disappearance of aggression and defensive reactions towards certain animals, such as snakes, or even unfamiliar people. Monkeys no longer reject as much food and especially no longer show food preferences. Macaque

---

[5] The ventral striatum (equivalent in humans to the accumbens in animals)—one of the critical centres of projection for the dopaminergic system—is activated. The orbitofrontal cortex and the caudal nucleus are activated in proportion to the general degree of excitation linked to risk taking when the subject had big winnings and a lot of capital, or a substantial loss and a large deficit, situations that may induce activity to evaluate the need to change strategy.

monkeys with lesions have trouble distinguishing reward from punishment and especially changing behaviour when the response is not appropriate to the situation. They continue with what they are doing. They make errors in go/no-go type tasks [23]: they cannot suppress action. When a reward is suppressed, a healthy monkey progressively weakens its response because it no longer believes in the reward. Monkeys with damage to area 13 cannot choose between two visual stimuli, which perhaps is due to the trouble they have inhibiting a response to a stimulus that is not followed by a reward [24].

Inhibition is a basic component of decision making. It is not the only one. For example, monkeys with lesions cannot memorize objects over a short period. But deliberating requires keeping in mind what is being deliberated. This working memory, which concerns detection of objects in the inferotemporal cortex (see Chapter 8), is different from that altered by a lesion of the dorsolateral prefrontal cortex, which is concerned with space. What a multitude of memory mechanisms need to be coordinated to come to a decision!

Salmon or sea bream?

If we prefer salmon to sea bream, salmon is a reward for us. Are there neurons in the orbitofrontal cortex that encode the value of a reward? Answering this question requires isolating a moment in an experiment when an animal estimates the value of a reward. It is not easy, but physiologists have succeeded in doing it by breaking tasks down into successive stages linked to the different aspects of such a moment. For example, suppose that you want to drink a glass of wine: you must first direct your gaze towards the glass, then initiate the action required to pick it up. The brain works so well that we must be able to record several types of neuronal activity distinguished by the intensity of their discharge or the number of action potentials they produce per second: some will be activated according to the specification of the goal, others with execution of the chosen action, finally others with the evaluation of the reward.

To solve this puzzle, a monkey is trained to drink in response to various signals during which the activity of its neurons is recorded. The monkey is first shown a visual stimulus, for example, the image of an apple on a screen (Fig. 11.2). A delay is introduced, during which the neuronal activity must be linked just to the visual perceptual properties of the neuron. Then the monkey is given a signal that requires it to evaluate whether it has the chance to obtain a reward by responding. Thus it must think before responding. The monkey must press a button. To observe the activity during which it prepares its response, another delay is introduced. This second activity is considered as reflecting the cognitive processes associated with the decision to respond. Finally, it is permitted to respond by pressing on a button. It responds and obtains a reward (fruit juice and so on). To distinguish *imagining the reward* from *actually getting* it, at times a stimulus is substituted that entails no reward.

The orbitofrontal cortex contains neurons that activate only when the monkey is engaged in evaluation and decision making (the 'appraisal' of the psychologists

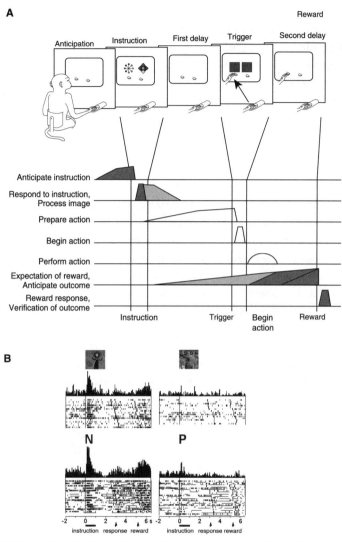

Fig. 11.2 **(A)** Types of activity found in the orbitofrontal cortex of the monkey in the context of the choosing task (activity represented in grey). The task could have shown activity linked to the preparation and execution of movement, but no activity of this type is present at the level of the orbitofrontal cortex (activity represented in white). **(B)** Activity of a neuron of the orbitofrontal cortex that responds specifically to stimuli representing the nonpreference of the animal, independent of the type of nonpreferred stimulus (letter N or picture by Miró associated with this reward). The neuron does not respond to images associated with the monkey's preferences (letter P or another picture by Miró associated with preference). After Tremblay and Schultz (1999)

mentioned in Chapter 2) concerning a possible reward. Recording these neurons and those of the basolateral part of the amygdala reveals that these two structures are involved in this learning but in different ways.[6] These results confirm the generally acknowledged hypothesis that the prefrontal cortex has an executive function, in part by representing ongoing events and the expected results of action [25].

The property that distinguishes the neurons of the orbitofrontal cortex from those of the amygdala does indeed seem to be the *reversal* of associations between a visual stimulus and a reward, as pathology suggests. Suppose that we associate the image of an apple with a piece of sugar and the image of a tomato with salt or vinegar. A monkey will enjoy seeing the image of the apple because it evokes sugar, but it will be disgusted on seeing the tomato because it evokes salt.

Next the relationships are reversed. The monkey is shown the apple on being given vinegar or salt; it will very quickly reverse its behavioural response. Rolls has shown that the neurons of the orbitofrontal cortex have the ability to change rules of association, which is not so much the case for the neurons of the amygdala. What is important in attributing a role for these neurons in decision making is that they are active when the animal evaluates the reward or the punishment *independent* of its response [26]. The neurons are not concerned with the action of consuming the preferred product as such.

So the orbitofrontal cortex is important in selecting a goal. If the animal is shown images of *two* foods at the same time, for example, an apple and a banana, the animal must decide. Certain neurons respond to the preferred food, others to the one the animal does not like. They participate in the neural basis of *aversion* to or preference for food, not to evaluation of the image itself.

If a monkey is offered three of its favourite foods in a certain order, three rewards with a hierarchical value, the neurons of the orbitofrontal cortex are activated by the stimulus in every case. By analogy, a preference neuron in your orbitofrontal cortex will be activated both by a Côtes-du-Rhône when you give it a choice between that and a table wine, and by a Clos-Vougeot when compared with the Côtes-du-Rhône. The system of preference is independent of the object or its consumption.

Motivation clearly comes into play in the activity of these neuronal constituents of a decision-making system. This correlation of their discharge with the development of preference does not mean that they belong to an 'executive or supervisory' system or that it is the neurons that decide. So it is not surprising that damage to these structures makes it difficult for subjects to make decisions, to organize their behaviour based on context or value, and so on.

---

[6] In a task of olfactory discrimination, the neurons of both regions were activated. Most of the neurons of the amygdala, considered to be critical for learning associations, were activated only by mistakes and changes in strategy to avoid being punished with a disagreeable taste. In contrast, the neurons of the orbitofrontal cortex were distributed equally between those that were activated when the choice was either positive or negative.

Chocoholism and gourmand syndrome

Our poodle Lolita is a gourmand. She adores chocolate. It is very bad for her, but we give it to her anyway! The first time, she was pleasantly surprised; by the third, she had developed a bad habit. Now, she comes begging for some as soon as the box appears. She expects a reward. Experiments suggest that her orbitofrontal neurons discharge systematically while she is waiting, wagging her tail. They are also involved in addictions.

The same food that pleases when one is hungry can be disgusting when one is sated. The technical term for this reversal of appetency is 'alliesthesia'. An amusing positron-emission tomography study concerns pleasure in eating chocolate [27]. Subjects ate chocolate until they were full. After each piece, they assessed the agreeable or disagreeable character of the sensation and the degree to which they desired another piece. Two different groups of structures were activated. When subjects want more chocolate and report subjective pleasure in eating it, the caudomedial orbitofrontal cortex, the operculum of the insula, the striatum, and the mesencephalon are activated. When they have had enough and are disgusted, the caudolateral orbitofrontal and prefrontal regions of the cortex and the parahippocampal gyrus are activated. And in all cases, the posterieur cingulate cortex is activated. The value of the food is thus indeed processed in the orbitofrontal cortex, where different regions encode opposite values: agreeable and disagreeable, sated or not. Two motivational systems resulting from this evaluation then induce either approach or avoidance.

It is possible that the *right cortex is particularly important in our taste for good things*. The literature describes a 'gourmand syndrome' [28] of patients with focal lesions of the right anterior cortex (corticolimbic region). These people can only think of one thing: eating fine food. It is not bulimia. One of them gave up his occupation as a political journalist to become a food critic [29]!

Other neurons of this cortex are active while a reward is being consumed. They may be involved in the validation of success. Still others are activated when the context changes, and some are sensitive to the level of satiety. So there may be competition at the cortical level between the regions that encourage approach and others that encourage avoidance.

At the end of this book (Chapter 13), we will consider another basic property of the orbitofrontal cortex: that of comparing past events with the present and preventing intrusion of these memories, that is, keeping past facts from invading the present.

These experiments provide access to mechanisms of thought and of judgement. They are obviously very simple, but they open uncharted territory to the science of reasoning. They are even fascinating, because we have only to hear the activity of the neurons of the orbitofrontal cortex over the laboratory loudspeaker to guess what the animal prefers! We know several moments before the animal does anything how it will choose. We can thus 'read' the monkey's mind, as we noted in the chapter on vision. The practical implications of these discoveries are imaginable today: in the future it will be possible to implant electrodes in the brain of a person and to know in advance, by connecting his brain to a computer, what he prefers. This is a very contemporary and formidable version of truth serum or the tests of skin resistance that are used in lie detection. We will have to establish ethical guidelines for the use of these techniques.

Intention to act: crossroads of the accumbens

How are the preferences developed in the prefrontal and limbic cortex translated into action? A little nucleus, the nucleus accumbens, probably plays a critical role in this transfer. In the rat, it is part of the striatum (Fig. 11.3).

The dorsal striatum, which contains the so-called caudal and putamen nuclei, and includes the accumbens, is distinct from the olfactive tubercule and the islands of Cajal. The accumbens is a key structure for handling the transition between the centres of emotion and those that control action. It receives projections from the (basolateral and lateral) amygdala, the orbitofrontal and entorhinal cortex, the associative temporal cortex, and the anterior cingulate. In the rat, the accumbens comprises two parts: the nucleus or centre and the 'shell' [30], which make up two separate circuits (Fig. 11.3). The nucleus belongs to the limbic module of the loops connecting the cortex to the basal ganglia and the thalamus.[7] In humans, its organization is still poorly understood. In the monkey, the striatum is clearly involved in assessing reward expectation [31].

The accumbens is influenced by all the major neuromodulatory systems (serotonin, dopamine, norepinephrin, acetylcholine). For example, increasing dopamine in the nucleus accumbens after a rat has made the association between light and food suppresses its behavioural response only to the light [32]. This suggests that the accumbens is involved in transfer of conditioned learning with reward towards the structures that control appetitive behaviour, and thus also involved in developing incentives to act or to consume, if not actual intentions. This function could also be extended to avoidance reactions. It is involved in choice of action by the major centres of sensory information processing (visual, auditory, and somatosensory cortex, but also the intralaminary thalamus). It plays a role both in integrating sensory information and attributing value by the limbic system and activates the dopaminergic system. The nucleus accumbens is thus a critical centre for preference and desire to consume.

Suppose that you are visiting a city for the first time. One day, you are very thirsty, and you find yourself in a square in the centre of the city. For example, the fountain of the four dolphins in Aix-en-Provence. Suppose, too (which is not the case at Aix), that there are four bistros in the four streets, and that you want to explore all of them.

---

[7] It receives projections especially from the prefrontal cortex and in turn projects towards the substantia nigra pars reticulata. The second part mainly receives projections from the hippocampus and has an output towards the main centre of dopamine production (the VTA nucleus), which loops towards the amygdala and the frontal cortex. Finally, it projects towards the substantia nigra pars compacta and the caudal and putamen nuclei, and thus towards the centres that control action.

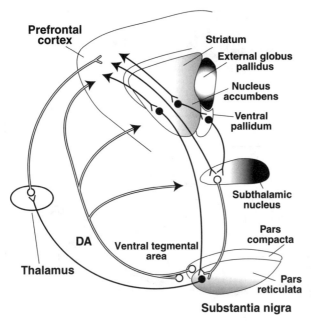

Fig. 11.3 Details of a circuit linking the prefrontal cortex, the basal ganglia, and the dopaminergic system in the rat. DA, dopamine. Courtesy A.-M. Thierry and J.-M. Deniau

Only one sells the pastis[8] that you like. The next day, you happen to be in the same place, and you say to a friend: 'Come! I want you to taste something really good.' You have to decide which street is the right one. To make this decision, you associate your memory of the bistro and its location with respect to landmarks—buildings, trees, stores—in the city. We have put a rat in an identical situation, in a labyrinth with four arms (Fig. 11.4). The rat was placed at the intersection and had to decide which 'bistro' to go to. Here, the bistros were represented by the ends of the arms where the rat could receive one to six drops of a liquid that it liked. It learned very quickly how to get to the 'best bistro'. It could orient itself based on cues outside the maze, located on the walls of the laboratory. We know that the hippocampus of the rat contains neurons called 'place cells' that are activated only when the animal is in a given place and whose spatial specificity appears with the ability to move in the young rat [33]. Other neurons called 'head direction' cells allow the rat to know the direction of its head with respect to distal cues [34]. This information about space converges on the accumbens with signals coming from the limbic structures that evaluate reward [35]. But when the neuronal pathway that links the accumbens to the hippocampus was severed, the rats became disoriented.

In the rat, the accumbens really does represent an intersection of space and motivation. That is why it is thought to play a decisive role in addiction, which we might think of as an error in our ability to regulate our decisions or to resist our desires in the face of powerful associations. Recent findings on the phenomenon of craving show that there is a ritual to the way addicts take drugs [36]. They consume by reproducing the same

---

[8] An aniseed-flavoured liqueur popular in France.

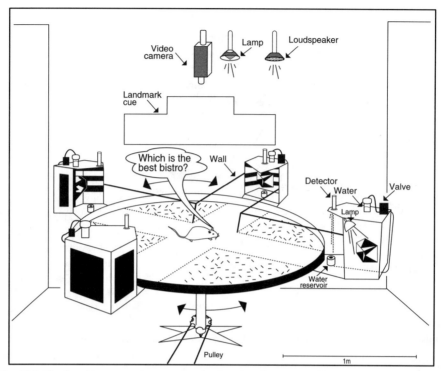

Fig. 11.4 The four-bistro experiment. A rat is placed at the centre of a four-armed maze. At the end of each arm is a 'bistro', that is, a station where, when the rat approaches, a photoelectric cell detects its presence and gives it a sip of liquid that it likes. It doesn't receive the same dose of juice at all of the 'bistros'. There are good and bad bistros. Because the arms of the maze are all identical, to remember the location of the best bistro, the rat must use landmarks outside the maze. For this purpose, large visual cues were placed on the walls (just as the Eiffel Tower helps us to navigate around Paris). To find the bistro, the rat has to make an association between the place and the reward. The nucleus accumbens plays a major role in this association because it links the information of the hippocampus and the neurons that indicate the direction of the head with the value of the reward and guide action. Damage to the pathways that link the accumbens to the hippocampus creates a marked deficit in the rat's capacity to decide which is the good bistro. Albertin et al. (2000)

gestures in the same places and so on. When they have kicked their habit, they may find themselves in the same place and feel an irrepressible desire for a fix. This supports the analysis of these mechanisms by a neobehaviourist approach based on the idea of instrumental conditioning, in which the amygdala, the orbitofrontal cortex, the accumbens, and possibly nuclei such as the ventral tegmentum are the principal relays [37].

Researchers have hypothesized that the accumbens acts as a gateway for integrating cognitive information (from the ventromedial prefrontal cortex and the hippocampus) and emotional information (from the amygdala) in the process of human decision making [38]. Although the Iowa Gambling Task is currently used to study the ventromedial prefrontal cortex, recent work has suggested that two other tasks—the Cambridge Gambling Task and the Risk Task—may be even more specific for the role of this brain region [39]. Accumbens activity correlates positively with the harm avoidance subscale of the Temperament and Character Inventory, suggesting that together with the other areas involved in decision making, the accumbens may be involved in risk taking [40].

Extrapolating this analysis suggests that many of our habits, not to mention our 'habitus'[9] in the sense of Pierre Bourdieu, stem from the same type of mechanism. Decision making is often the ability to escape all our acquired habits, modes of thinking, or customary actions to create a new solution. The physiology of preference naturally entails a physiology of will.

A supervisor: the anterior cingulate cortex

So the amygdala is involved in building associations, the orbitofrontal cortex in the ability to manipulate these associations according to rules and based on context and various cognitive factors, and the accumbens occupies a key position in the interface with the basal ganglia and the networks for choosing action. But we still need to consider another region of the brain that contributes to relations between emotion and decision making: the anterior cingulate cortex [41]. When James Papez reconstructed the circuit that, according to him, governs the emotions, he included the anterior cingulate cortex because lesions caused by callosal tumours spilling into this part of the brain cause emotional disturbances. In all, the cingulate cortex extends from the parietal cortex to the prefrontal cortex. It encompasses a posterior region that is especially involved in evaluation and an anterior region that participates in executive functions.

The anterior cingulate cortex does not only encompass the anterior horn. It contains very large pyramidal neurons in layer V. The details of its organization are critical, because the composition of the layers is extremely heterogeneous. The anterior cingulate cortex itself is divided into two areas with different functions: the 'affective' region and the 'cognitive' region[10] (Fig. 11.5).

---

[9] According to a personal communication from Richard Nice, who translated several of the works in which Bourdieu most fully develops the concept, 'the habitus is a long-lasting, deeply embedded set of "dispositions", "ways of doing things, of feeling, thinking and judging" '. See, for example, P. Bourdieu, *The logic of practice*, trans. R. Nice (Cambridge: Polity, 1990).

[10] An initial so-called affective region is located at the periphery of area 32, including most of rostral areas 24 a–c and 32, and ventral areas 25 and 33. This affective region is connected to the amygdala, the periaquaductal grey matter, the nucleus accumbens, the hypothalamus, the anterior insula, and the orbitofrontal cortex. It projects to the autonomic, visceromotor, and endocrine systems. The second 'cognitive' region of the anterior cingulate cortex includes areas 24b', c', and 32'.

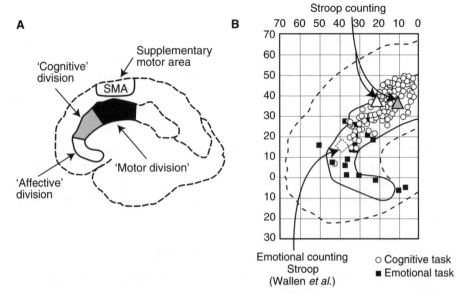

Fig. 11.5 Emotion, cognition, and action: three components of decision making. **(A)** The anterior cingulate cortex probably plays a major role in the combination of emotional, cognitive, and motor aspects of decision making. This scheme shows its localization in the brain. The cingulate cortex extends from the prefrontal cortex to the parietal cortex, but it is the anterior part that is considered here. Three connected regions have been identified. They occur one after the other in the structure: the anterior-most portion is involved in emotional evaluation. It forms part of the circuits that involve the orbitofrontal cortex. Just behind is a region that is the seat of operations and cognitive decision making (evaluation of conflicts and errors). Finally, a third part called the cingulate motor field is situated under the supplementary motor area, which is very important in initiation and coordination of complex actions, and serves as an interface with the executive systems of action. **(B)** Synthesis of the areas activated during several emotional or cognitive tasks. For the definition of the emotional or cognitive Stroop task, see the text. After Bush et al. (2000)

Behind these two anterior regions, in what one might call the medial cingulate cortex, motor activities have been located that bring to the fore the cingulate oculomotor areas. This motor portion of the cingulate cortex is involved in the control of saccades [42]. It probably also serves as an interface between cognition and control of movement [43].

Lesions of the anterior cingulate cortex produce apathy, inattention, dysregulation of the functions of the autonomic system, akinetic mutism, and emotional instability or lability, for example, absence of distress or compulsive weeping.

Patients also present with cognitive deficits, such as in executing Stroop attentional conflict [44] tasks, which cause two different interpretations of the same reality to conflict: for example, writing the word 'red' in green. A subject reading the word says 'red', but his brain tells him that the word itself is green. A mathematical type [45] of cognitive conflict can also be provoked, for example by asking a subject to read this series of words:

three
three
three
three

then asking him *how many* words there are ('mathematical' counting task). There is obviously a cognitive conflict between the meaning of the word (three) and the number of words (four). But it is also possible to create emotional conflict by showing a subject as in the following list:

murder
murder
murder
murder

and asking him *how many* words there are ('emotional' counting task). Someone reading words that have no emotional content serves as a control.

The (anteriormost) affective division of the cingulate cortex is activated in tasks that involve emotional content, but in psychiatric patients it is also involved in provocation of symptoms related to their ailment, anxiety, simple phobia—for example, fear of spiders—or obsessive-compulsive disorders. It is also activated during bouts of sadness and in depressed subjects.

The region just behind this affective area—more dorsal—of the anterior cingulate cortex is activated during conflictual tasks of a purely mathematical nature [46], whence the idea of a more affective region and a more cognitive region. So from back to front, the regions are affective, cognitive, and motor [47]. It is tempting to see in that order a hierarchy going from evaluation right up to action! The two portions of the anterior cingulate cortex may compete for tasks that are predominately emotional or cognitive, respectively, and they are involved in a complex circuit that is also involved in the pathology of depression [48] (Fig. 2.4). When one is activated, the other is inactived. In Chapter 13 we will see another case of reciprocal interaction between emotion and cognition. This interaction shows, once again, how emotion can be important for cognition. Too intense an emotion can have a powerful influence over decision making—we have trouble making decisions when we are overwhelmed—by blocking cognitive activity, whereas an intense cognitive task can suppress emotion [49]. Thus emotion is useful for cognition under very particular conditions that are still not well understood.

Arbitrating conflict

More generally, the anterior cingulate cortex is involved in detecting conflicts and in delayed processing. For example, the activity of the affective subdivision is modified (in fact, inactivated) during tasks of attention, in which a subject is asked to divide her attention between two targets (divided attention); antisaccades (looking in the direction opposite to a target); working memory (remembering a telephone number for a few moments); sequence learning; and visuospatial tasks. Many examples of this reciprocal exclusion are to be found in everyday life: intellectual work can overcome strong emotion, and vice versa, strong emotion can block intellectual activity and the ability to think ahead. On the contrary, brain imaging has shown that subjects suffering severe depression or healthy subjects deeply moved by a film or by pain exhibit inactivation of the cognitive subdivision. Emotion brings us back to the present.

Showing subjects pictures of mutilated bodies, naked women, or babies with eye tumours and asking them to direct their attention to the emotional feelings caused by the images rather than to their spatial aspects activates the anterior cingulate cortex.

This part of the brain is believed to play a role in emotional consciousness. Generally speaking, around 10 per cent of the general population has difficulty expressing their emotions. This symptom of 'alexithymia' consists in difficulty in verbally describing one's own feelings, distinguishing them from their somatic expression, and a tendency to focus on external events rather than on personal experiences, and so on. For example, a person who has recently buried a loved one may recall the colour of the clothes she was wearing at the funeral but not her emotional state. Richard Lane has demonstrated a correlation between cerebral activation of the anterior cingulate cortex and the level of emotional consciousness.

Others in a state of posttraumatic stress are alexithymic (more than 70 per cent of patients). A recent study measured the brain activity of eight patients with alexithymia and subjects without [50]. Subjects were shown images of an emotional nature, and the activity induced by the images was compared with that caused by controls. Another comparison involved so-called neutral pictures and controls. The results of functional magnetic resonance imaging reveal an interesting dissociation. Showing alexithymic patients images intended to provoke 'positive' emotions (for example, erotic images) increased activity in the anterior cingulate cortex, the mediofrontal cortex, and the mediofrontal gyrus. In contrast, showing them strongly negative images reduced activity in the mediofrontal and paracingulate cortex. These disturbances are thus clearly associated with structures involved in evaluating emotions at an elevated level of what Lazarus would call 'appraisal' (see Chapter 2). The problem is that we do not know whether this structure is inhibitory, and much remains to be done in understanding its physiology. What matters is that there exists an area specialized in processing emotions so close to a system involved in anticipating action, evaluating perceptual data, and choosing actions, including suppressing them. A particular form of suppressive action, often described as psychomotor slowing and lessening of desire, is the pathological form of depression. Recent work in brain imaging suggests an important role for the orbitofrontal cortex and anterior cingulate cortex in the network of brain regions involved in depressive states.

The anterior cingulate cortex and error detection

When subjects are asked to respond rapidly to a visual stimulus, the corresponding brain activity is reflected at the surface of the skull by variations in electrical potential—known as evoked potential—that can be recorded. One particular evoked potential associated with error detection in cognitive tasks, consisting of a negative wave, was localized in the region of the anterior cingulate cortex [51]. This negative wave, which reflects the activity of the neurons of the underlying cortex, is indeed linked to assessing error but not its correction, as well as to the motivation of the subject.[11] The cingulate cortex is also activated during reversal of learning, and is sensitive to the probability of a conditioned stimulus appearing. It is activated proportionally to the number of false alarms during a go/no-go task in 7- to 12-year-old children and in adults [52]. A high-density electrical mapping study of adults revealed different patterns of activation in changing plans; switching from a no-go to a go situation primarily involves the parietal area, whereas switching from a go to a no-go involves the anterior cingulate and prefrontal cortex [53].

In short, the anterior cingulate cortex is located in the medial brain regions, such as the orbitofrontal cortex, which evaluate stimuli and information about the body and the world, while the lateral regions hold that information in working memory. It is part of a long pathway involved in evaluating situations, guiding or interrupting action as a function of context, intent, and comparison with the recollection of past events, in contrast with short, automatic pathways (for example, the amygdala or insula). It is involved in detecting conflicts and error and in conscious feelings of satisfaction or distress. The anterior cingulate cortex also comes into play in evaluating how much effort is required to obtain a specific reward [54]. When we are faced with a problem of monetary gain or loss, the dorsal anterior cingulate cortex is activated, and this activity is related to both internal and external sources of information concerning the errors we make [55].

Morever, the anterior cingulate cortex may be part of a system of 'attentional supervision' [56] such as that proposed by Norman and Shallice (see the Introduction). These ideas constitute the supervisory theory of competition: when there is conflict between different lateral cortical regions, confusion is avoided by arbitration of the medial areas.

Selecting action: the basal ganglia

How are decisions developed in the frontal cortex and the centres of emotion and motivation translated into action? Obviously, we have to be very cautious. Our knowledge is

---

[11] For example, if a subject is asked to carry out a task of increasing precision (rather than speed), the amplitude of the potential increases. It diminishes with learning when the error rate is less than 30 per cent.

still very fragmentary. The basal ganglia possess mechanisms that are especially well adapted to choose responses that offer the best reward after a competition at the level of the neuronal synapses [57].

As we have seen several times, the basal ganglia play an essential role in decision making, in particular in suppressing undesired movements. But in both humans and monkeys, collections of neurons are recruited differently depending on the nature of the task [58]. In humans, for example, decisions involving short-term monetary rewards activate the caudate nucleus [59], whereas switching tasks involves both the caudate and parietal cortex [60]. Diseases that affect the basal ganglia cause obsessional behaviours or endless repetition of an action (Chapter 3). To better understand how these structures are involved in halting or suspending action, it is necessary to examine their anatomy and physiology a little more in detail [61].

One of the most amazing aspects of the organization of the basal ganglia is that they are part of parallel systems (loops) that link them to the thalamus and to the cerebral cortex. These loops are modular: they concern control of gaze, motor movements of the limbs, the limbic system, motivation, and so on, in a very specific way[12] (Fig. 3.3).

So much for the summary, but what does it have to do with decision making? If there are actually two pathways with two different effects, there should be populations of neurons that are activated, respectively, on initiation of movement and on its suppression. Giving monkeys go or no-go tasks [62] reveals that half the neurons of the putamen (one of the basal ganglia) are activated specifically in a go or no-go task, whereas the rest are activated during both tasks. But that is only a preliminary observation. The workings of these nuclei are much more complex than a simple dichotomy would indicate, and further work is needed to understand the mechanisms of functional segregation in these structures.

One could imagine a mechanism comprising circuits that sum, at the level of a postsynaptic neuron, the contribution of several thousands of independent inputs, each weighted according to the force of its synaptic connection [63]. Rolls proposes a hierarchical theory. To establish a boundary, he distinguishes structures that have back

---

[12] The circuit that controls gaze movement, for example, consists of two main pathways. One is *direct*: it joins the striatum (Str) and the output nuclei of the basal ganglia, namely, the internal portion of the globus pallidus and the substantia nigra pars reticulata (SNpr). The other is *indirect* and connects the striatum to the internal globus pallidus (Gpi) and to the substantia nigra through a sequence of connections that encompass the external globus pallidus (Gpe) and the internal subthalamic nucleus (STNi). These two projections are believed to be the product of distinct populations of neurons in the striatum. In both cases, the final output of the basal ganglia is inhibitory and assumed to suppress motor responses via the thalamus, the colliculus (as we have seen in the case of gaze control), or the brain stem. Activation of the direct pathway diminishes Gpi and SNr output, and consequently disinhibits motor commands. Activation of the indirect pathway increases inhibition and thus suppresses motor commands. It is this last property that may be of use in stimulating the subthalamic nucleus to reduce (in a more or less lasting way) the motor symptoms of Parkinson's caused by the lack of dopamine that is the cause of this disease.

projections to the regions of the cortex that project into those structures, such as the hippocampus, the orbitofrontal cortex, and the amygdala. He suggests that choice and decision making carried out in the basal ganglia constitute an 'implicit route to action', whereas the route through the orbitofrontal cortex is explicit and conscious and thus might be the seat of deliberation in humans, notably through its link with structures related to language. He also wonders whether choice and selection might be carried out by hierarchically inferior mechanisms, at the level of the brain stem or by the short pathway of the amygdala, which allows rapid responses to aversive stimuli. But in primates and in humans, these mechanisms are probably not involved in evaluating reward. So they would not qualify as genuine mechanisms of decision making.

This theory is debatable, but also very interesting. It reinforces the underlying thesis of this book, namely, that decision making is a fundamental property of the nervous system. But it also introduces the idea of a hierarchy in the systems where deliberation becomes ever more complex and the encoding of values increasingly sophisticated.

Dopamine and reward

Estimating errors is thus essential for adapting to conflict situations and also for decision making. Is it possible to find in the nervous system a quantity that varies with the estimated error? Yes: changes in production of dopamine could be linked to errors in predicting the amount of the expected reward [64]. In short, an increase in the level of dopamine signifies 'better than expected' and a decrease 'less than expected'. Dopamine is a neurotransmitter produced mainly by two nuclei located in the depths of the brain, in the ventral tegmentum, and in the compact part of the substantia nigra (Fig. 11.3). The neurons of these centres are activated when a monkey makes errors in performing a task that requires prediction. The amazing thing is that these neurons signal the error in a prediction. In other words, if the animal correctly predicts that a reward is associated with a stimulus, they are not activated. If a surprise reward arrives, they are activated!

If the animal is expecting a reward and it does not arrive, the neurons stop discharging; if an unexpected reward arrives or if punishment occurs, the dopaminergic neurons are activated. It is thus a transitory activity, linked to a prediction error. These cells are also sensitive to familiarity: because they are activated by new stimuli, if the presentation is repeated, they quiet down and discharge less and less frequently.

So the brain works by detecting differences between *expected states* and *actual states*. It is not so much the *absolute value* of a reward that brain cells are concerned with, but its *relative value*. Which, incidentally, was predicted by researchers such as Kahneman and Tversky in prospect theory (see Chapter 1).

Dopaminergic neurons modulate the activity of the structures of the cerebral cortex involved in processes of evaluating or motor or perceptual decision making, for example, the amygdala, the orbitofrontal cortex, the entorhinal cortex, the cingulate cortex, the frontal oculomotor fields, and so on. It has been suggested that dopamine acts directly at the level of the synapses of the targeted centres. For example, in the nucleus accumbens, the dendrites of the neurons receive signals from the cerebral cortex and the dopaminergic neurons. If a dopaminergic neuron is activated at the same time as a cortical neuron, the synapse is reinforced, as predicted by Hebb's rule ('cells that fire together, wire together').

Despite the very seductive character of this theory of dopamine as an error detector, one must be careful not to misconstrue its role. Its function could be more global than computational models let on. In any case, the point is that we are dealing with a high-level process, a cognitive evaluation of familiarity, and not only 'habituation', that is, a diminution of response with repetition.

The idea's novelty consists in using *error to predict*. The brain does not process data from the world, but biochemical or neuronal variables that signal what is important in the world, that is, *intermediary* variables. Roboticists also create variables that have little direct relation to the variables controlled by the robot, because these intermediary variables—which I call 'working variables'—make it possible to simplify the calculations, for example, to process nonlinear problems in a linear fashion. We have suggested that the brain uses composite variables to control movement [65]. It is easier for the calculations to control a composite variable, which is a combination of position, speed, and acceleration. Here the complexity makes it possible to simplify the neurocomputation. That is not at all intuitive. The brain is a genuine emulator of *new* variables useful in working out decisions. It retranscribes these *variables* into action via specialized circuits. Amazing!

Modelling the role of dopamine in decision making

Decision making thus involves evaluating errors that we have made in previous choices. Practically all learning models of decision making, represented by equations, contain at least one error term. Alteration in the production of dopamine, measured during decision making, represents in the brain precisely this error term of predicting reward. Consider an example. Suppose that I must decide whether I want to drink green tea (A) or coffee (B). Based on my preference, I will attribute to A or to B a certain probability of reward: for example, suppose that I don't want to fall asleep while I am writing a book on decision making or that I want to calm my anxiety. If I try the green tea and I fall asleep, I will have made a mistake (although, if it manages to alleviate my fears, basically, I will have been rewarded). The accumulation of these errors and rewards enables me somehow to estimate the rules that govern the value that green tea (A) and coffee (B) have for me.

The idea that production of dopamine is linked to evaluation of reward during motor learning has been applied both to interpretation of neuronal recordings in primates and to describe bees' behaviour in searching for food. Several mathematical models were recently proposed to simulate these processes and the role of dopamine [66]. Because they are formulated in a very general fashion, they do not explicitly mention the accumbens, though it is part of the striatum.

One recent model [67] works in the following way: A subject has two possible behaviours for making a decision, associated with two different groups of neurons in the cortex, which result in different configurations of neuronal activity. The neuronal model is a classical, three-level model: a superficial layer that receives information from the cerebral cortex or other sources; an output layer that produces the effects; and a so-called intermediate level that is the principal locus of learning where neuromodulation by dopamine takes place. These cortical activities weight the activity of the intermediary layer of the model, which corresponds to structures that project, for example, from the cortex into the basal ganglia. This weighting regulates the selection of such-and-such behaviour, the decision to take this or that action[13]. This model has been tested in experiments involving decision making with human subjects. Obviously, that does not constitute proof of its validity, but it suggests that it is possible to have going simultaneously both an experimental approach and a theoretical approach inspired by biological mechanisms.

The parable of familial decision making

How is the basal ganglia's mechanism for choosing action encoded into the collection of processes that link the cerebral cortex to the centres of movement and action? On the one hand, we have found preformed automatic mechanisms—nearly automatic selection, decision without deliberation—that take into account neither affective value nor utility, nor context, nor memory of the past. On the other hand, we have found mechanisms, some of which may be conscious, that exert on the first group a basic role of signalling, selection, modulation, suppression, delay, exclusion, and so on. The problem is finding the interface among these different mechanisms. Here is an example

---

[13] The model introduces several variables: P is a linear unit (neuron or group of neurons) whose output activity can be represented by equations where δ is the fluctuation of the production of dopamine and V is a signal that arrives from the cortex and is assumed to be a temporal derivative of cortical excitatory activity:

$\delta(t) = r(t) + V(t) + b(t)$, where r is the input that represents the stimuli resulting in a reward and $b(t)$ is the basal activity that here is assumed to be zero;

$V(t) = \Sigma_i x(i,t) w(i,t)$, where $x(i,t)$ is the activity associated with choice i at time t, and w the synaptic weight in the intermediary layers.

In this case, there are only two $x(t)$ that represent the two possible choices (tea or coffee), each being assumed to represent only one binary action: 1 if the behaviour is chosen, and 0 if it isn't. $V(t)$ is positive if the level of dopamine is above the basal level and negative if it is below. Under these conditions, δ is interpreted as a function of *error prediction* between the value of the expected reward and the value actually received. This value is used to direct the selection of synaptic weight w in the intermediary layers.

of a recent summary of the problem of functional architecture of the decision-making processes for controlling action:

> The ethological and neurobiological analysis of behaviour suggests there are indeed multiple levels of selection within the vertebrate nervous system . . . At the highest level, selections are required that decide the current general course of action. At intermediate levels, selection specifies appropriate patterns of coordinated movements in the context of the current high-level aim. Finally, at the lowest levels, selection determines patterns of appropriate muscular activity that can deliver the currently selected action. This hierarchical decomposition of selection makes decision making a tractable enterprise since, at any given moment, it restricts lower-level competitions to just those competitors capable of implementing current higher-level objectives [68].

Note the importance of *competition*. If decision making is a fundamental property of the nervous system, it is because it has available so many mechanisms in competition. The most elementary are the systems of reflexes. In *The Brain's Sense of Movement* I showed how, for example, inhibition helps to avoid compensating gaze movements (the vestibuloocular reflex) from preventing gaze-orienting movements. But here we are talking about competition at a higher level of complexity, such as financial conflicts of interest, emotional or career choices, and so on. Imagine suppressing a bad memory associated with an action to be able to begin again, for example, when we encourage a horse rider to get back on his mount directly after a fall, or when during a lover's quarrel we make an effort to suppress our resentment so as to regain a level of understanding where affection quells anger. What solutions has nature found to resolve these problems of competition?

Three types of architecture have been suggested for choosing actions (Fig. 11.6). The first, *subsumption*, assumes that these actions are organized in a hierarchical way that allows them to be selected automatically according to a fixed order of priority. The disadvantages of this organization are obvious. In the second, more *distributed* architecture, actions are networked in a series of reciprocal inhibitory connections. One or the other action will automatically be chosen according to links endowed with certain sensory configurations. (This recalls the toad's mechanisms of fight or flight that we saw in Chapter 4, except that the toad has much more sophisticated ways of taking context into account.) Yet this circuit, too, has serious limitations, a fact that becomes obvious when the behaviours multiply and require reciprocal inhibition to govern complex interactions.

The third architecture encompasses a *supervisor* or central *interruptor* that activates circuits selectively. The advantage of this circuit is that it requires fewer connections and allows more flexibility. It combines the advantages of modularity and centralization.

Everyone knows that the family is the locus of love and devotion and at the same time all sorts of conflicts. 'Family, I hate you!' wrote André Gide, and one of the reasons is that it engages in perpetual, often pointless decision making. I have always been struck by the emotional violence that seizes members of a family over decisions that are very simple and, basically, of no great consequence. This decision-making

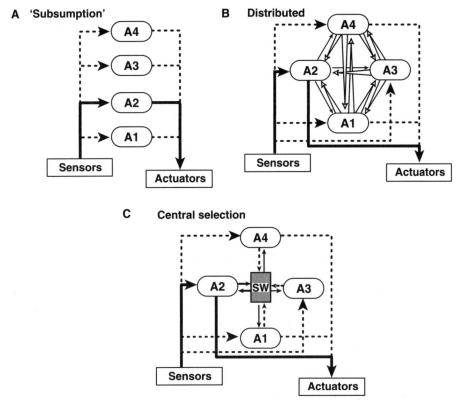

Fig. 11.6 Proposed architectures for resolving the problem of selecting action. **(A)** A 'subsumption' architecture. The competition between possible actions (A1 and so on) represented in the different layers is solved by a predefined priority order. This model is obviously too rigid, except perhaps for very simple animals. **(B)** Distributed network of neurons in which each module is in competition (A1, and so on) with a reciprocal inhibitory connection with each of the other modules and an excitatory link with the mechanism of the motor output. **(C)** A central selection mechanism (reminiscent of Norman and Shallice's supervisor) can be used to determine access to the shared output resource with fewer connections than in the preceding model. This model includes a switch (SW) that detects the fact that A2 has a greater weight and 'supports' this element. The less-supported components (in the case of deliberation for a decision, we would say the less convincing arguments), and thus losers, are inhibited by the central switch. After, Redgrave, Prescot, and Gurney (1999)

battlefield illustrates very well a problem of physiology. Imagine a competition between three cousins (behavioural circuits) during a family reunion. They are bored and want to use the only available car. They receive phone calls from their pals (cerebral cortex) inviting them (excitatory input) to do three different things: one, Peter, is invited to go swimming; the other, Annie, to go to the movies; the third, Roger, to go fishing. Their parents (basal ganglia) are all sitting in one room, talking to each other.

They know about their children's plans because they have a telephone service that informs them of the communications with the friends (copy of the motor cortex commands sent to the basal ganglia). Each of the parents hopes that his child stays at the party (inhibitory output of the basal ganglia) and thus seeks to dissuade the child from leaving (everyone knows that the only thing parents are good for is keeping their children from having fun!).

Yet an important principle governs this little society: If a child wins and obtains the permission of his father, P, not only does P not keep him from going out (represented in the model by inhibitory feedback to the basal ganglia, inhibitor of inhibition), but P must persuade the other parents that his choice is the best and that he would like the car for his own child. This mechanism—called 'winner takes all'—is believed to be the basis for decision making among the neurons of the basal ganglia. To keep this privilege for his child, P stimulates the other parents to prevent their children from leaving. Thanks to this mechanism, if one of the parents accepts or if the host (which in the model represents the external motivational influences of the limbic system on the operation of the basal ganglia) intervenes, or if one of the children is particularly persuasive, and so on, only that child may leave, and that choice-decision will take effect.

This mechanism has a triple secret: first, the maintenance of specialized modules each of which plays its role and directs a behaviour (the children); second, a system that produces intention to act (the friends; the cortex); third, the selection system of the basal ganglia that controls intiation of action through double inhibition (see Fig. 11.7).

This research has stimulated various working models for computers and robots inspired by biology, incorporating some modification to resolve problems related to action selection [69].

The physiology of hesitation

Another theoretically important aspect to take into account is the problem of hesitation. When we are deliberating over a decision, at least two solutions are available to us. This state of uncertainty is very unpleasant, and we often perceive it as an instability: hesitation is common. The less clear-cut competing solutions are, the more we hesitate, like Buridan's ass. So it is essential that the brain choose cleanly between two competing solutions: it must resolve this problem at all levels of its operation, constructing unbreachable boundaries. In the visual system, lateral mechanisms of inhibition perform that function. In the basal ganglia, reciprocal or even anticipatory (feedforward) mechanisms of inhibition accelerate the cascading phenomenon that leads to victory of one of the solutions in the winner-takes-all [70] process. Analysis of the circuits in the basal ganglia and the loops linking them to the cerebral cortex and the thalamus suggest how the elements of the motor repertoire are selected. But how then is the brain able to use these mechanisms to govern behaviour more generally?

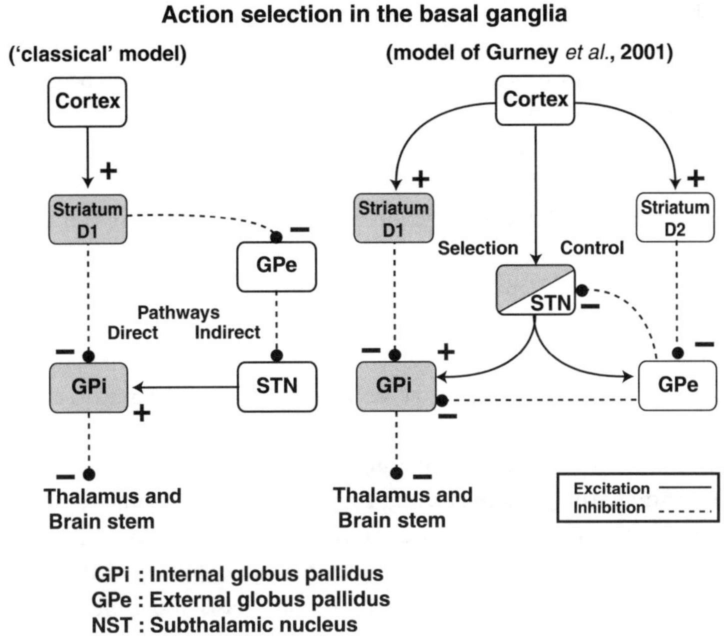

Action selection in the basal ganglia

Fig. 11.7 Two conceptions of the mechanisms for selecting action in the basal ganglia. A classical model assumes that a motor order is produced in the cerebral cortex and activates the striatum, which inhibits the internal globus pallidus (Gpi). Because the projection of the internal globus pallidus on the motor thalamus is inhibitory, this double inhibition cascade activates a motor program. This 'direct' pathway is backed up by an 'indirect' pathway that, through the intermediary of the external globus pallidus (GPe) and the subthalamic nucleus (STN), modulates this action. The two parts of the striatum involved in this process contain neurons sensitive to receptors other than dopamine receptors (D1 and D2). After Humphries and Gurney (2002)

Heterarchy and hierarchy

One solution, perhaps found through evolution, is precisely a double mechanism of modularity (heterarchy) and hierarchy: heterarchical because the loops of the thalamo-cortical basal ganglia are specialized in the control of movements related to gaze, gesture, memory, and so on; and hierarchical because these 'parallel' (if their anatomy is to be believed) modules may actually show a transversal hierarchical connection [71]. Since each region is regulated by a closed circuit and the different regions are hierarchically organized, competition could occur first, at the highest level, between the different motivational systems through the approach of the limbic system and the control of emotions. If the motivational system wins, it could selectively facilitate this or that modular implementation circuit of the basal ganglia, which would then themselves have to resolve the problems of actually carrying out the action. However, we need more evidence that this organization is really the prevailing one.

The beauty of this idea is that it avoids the classical scheme of a formal concept, cortical in origin, which dictates its orders to the lower levels. The notion of *global* to *specific* replaces that of *abstract* to *concrete*. But the concrete can be global. Mood involves the whole body, including posture, attitude, gaze, and the viscera, but it does not guide the hand or the tongue towards an object. Mood prepares the senses and the brain to examine the world with a certain quality of purpose—Husserl would have said, 'with a certain directedness'. Mood organizes reality before it is the object of action, but it organizes it while also preparing the body, memory, and so on. For reality constructed by the brain depends fundamentally on the preparation and state of the sentient body. Decision making is thus hardly a simple calculation of utility or probability, but engagement with the world, grasping it for a certain purpose, the world itself transformed by decision making, since decision making alters our perception of the world. And perceiving is deciding.

The projecting brain is a machine that makes decisions at all levels of operation, for predicting, anticipating, and proposing hypotheses is essentially deciding how the world should be. It is said that desires should not be taken for reality, but that is precisely what deciding is. The error economists make is to believe that deciding is a secondary rather than a primary process. It is the very basis of our life, and probably of our awareness, since to be aware is to decide that the world is this way and not some other way. And that is a matter of desire and belief, not of utility.

Emotion is to decision making what posture is to gesture

Emotion prepares us for action. It establishes the context in which action is experienced. It creates a world that resolves conflicts, a possible world, acceptable to our brain, its desires, its constraints, its hopes. Fundamentally, emotions are like colours: they help to categorize the world and simplify neurocomputation. In the infinite complexity of the physical world, they aid the brain in sorting things out.

Perception is decision making. I will say it again: Perception is deciding, and emotion is the ultimate judge of those decisions. We have seen that decisions are made from the earliest sensory relays: Is an object in front or behind? Is it hard or soft? Rigid or pliable? Is the world moving, or is it my body? These decisions classify the world in ways that simplify further analyses and enable the construction of constants for behaviour to organize around.

Faced with a collection of possible choices, preference prepares our actions just as posture prepares our movements. Nicolai Bernstein said that posture is preparation for moving. *Emotion is preparation for acting.* But its scope goes beyond unitary action; it becomes fuller in time because its field is the entire action and its expected consequences. Preference means imagining the future, projecting the consequences of action, remembering the past to predict the future. It is a subtle mix of archaic responses and fertile, living imagination. Like the water of a fountain, preference is the desire both to jump and to be still, evoking the source as well as the hope to return to the earth from which it sprang. It is a reason for joy. Unable to decide, Buridan's ass was indeed very unfortunate!

References

1. E Cassirer, *The philosophy of symbolic forms*, trans. R Manheim (New Haven, CT: Yale University Press, 1957), p. 72.
2. D Hermans, J De Houwer, and P Eelen, 'Evaluative decision latencies mediated by induced affective states', *Behaviour Research and Therapy*, **34** (1996): 483–8. For more about decision making and reward, see the special issue of *Neuron*, 'Reward and decision', **36** (2002): 189–332.
3. E Rolls, *The brain and emotion* (Oxford: Oxford University Press, 1999).
4. RL Lane and L Nadel, *Cognitive neuroscience of emotion* (Oxford: Oxford University Press, 1997).
5. JP Aggleton, *The amygdala: a functional analysis*, 2d edn (New York: Oxford University Press, 2000).
6. P Gloor, *The temporal lobe and the limbic system* (Oxford: Oxford University Press, 1997).
7. J LeDoux, The *emotional brain: the mysterious underpinnings of emotional life* (New York: Touchstone, 1996).
8. JV Haxby, EA Hoffman, and MI Gobbini, 'The distributed human neural system for face perception', *Trends in Cognitive Science*, **4** (2000): 223–33.
9. A Bechara, D Tranel, H Damasio, R Adolphs, C Rockland, and AR Damasio, 'Double dissociation of conditioning and declarative knowledge relative to the amygdala and hippocampus in humans', *Science*, **269** (1995): 1115–18; R Adolphs, D Tranel, H Damasio, and AR Damasio, 'Fear and the human amygdala', *Journal of Neuroscience*, **15** (1995): 5879–91.
10. R Kawashima, M Sugiura, T Kato, A Nakamura, K Hatano, I Kengo, H Fukuda, S Kojima, and K Nakamura, 'The human amygdala plays an important role in gaze monitoring: a PET study', *Brain*, **122** (1999): 779–83; N George, I Dreter, and R Dolan, 'Seen gaze: direction modulates fusiform activity and its coupling with other brain areas during face processing', *Neuroimage*, **13** (2001): 1102–12.
11. H Fischer, M Tifors, T Furmark, and F Fredrikson, 'Dispositional pessimism and amygdala activity', a PET study in healthy volunteers', *Neuroreport*, **12** (2001): 635–8.
12. JP Aggleton, 'The enigma of the amygdala', in Lane and Nadel, *Cognitive science of emotion*, p. 106.
13. Rolls, *The brain and emotion*.
14. RD Rogers, AM Owen, HC Middleton, EJ Williams, JD Pickard, BJ Sahakian, and TW Robbins, 'Choosing between small, likely rewards and large, unlikely rewards activates inferior and orbital prefrontal cortex', *Journal of Neuroscience*, **20** (1999): 9029–38.
15. S Berthoz, JL Armony, RJ Blair, and RJ Dolan, 'An fMRI study of intentional and unintentional (embarrassing) violations of social norms', *Brain*, **125** (2002): 1696–1708.
16. LK Fellows and MJ Farah, 'Ventromedial frontal cortex mediates affective shifting in humans: evidence from a reversal learning paradigm', *Brain*, **126** (2003): 1830–7.
17. A Pears, JA Parkinson, L Hopewell, BJ Everitt, and AC Roberts, 'Lesions of the orbitofrontal but not medial prefrontal cortex disrupt conditioned reinforcement in primates', *Neuroscience*, **23** (2003): 11189–201; K Happaney, PD Zelazo, and DT Stuss, 'Development of orbitofrontal function: current themes and future directions', *Brain and Cognition*, **55** (2004): 1–10; A Bechara, 'The role of emotion in decision-making: evidence from neurological patients with orbitofrontal damage', *Brain and Cognition*, **55** (2004): 30–40.

18 B Milner, 'Some cognitive effects of frontal lobe lesions in man', *Philosophical Transactions of the Royal Society B*, **298** (1982): 211–26.

19 R Elliott, K Friston, and R Dolan, 'Dissociable neural responses in human reward systems' *Journal of Neuroscience*, **20** (2000): 6159–65.

20 WC Drevets, JL Price, JR Simpson Jr, RD Todd, T Reich, M Vannier, and ME Raichle, 'Subgenual prefrontal cortex abnormalities in mood disorders', *Nature*, **386** (1997): 824–7.

21 NH Frijda, *The emotions* (Cambridge: Cambridge University Press, 1986).

22 RJ Davidson and W Irwin, 'The functional anatomy of emotion and affective style', *Trends in Cognitive Sciences*, **3** (1999): 11–20.

23 L Tremblay and W Schultz, 'Relative reward preference in primate orbitofrontal cortex', *Nature*, **398** (1999): 704–8; L Tremblay and W Schultz, 'Reward-related neuronal activity during go/no go task performance in primate orbitofrontal cortex', *Journal of Neurophysiology*, **83** (2000): 1864–76.

24 JM Fuster, *The prefrontal cortex: anatomy, physiology, and psychology, of the frontal lobe*, 3rd edn (Philadelphia, PA: Lippincott-William & Wilkins, 1997).

25 G Schoenbaum, A Chibqa, and M Gallagher, 'Orbito-frontal cortex and basolateral amygdala encode expected outcomes during learning', *Nature Neuroscience*, **1** (1998): 155–9.

26 L Tremblay and W Schultz, 'Relative reward preference in primate orbitofrontal cortex', *Nature*, **398** (1999): 704–8; Tremblay and Schultz, 'Reward-related neuronal activity'.

27 DM Small, RJ Zatorre, A Dagher, AC Evans, and M Jones-Gotman, 'Change in brain activity related to eating chocolate: from pleasure to aversion', *Brain*, **124** (2001): 1720–33.

28 M Regard and T Landis, ' "Gourmand syndrome": eating passion associated with right anterior lesions', *Neurology*, **48** (1997): 1185–90.

29 C Davis, RD Levitan, P Muglia, C Bewell, and JL Kennedy, 'Decision-making deficits and overeating: a risk model for obesity', *Obesity Research*, **12** (2004): 929–35.

30 A-M Thierry, Y Gioanni, E Gégénétais, and J Glowinski, 'Hippocampo-prefrontal cortex pathway: anatomical and electrophysiological characteristics', *Hippocampus*, **10** (2000): 411–19.

31 L Tremblay, JR Hollerman, and W Schultz, 'Modification of reward expectation-related neuronal activity during learning in primate striatum', *Journal of Neurophysiology*, **80** (1998): 964–77.

32 B Everitt and T Robbins, 'Amygdala-ventral striatal interactions and reward related processes', in JP Aggleton, ed., *The amygdala: neurobiological aspects of emotion, memory, and mental dysfunction* (New York: Wiley-Liss, 1992), pp. 401–29.

33 P Martin and A Berthoz, 'Development of spatial firing in the hippocampus of young rats', *Hippocampus*, **12** (2002): 465–80.

34 MB Zugaro, A Berthoz, and S Wiener, 'Background, but not foreground, spatial cues are taken as references for head direction responses by rat anterodorsal thalamus neurons', *Journal of Neuroscience*, **21** (2001): 1–5.

35 R Shibata, AB Mulder, O Trullier, and SI Wiener, 'Position sensitivity in phasically discharging nucleus accumbens neurons of rats alternating between tasks requiring complementary types of spatial cues', *Neuroscience*, **108** (2001): 391–411; AH Tran, R Tamura, T Uwano, T Kobayashi, M Katsuki, G Matsumoto, and T Ono, 'Altered accumbens neural response to prediction of reward associated with place in dopamine D2 receptor knockout mice', *Science*, **99** (2002): 8986–91.

36 B Everitt, 'Craving cocaine cues: cognitive neuroscience meets drug addiction research' *Trends in Cognitive Sciences*, **1** (1997): 1–2.

37 J-P Changeux, *The physiology of truth: neuroscience and human knowledge* (Cambridge: Belknap Press, 2004).

38 BM Wagar and P Thagard, 'Using computational neuroscience to investigate the neural correlates of cognitive-affective integration during covert decision making', *Brain and Cognition*, **53** (2003): 398–402; BM Wagar and P Thagard, 'Spiking Phineas Gage: a neurocomputational theory of cognitive-affective integration in decision making', *Psychology Review*, **111** (2004): 67–79.

39 L Clark, F Manes, N Antoun, BJ Sahakian, and TW Robbins, 'The contributions of lesion laterality and lesion volume to decision-making impairment following frontal lobe damage', *Neuropsychologia*, **41** (2003): 1474–83.

40 SC Matthews, AN Simmons, SD Lane, and MP Paulus, 'Selective activation of the nucleus accumbens during risk-taking decision making', *Neuroreport*, **15** (2004): 2123–7.

41 OD Evinsky, MJ Morell, and BA Vogt, 'Contributions of anterior cingulate cortex to behaviour', *Brain*, **118** (1995): 279–306; G Bush, P Luu, and M Posner, 'Cognitive and emotional influence in anterior cingulate cortex', *Trends in Cognitive Sciences*, **4–6** (2000): 215–22.

42 L Petit, C Orssaud, N Tzourio, G Salamon, B Mazoyer, and A Berthoz, 'A PET study of voluntary saccadic eye movements in humans: basal ganglia-thalamo-cortical system and cingulate cortex involvement', *Journal of Neurophysiology*, **69** (1993): 1009–17; B Gaymard, S Rivaud, JF Cassarini, T Dubard, G Rancurel, Y Agid, and C Pierrot-Deseilligny, 'Effects of anterior cingulate cortex lesions on ocular saccades in humans', *Experimental Brain Research*, **120** (1998): 173–83.

43 T Paus, 'Primate anterior cingulate cortex: where motor control drive and cognition interface', *Nature Reviews Neuroscience*, **2** (1999): 417–24; AU Turken and D Swick, 'Response selection in the human anterior cingulate cortex', *Nature Neuroscience*, **2** (2001): 920–4.

44 JV Pardo, PJ Pardo, KW Janer, and ME Raichel, 'The anterior cingulate cortex mediates processing selection in the Stroop attentional conflict paradigm', *Proceedings of the National Academy of Sciences of the USA*, **87** (1990): 256–9.

45 G Bush, P Luu, and MI Posner, 'Cognitive and emotional influences in anterior cingulate cortex', *Trends in Cognitive Sciences*, **4** (2000): 215–22.

46 G Bush, JA Frazier, SL Rauch, LJ Seidman, PJ Whalen, MA Jenike, BR Rosen, and J Biederman, 'Anterior cingulate cortex dysfunction in attention deficit/hyperactivity disorder revealed by fMRI and the counting Stroop', *Biological Psychiatry*, **45** (1999): 1542–52.

47 PJ Whalen, G Bush, RJ McNally, S Wilhelm, SC McInerney, MA Jenike, and SL Rauch, 'The emotional counting Stroop paradigm: a functional magnetic imaging probe of the anterior cingulate affective division', *Biological Psychiatry*, **44** (1998): 1219–28.

48 WC Drevits, 'Neuroimaging and neuropathological studies of depression: implications for the cognitive-emotional features of mood disorders', *Current Opinion in Neurobiology*, **11** (2001): 240–9.

49 FG Ashby, AM Isen, and AU Turken, 'A neuropsychological theory of positive affect and its influence on cognition', *Psychology Reviews*, **6** (1999): 529–50.

50 S Berthoz, E Artiges, P-F Van de Moortele, J-P Poline, S Rouquette, SM Consoli, and J-L Martinot, 'Effect of impaired recognition and expression of emotions on frontal cingulate cortices: an fMRI study of men with alexithymia', *American Journal of Psychiatry*, **159** (2002): 961–7.

51 H Gemba, K Sasaki, and VB Brooks, ' "Error" potentials in limbic cortex (anterior cingulate area 24) of monkeys during motor learning', *Neuroscience Letters*, **70** (1986): 223–7; P Luu, T Flaisch, and DM Tucker, 'Medial frontal cortex in action monitoring', *Journal of Neuroscience*, **20** (2000): 464–9.

52 BJ Casey, RJ Trainor, JL Orendi, AB Schubert, LE Nystrom, JN Giedd, FX Castellanos, JV Haxby, DC Noll, JD Cohen, SD Forman, RE Dahl, and JL Rapoport, 'A developmental functional MRI study of prefrontal activation during performance of a go/no go task', *Journal of Cognitive Neuroscience*, **9** (1997): 835–47.

53 EC Dias, JJ Foxe, and DC Javitt, 'Changing plans: a high density electrical mapping study of cortical control', *Cerebral Cortex*, **13** (2003): 701–15.

54 ME Walton, DM Bannerman, K Alterescu, and MF Rushworth, 'Functional specialization within medial frontal cortex of the anterior cingulate for evaluating effort-related decisions', *Journal of Neuroscience*, **23** (2003): 6475–9.

55 CB Holroyd, S Nieuwenhuis, N Yeung, L Nystrom, RB Mars, MG Coles, and JD Cohen, 'Dorsal anterior cingulated cortex shows fMRI response to internal and external error signals', *Nature Neuroscience*, **7** (2004): 497–8.

56 CS Carter, TS Braver, DM Barch, MM Botvinick, D Noll, and JD Cohen, 'Anterior cingulate cortex error detection and on line monitoring of performance', *Science*, **280** (1998): 747–9.

57 O Hikosaka, 'A new approach to the functional systems of the brain', *Epilepsia*, **43** (2002): 9–15; B Coe, K Tomihara, M Matsuzawa, and O Hikosaka, 'Visual and anticipatory bias in three cortical eye fields of the monkey during an adaptative decision-making task', *Journal of Neuroscience*, **22** (2002): 5081–90; PM Blazquez, N Fujii, J Kojima, and AM Baybiel, 'A network representation of response probability in the striatum', *Neuron*, **33** (2002): 973–82.

58 SV Afanes'ev, BF Tolkunov, AA Orlov, and EV Selezneva, '[Collective reactions of neostriatum (putamen) neurons to alternative behaviour in monkeys]', *Rossiiskii fiziologicheskii zhurnal imeni I. M. Sechenova*, **83** (1997): 19–27; SC Tanaka, K Doya, G Okada, K Ueda, Y Okamoto, and S Yamawaki, 'Prediction of immediate and future rewards differentially recruits cortico-basal ganglia loops', *Nature Neuroscience*, **7** (2004): 887–93.

59 M Haruno, T Kuroda, K Doya, K Toyama, M Kimura, K Samejima, H Imamizu, and M Kawato, 'A neural correlate of reward-based behavioral learning in caudate nucleus: a functional magnetic resonance imaging study of a stochastic decision task', *Journal of Neuroscience*, **24** (2004): 1660–5.

60 SP Verney, GG Brown, L Frank, and MP Paulus, 'Error-rate-related caudate and parietal cortex activation during decision making', *Neuroreport*, **23** (2003): 923–8.

61 Y Smith, MD Bevan, E Shink, and JP Bolam, 'Microcircuitry of the direct and indirect pathways of the basal ganglia', *Neuroscience*, **86** (1998): 353–87.

62 M Inase, BM Li, and J Tanji, 'Dopaminergic modulation of neuronal activity in the monkey putamen through D1 and D2 receptors during a delayed go/no go task', *Experimental Brain Research*, **117** (1997): 207–18.

63 Rolls, *The brain and emotion*.

64 W Shultz and A Dickinson, 'Neuronal coding of prediction errors', *Annual Reviews of Neuroscience*, **23** (2000): 473–500.

65 S Hanneton, A Berthoz, J Droulez, and JJ Slotine, 'Does the brain use sliding variables for the control of movements?' *Biological Cybernetics*, **77** (1997): 381–93.

66 GS Berns and TJ Sejnowski, 'How basal ganglia make decisions', in AR Damasio, H Damasio, and Y Christen, eds, *Neurobiology of decision making* (Berlin: Springer, 1996), pp. 101–13.

67 DM Egelman, C Person, and PR Montague, 'A computational role of dopamine delivery in human decision making', *Journal of Cognitive Neuroscience*, **10** (1998): 623–30.

68 P Redgrave, TJ Prescott, and K Gurney, 'The basal ganglia: a vertebrate solution to the selection problem', *Neuroscience*, **89** (1999): 1009–23.

69 B Girard, V Cuzin, A Guillot, KN Gurney, and TJ Prescott, 'Comparing a brain inspired robot action selection mechanism with "winner-takes-all" ', in B Hallam, D Floreano, J Hallam,

G Hayes, and J-A Meyer, eds, *From animals to animats 7*, Proceedings of the Seventh International Conference on Simulation of Adaptive Behavior (Cambridge, MA: MIT Press, 2002), pp. 75–84.

70 **J Wickens**, *A theory of the striatum* (Oxford: Pergamon, 1993); J Wickens, 'Basal ganglia: structure and computations', *Computation in Neural Systems*, **8** (1997): R77–R109; JW Mink, 'The basal ganglia: focused selection and inhibition of competing motor programs', *Progress in Neurobiology*, **50** (1996): 381–425.

71 **D Joel and I Weiner**, 'The organisation of the basal ganglia thalamo-cortical circuits: open interconnected rather than closed segregated', *Neuroscience*, **96** (2000): 451–74.

Chapter 12

'I think, therefore I suppress'

The suspended kiss

Everyone is familiar with the magnificent attitude of a pointer dog stopped, frozen before its prey, fascinated by the idea of devouring it, ready to go, no doubt already imagining its satisfaction. The dog is trained to perfect its gesture, adopting a posture that terrorizes the prey, leaving the hunter, its master, the enjoyment of vanquishing the pheasant or the hare. Arrested action is expressed by the card player staying his hand an instant before laying his cards on the table, the diver pausing just before diving, the writer lifting his pen to search for a word, the sigh in the musical phrase, the kiss momentarily suspended to enable the lips to resume their soft encounter, the hand raised in anger that hesitates to strike. How many times during the day do we interrupt an action, sometimes to enjoy the pleasure to come? Deciding also means deciding not to do something.

Inhibition is evident when we link several actions together. An interval of time, which depends on the complexity of the gesture, the context, and so on, separates them. This phenomenon is called the 'psychological refractory period'. The role of this delay is unknown. Does it enable us to restart command systems from zero to synchronize them? The physiological literature is filled with such unexplained inhibition phenomena that hint at the brain's need to use inhibition to control action. For example, just before making a movement, the motor neurons experience a brief 'period of silence' whose function remains a mystery. In *The Brain's Sense of Movement*, I mentioned that almost all the major structures of the brain that have a regulatory role and a function of selection or learning are structures for which action is inhibitory. The neurons that project the results of processing from these structures are inhibitory neurons. I have come up with the following aphorism: 'The brain is a fiery steed controlled by the reins of inhibition.' That is the case of the cerebellum, all of whose output neurons are inhibitory (the lovely Purkinje cells!); it is also the case for the frontal and prefrontal cortex, the basal ganglia, the striatum, and so on.

In Japan, as much importance is accorded to the absence of action as to action itself. No and Kabuki are above all arts of inaction awaiting future action, reflecting feelings and the storms of mood, which register both the fortune and misfortunes of the past and hopes for the future. Threat and tenderness, fear, suffering, and happiness are expressed by poses in which suspended movement creates a dynamic form even more powerful than many gestures. In contemplating this expressive immobility, our brain is led into mental movement simultaneously free and constrained that instead of being the movement of others becomes our own movement and for that reason has a greater effect on us than if it had been carried out. It lets us decide ourselves the fate of the suspended movement, of this contained attitude. The decision not to make the gesture leaves us free to decide which gesture we would have made ourselves. The musical sigh is an exquisite moment, a pause that suggests anything could happen.

The concept of '*ma*' epitomizes the Japanese love of the suspended act, the peak of the salmon's leap in the rapids. It is an instant where movement, though continuous, is arrested, where the mind can deliberate while following the flow of time. 'O time! suspend your flight,' wrote the poet Lamartine.

But suppressing an action does not happen only during this immobilization; it is an essential part of its very initiation. Indeed, the decision to act is always a choice among several possible actions. Here it is worthwhile reintroducing the concept of 'repertoire of actions' (see the chapter on walking and balance). We possess a repertoire of actions comprising both the genetic baggage of our species and the knowledge acquired during our life. So initiating an action means inhibiting many others. It also always entails making a choice between one action and its opposite. It is therefore accompanied by suppression of nonselected actions. Moreover, if the decision is the result of a deliberation, the period of deliberation includes holding back an imagined action. The delay occasioned by reflection is an active process that restrains the action just as one restrains a dog with a lead, a spirited horse with reins.

What parts of the brain are involved in blocking movement before it starts?

The triad: flee, fight, freeze

Distinguishing prey from predator is a basic function of the brain. In Chapter 4 we saw that the toad chooses among three behaviours: approaching, fleeing, or staying put. This triad is also present in the most complex decision-making behaviours. For example, these days, one of the places where the greatest number of decisions are made in the least amount of time is the dealing rooms of big banks where decisions that involve gains and losses of billions of euros must be made in a few seconds. The same goes for the stock exchange. There, too, telling 'prey' from 'predator' is essential! These operators must decide not to buy or not to sell as often as the converse. The danger here is not of being eaten, it is of losing everything. It eventually comes down to the same thing, but less directly!

Evolution thus chose several essential strategies: flight, combat, immobilization. Avoidable danger triggers a flight or avoidance response. On the other hand, if the danger is not avoidable, we must confront it and fight; finally, we could also just stand still. It is this last strategy—very widespread—that I would like to discuss in this chapter. It often happens that gripped by fear or worry, we freeze, senses concentrated, waiting for a cue, a sign, ready to jump or to flee, to scream or to hide.

Several types of immobilization reactions need to be distinguished. Jeffrey Gray [1] proposes a hierarchical organization for the control of movement: for example, the lowest level contains the triad 'flee, fight and freeze', useful when the danger is near or the response must be very fast. A more distant yet certain danger allows 'directed flight' and a choice among a variety of possible responses and consequences. Here, the analysis and motor programming must reach high levels of abstraction, which often involves conditioning. Finally, 'behavioural inhibition', that is, not doing anything, would entail an even higher level when the group of constraints demands processing complex information.

The repertoire of our actions is immense and has been refined over the course of evolution thanks to the long development that characterizes the human baby. Thus undesirable actions need to be eliminated from the motor repertoire. Several brain centres participate in this selection.

Let us consider a few examples. The periaqueductal grey matter is located in a centre deep in the brain that participates in working out responses to a proximal danger that induces the triad flee, fight, freeze. The medial thalamus, a nucleus that transmits sensory information to the cortex, is involved in the choice of behaviours when a distant danger appears to which one responds by fleeing or suppressing aggression. It is part of the loops linking the basal ganglia to the cortex. The anterior cingulum, involved in attentional processes, is strongly linked to the prefrontal cortex and the amygdala, a major centre for initiating aversive emotions such as fear. All these structures are important for avoidance behaviours. Finally, the posterior cingulum as well as the septohippocampal system, temporal lobe centres involved in episodic memory and spatial memory as well as detecting novelty for evaluating risk and inhibition behaviour, are involved in the case of potential danger that one chooses to approach!

The prefrontal cortex plays an important role in suppressing and suspending movement during deliberation. Electroencephalogram recordings in humans reveal, for example, electrical and magnetic (taken by magnetoencephalogram) activity associated with suppression of action. This potential is called no-go potential. The potentials appear to represent cortical activity linked to judgement and to a decision, for example, not to move one's hand when presented with a stimulus for stopping movement [2]. A negative wave appears in 200 to 300 ms on the frontocentral scalp. This 'N200' wave is greater in the case of no-go than in the case where the motor movement is executed. It is also increased in the case of more cognitive suppression, for example, during response to a mental counting task [3]. The next wave to appear has a latency of 300 to 600 ms (P3) that is also increased during suppression in the frontocentral regions, particularly in the left hemisphere in subjects who respond with their right hand[1].

Pathologies specific to the prefrontal cortex suggest a critical role for this part of the brain in suppressing undesired action, for example, in children afflicted with phenylketonuria. This deficit is caused by inadequate conversion of one amino acid, phenylalanine,

---

[1] This wave is often called 'P3 no-go' owing to its special topography, namely, maximal in the frontocentral region, whereas another P3 wave—P3 go—of identical latency in the centroparietal regions is maximal during execution.

into another, tyrosine, which is the precursor to dopamine. These children have an impaired ability to inhibit their actions, typically a function of the dorsolateral prefrontal cortex. This disorder particularly affects this structure, whose neurons are very sensitive to any variation in tyrosine.

Detecting novelty: a distraction?

Konrad Lorenz said that we are 'neophiles': we are always looking for new things. An innate biological force pushes us ineluctably in that direction. Not responding to it is an intriguing form of inhibiting action. Many brain mechanisms detect it. For example, in the 1950s the Russian school of neurophysiology suggested that the hippocampus is a detector of novelty [4]. Indeed, the hippocampus is the most highly placed brain structure in the hierarchy of analysers of the environment; it is one of the essential centres of memory of actions (called episodes). For 'new' can mean 'danger' or, on the contrary, 'pleasure', 'competition' or 'help'. Curiosity is one of our qualities; it is part of the system that Panksepp calls 'seeking' and is very developed in certain animals, such as the panda—charming, pacific, and deeply inquisitive—and in certain humans, such as researchers, who always want to know more.

However, responding to a new stimulus presents disadvantages. When we are deliberating over a decision, a new event can distract us and disturb our decision making. This depends critically on the way we select our goals, and if we are too sensitive to temptations that divert us from our purpose, our decisions become confused [5]. So we need to be able to suppress the orientation response towards the newcomer, the 'distractor', whether it is a person, an object, or even a thought or a recollection.

Inhibition of novelty can be studied experimentally. For example, I can ask you to detect a new sound stimulus in a series that you are familiar with (say, a false note in a popular melody) and to respond by pressing a button each time you detect this new sound. But I can also ask you sometimes to abstain from responding. For example, a subject is asked to listen to some sounds—bip, bip, bip, bip, and so on—made in a regular way (roughly every one and a half seconds). The frequency of the sounds is 1000 Hz. But from time to time, sounds of a higher frequency (1010, 1012, 1015, 1100 Hz) are inserted in the series of basic sounds. The sequence thus becomes bip, bip, bip, *bap*, bip, bip, bip, bip, *bap*, bip, bip, *bop*, bip, bip, bip, bip, *bap*, and so on. When the subject detects the higher-frequency sound (bap), he must respond and, sometimes, not respond (no-go) even though he has identified it.

This situation is very common in daily life: we are used to living amid familiar sounds and to detecting new noises. Sometimes we pay attention to them and respond to them; other times, we ignore them. This happens to us in the car as well as at home or work. Detection of unfamiliar noises is part of the normal process of decision making.

When a subject is asked to perform these tasks, the potentials evoked differ clearly among the three regions of the cerebral cortex—inferior frontotemporal, frontocentral, centroparietal—following the instructions given. Unfortunately, electroencephalograms do not enable more precise resolution. After about 260 ms, the effect of inhibition is clear in the inferior prefrontal regions. Later, around 300 to 600 ms, an evoked potential linked to suppression appears in the left frontocentral region that is distinct from an effect caused by difficulty in executing the task and thus from an effect in the level of attention. These potentials recorded on the surface of the scalp suggest activity located in the prefrontal cortex and in the anterior cingulum.

The variety of structures involved in inhibiting action can also be demonstrated in the case where several types of action inhibition are being compared. For example, one can give a subject three kinds of tasks, suppressing all action (stop), stopping one action and choosing another (stop-change), and stopping only one of two actions one is carrying out simultaneously, for example, hanging up the phone while driving (selective stop). Recordings of the cortical potentials that precede the movement or the no-go reveal that for both tasks, inhibition is carried out in different centres of the brain [6]. So decision making is a distributed mechanism, and we still need to figure out its physiology to understand its organization. But we are on the right track.

Tics and jitters

We all have tics: some people relentlessly scratch their chin, others twist their hair furiously, still others wink, flare their nostrils, wrinkle their eyebrows. These tics are not obsessive actions, like those described in the chapter on pathology. Patients afflicted with Gilles de la Tourette syndrome exhibit these tics in an abnormal way, along with vocal tics. The cause of Tourette's is still unclear. The disorder has been attributed to abnormalities in the loops linking the basal ganglia, the thalamus, and the cerebral cortex. Transcranial magnetic stimulation suggests a deficit at the level of the motor cortex with reduced intracortical inhibition. Tics might thus be a movement of our repertoire that slips through the inhibitory gateway, as happens sometimes in a political demonstration when some clever person manages to cross the police barricades.

But tics might also have a deeper significance. There must be a reason that one person twists the strands of her hair while another blinks and yet another coughs each time he has to speak. Tics are a way of reassuring ourselves by associating an action with a thought. This type of tic appears precisely when people are thinking or have to make a difficult decision.

Inhibition also prevents us from jumping every second, as soon as someone surprises us or we hear a loud noise. This reflex, which is basic to our survival, has its origin in short neural circuits, such as the fear reaction. But the second time the noise occurs, we don't jump. We are thus capable of inhibiting the reflex, powerful though it is, but not always—as in the case of patients with Huntington's disease[2].

These undesired actions remind us above all that decision making is about disinhibition as much as excitation, or a combination of the two. But, you will say, why insist so

---

[2] The origin of this disease, which manifests through involuntary intrusive and uncontrollable movements (called 'choreic'), is degeneration of the striatum, that is, mechanisms of behavioural inhibition of the basal ganglia.

much on the distinction between excitation and inhibition? Is it to prove to us that the brain is complicated? No, it is to show that the brain is subtle. Indeed, inhibition is not the mirror image of excitation. In mathematics, it suffices to change the sign to go from inhibition to excitation. In biology, these mechanisms of inhibition are incredibly delicate. I think that the greatest discovery of evolution is without doubt the exceptional flexibility the excitation-inhibition combination offers. If dopamine is so decisive in the processes of decision making and learning, it is probably also because its action is exerted preferentially on structures such as the basal ganglia, whose control encompasses many inhibitory mechanisms.

The role of the prefrontal cortex: monkeys playing cards

As suggested in the chapter on the pathology of decision making, the prefrontal cortex plays a fundamental role in selecting action and in inhibiting undesired responses [7]. In addition, the prefrontal cortex may link past rewards to future outcome and thus help to optimize decision-making strategies. A study of monkeys playing a game against a computer concluded that neurons in the dorsolateral prefrontal cortex appear to guide the animals' decisions based on prior choices and rewards. Brain imaging has recently revealed that different areas of the prefrontal cortex are differently involved in various aspects of decision making and evaluating advantages and disadvantages [8]. Decision-making abnormalities observed in patients with lesions and data reported below concerning electrical stimulation have encouraged neurophysiologists to record the activity of neurons in the prefrontal regions of the brain in the monkey. For example, the first Japanese authors who worked on this question initially recorded the activity of the prefrontal and prestriate cortex in monkeys, which had to press on a lever in response to light [9]. These areas are involved in the preparation and temporal organization of gestures: so potentials associated with this preparation appear in them [10]. Moreover, when the monkey suspends its gesture, other potentials appear[3].

The involvement of the prefrontal cortex in tasks of suppression is now well established [11]. In these experiments, it is always important to carefully describe the task to make clear what exactly was tested. In this case, a monkey is seated in front of a computer screen. First it sees a fixed point and must successively maintain its ocular fixation and carry out a task discriminating colour, shape, or position of a target that appears at the moment indicated on the computer screen by the word 'clue'[4].

---

[3] Of course, that is not to say that suppression is produced only in these areas, but they are at least involved in the circuits that suspend gesture. Electrical stimulation of this region leads to suppression of response to a visual signal when it occurs roughly 100 ms after the appearance of the stimulus.

[4] For each of these three conditions, the target could take different positions (left and right), or colours (red and green), or shapes (square or circle), and the monkey had to decide whether to press a lever depending on the position, colour, or shape of the target. For example, for the condition 'colour', the monkey had to press (go) if the target was red and not press (no-go) if it was green.

The point of this experiment is to space events in time. If a neuron is activated in the interval between the presentation of the target and the response or nonresponse, the neuron is assumed to have a role in the decision to respond or to abstain. Recordings show clearly that neurons respond differently during preparatory periods for tests of the type go and no-go and that their response is not the same, depending on the task. Of course, this does not prove that these neurons are involved in motor suppression itself, but it does at least suggest that they participate in the process of analysis and memorization of the properties of the target that leads to the decision to press or not to press[5]. Although, as we have seen, some decisions are even made in the parietal cortex [12], this prefrontal network may represent a final element of the analytical process—on the one hand, visuospatial properties, on the other hand, object identification.

The hierarchical character of the organization of decision making in the prefrontal cortex of primates was suggested by recording a monkey's neurons while it was observing bright spots and had to make a decision to respond based either on colour or movement. Some of these neurons respond differentially to such changes, and thus belong to a higher level of the hierarchy. They are independent of sensory contexts and are closer to genuine neurons for deciding whether to act or not to act. On the other hand, some neurons responding later have been found in the anterior part of the arcuate sulcus, which is known to be involved in developing motor responses. The discharge of these neurons is unrelated to the colour or motion of targets. This suggests a hierarchical succession of the process beginning with classification of sensory priorities up to organization of the motor response in the prefrontal cortex.

Another argument concerning the essential role of the prefrontal cortex (dorsolateral in particular) in suppressing undesired action is given by subjecting monkeys to the Piagetian test called 'object permanence', also called 'A not B'. Babies are taught to make a gesture to grab object A, and then are asked to make a gesture towards object B. This test assumes that they suppress the gesture learned towards A, whence the test's name. From 12 months onwards, babies succeed at this test, that is, making a gesture towards B and suppressing the gesture made towards A, whereas between 7 and 12 months, they make the A not B mistake. They persist in making a gesture towards A. Monkeys with lesions of the dorsolateral prefrontal cortex closely resemble 7- to 12-month-old babies. Development of the dorsolateral prefrontal cortex is thus integral to the capacity to hold a goal in working memory in the absence of any external cues, and to use this goal to guide behaviour and prevent undesired action [13]. 'Cognitive development can be conceived of not only as the progressive acquisition of knowledge, but also the enhanced

---

[5] These neurons are located in the inferior prefrontal cortex, thus in the pathways that analyse and memorize properties of objects and, here, their spatial properties. They are probably located at the confluence of activities encoded by the neurons of the temporal lobe (IT, TE, and TEO) (see Chapters 7 and 8), which analyse the visual properties of objects and images, and the pathways issuing from the parietal cortex in the dorsal pathway.

inhibition of reactions that get in the way of demonstrating knowledge that is already present [in this case, the permanence of the object]' [14].

There is a relation between the results obtained with monkeys and with humans. One finds, among other areas, activation of homologous regions in the ventrolateral prefrontal cortex (the inferior frontal sulcus in the monkey and the posterior part of the ventrolateral prefrontal cortex in humans) during a game of cards where the subjects are required to change the rules [15]. Cards of varying colour and pattern are presented in succession on a screen, and the monkey must establish the connection between the cards according to criteria that change (shape or pattern). The activity produced in the prefrontal cortex when the rules are changed is attributed to inhibition of the preceding rule.

I think, therefore I suppress: decision and competition

The mechanisms of suppression described at the beginning of this chapter would seem to form the core of a fundamental mechanism of deliberation and decision making in the most complex mental processes. Being able to choose the best cognitive strategy also entails inhibiting inappropriate ones. I am now going to show that this game of competition between activation and inhibition of cognitive strategies is present during an infant's development and thus perhaps constitutes (in any case, it's what I think) a basic component of our ability to reason.

'I think, therefore I suppress,' declared Olivier Houdé in a lecture at the Collège de France [16]. 'Memory is a jungle in which one finds appropriate strategies, but also dangerous ones. There are prey and predators, and intelligence requires processes of activation and coactivation, as well as inhibition.' Motor and cognitive developments are thus tightly linked, to the extent that even the cerebellum, that essential organ of coordination of movements that for a long time was assumed to have no cognitive role, turns out to be critical to development of cognitive function. I like this theory that 'incarnates' cognitive processes and, in particular, establishes decision making among the processes that involve action or, more specifically, acts.

Psychological theories about managing decisions do not always take this point of view into account. They assume the existence either of a 'central processor', a veritable conductor that has under its direction either specialized systems such as the visuospatial system, or of a 'central executive' [17], manager of an organization of specialized modules that ordinarily resolve ongoing problems. If uncorrected errors persist, a so-called supervisory attentional [18] system takes over and solves the problem. There is a certain similarity to ideas developed in the chapter on walking and balance where superimposed loops send back to a higher level errors whose origin is poorly understood and that are inadequately corrected by the lower level.

As well as recognizing the hierarchical organization of the decision-making process, we must imagine the possibility that there is no central supervisor but a *dynamic equilibrium* comprising winner-takes-all-type processes or, according to a model that is currently very fashionable, 'attractors'. Decision making is thus the result of unsupervised competition between inhibitory and excitatory processes. We know, too, that certain processes orchestrate the order in which we explore the world around us. For example, we sometimes have trouble looking at the same object in space two times in a row. This difficulty is in fact probably an advantage, because it enables us to be very economical in our exploration. There is a limit to our capacity to decide what we want to see, which psychologists call 'inhibition of return' [19]. Here, the workings of the brain give us an example that inspires educators and politicians without their realizing it: the limits of our freedom to decide are sometimes useful in better locating the right pathways. Somewhere there must be a proverb that goes: 'Don't look twice in the same spot if what you are seeking isn't there!'

Once again, the origin of these mechanisms may be found in the simplest animals. For instance, the cockroach responds to the approach of a predator by turning and fleeing. Three pairs of giant neurons (a bit like the Mauthner cell described in Chapter 4) are involved in determining the direction of the predator and organizing the direction of flight. Each neuron is responsible for one direction. How does the network decide which direction to take? If it is working in winner-takes-all mode, the discharge of each cell must inhibit that of all the others. This device, which exists in some animals, does not exist in the cockroach: it is the complex properties of interactions between the neurons that solve the problem and lead to a decision [20].

I am not saying that the human brain is identical to that of the cockroach. I am saying that the fundamental principle of decision making could already have been invented in simple animals. The complex capacity of the human brain may reside in the interaction between the modules, a greater contribution of suppression and especially of very varied and complex inhibitory mechanisms that offer an extraordinary diversity of possibilities [21], still mostly unknown, but that the integrative neurosciences will reveal—many retroactive loops, and so on. The idea of selective inhibition of competing processes is proposed by Alvaro Pascual-Leone's theory of 'constructive operators', which effectively describes working memory as a form of selective filter such that when a certain number of inadequate cognitive strategies are inhibited, one can then via a process of activation, decide—choose—the appropriate strategy from a repertoire of schemas [22]. We are going to examine the development of cognitive functions and show that the capacity of deciding and choosing, even for logico-deductive operations, depends on inhibiting undesired solutions [23].

Competition between numerical judgement and spatial assessment

The economic theories we briefly described in the first chapter rely on the capacity for numerical judgement. Probabilistic theories of utility are essentially founded on judgements that are related to gains expressed in numbers, and psychological research has shown the gap between the assumptions of the normative model and the reality of human brain processes. It is thus interesting to examine a few ideas concerning the way the brain carries out judgements having to do with numbers.

Suppose that a child is shown five tokens aligned horizontally in front of him, evenly spaced, like soldiers on parade. The child is asked, as Piaget did [24], to dip into a pile of tokens and to reproduce what he sees, so there is 'just as much'. The infant can interpret this ambiguous instruction either as reproducing the length of the virtual line along which the tokens are distributed, or using the same number of tokens. This test introduces competition, or conflict, between the concepts of number and length. In fact, on a first try, a 4- or 5-year-old child will line up more tokens, but against the same overall length as the model. He has taken space as a cue and likened quantity to length. Now the psychologist tries to catch him up. He moves the last token of the model line farther away, and asks the child to continue to make sure he has just as many of them in his own line. A child younger than 7 years will add tokens so that the line is longer: he remains a prisoner of the visuospatial arrangement and cannot put together the basic logical reasoning needed to estimate a number independent of the visuospatial context. After 7 years, the child lines up exactly the same number of tokens in the right arrangement parallel to that of the experimenter.

The interpretation of this failure of children younger than 7 to solve this problem is not easy. We do not really know whether it is due to the fact that they have no concept of number in working memory and thus cannot solve the problem and break away from the visuospatial context, or whether they do possess this cognitive strategy but are incapable, as Pascual-Leone suggests, of inhibiting the visuospatial strategy and replacing it with a numerical strategy. There could be cognitive competition between 'length equals number' and an 'abstract logical concept' of number. So the problem is to find out whether the child lacks a faculty or whether he has it but is incapable of inhibiting an opposing faculty. It is a bit like when people used to driving on the right arrive in England and have to drive on the left: we make mistakes not because we are incapable of driving on the left but because suddenly we are not suppressing our customary behaviour!

Recent work in psychology has focused on the calculating ability of children. At 4 to 5 months, a baby can do simple calculations, that is, ones that do not place several cognitive strategies in competition. The principle of the experiments is that of magic. The basic rules of reality are transgressed. A baby is shown numerical events that are possible or impossible ('magical'). If the baby looks especially at the magical events, it is because he has detected their unusual character. For example, a baby is shown a small theatre with a screen and a doll that comes on stage and hides. A second doll is introduced into the scene in sight of the baby, and that one hides, too. If the baby has mastered the concept of number, he should 'know' that there are two dolls behind the screen. If the screen is taken away and there is only one doll there (the other has 'magically' disappeared), the surprise of the baby shows that he possesses the concept of number. On the other hand, if he exhibits no surprise, one concludes that he has no such concept [25].

An older child, between 3 and 4 years, handles numerical information perfectly, but later, around 7 years, he suddenly starts having trouble. He is blocked in the task of competition between number and length. He cannot make a decision about the number owing to competition between visuospatial and numerical information.

More generally, the development of cognitive competence can now be studied using the new tasks designed for the study of decision making. For instance, the Iowa Gambling Task has now become a classic task for evaluating prefrontal competence along with other tasks such as the Tower of London or Wisconsin Card Sorting tasks. A reason to use this task is that the ventromedial cortex is one of the areas of the brain that develops late in children. Indeed, young children share with ventromedial prefrontal patients the inability to anticipate future outcomes. When tested with the Iowa Gambling Task, they rely more on immediate prospects [26]. The prefrontal cortex is clearly involved in the influence exerted by rules on perceptual decisions, as demonstrated in a study that required subjects to obey decision rules based on the information content—such as colour and shape—of a stimulus [27]. A familiar example of such a situation is that of driving a car and having to obey traffic lights. In addition, the Iowa Gambling Task shows that affective decision making develops rapidly during the preschool period, and is especially pronounced for girls [28].

How to beat a rival at love

Now imagine two rivals in competition, for example, two young men both in love with the same woman. At long last, the damsel chooses, and one of the two suitors wins her favour. How did he do it? There are two solutions: the winner might be more persistent or the loser might have abandoned the fight. But the winner could also have actively countered the loser by a variety of blocking mechanisms. The same is true of competition between two cognitive strategies for solving a problem.

Which of these processes—domination by force of argument or inhibition of the loser's solutions—is the work of a child's brain? A complex manipulation called 'negative priming effects' [29] makes it possible to solve the problem. Consider an example: A child is given two tasks in succession. The first serves as a primer for the second. In the first, the child is shown two lines, one above the other, each containing four tokens, but the upper line is longer, and consequently the tokens are more widely spaced. The child is asked which one has more. He must judge between visuospatial evidence (there is more above because it is longer) and numerical evidence (there is the same number above and below). To answer, the child must inhibit the cognitive strategy, which consists in taking into account the visuospatial evidence. In the second task, the child is shown a screen bearing three tokens above and two below, and is asked the same question. If, during the first task, he inhibited the visuospatial strategy, one should observe a negative priming effect in the second task, which is verified by measuring the time to reaction. Findings show that the child does inhibit one of the strategies. It is the second solution that seems to be used by the brain.

These data confirm the hypothesis of Frank Dempster, who wrote: 'Conservation and class inclusion have more to do with the ability to resist interference than they do with the child's ability to grasp their underlying "logic" ' [30]. The world is full of deceptions, such that intelligence is perhaps as much the ability to resist the sirens of false evidence as an aptitude for constructing complicated reasoning.

Inhibition and logical reasoning

Are the mechanisms of cognitive competition and reciprocal inhibition between mental strategies mentioned above involved in cognitive tasks? Reasoning implies deliberating and deciding, but the limits of human reasoning are many [31]. I covered them in the first chapter of this book, as well as critiques of most of the current theories of decision making, in particular those addressed to the prevailing theory of utility. The interest in a psychological and neurocognitive perspective on logicodeductive processing is heightened by the confusion that appears to reign among theoreticians of decision making. Recent research in psychology suggests new approaches [32]. An example will give an idea of the way in which one can pose the question: A subject is invited to play a game with conditional rules. The subject is shown geometrical shapes (circles, squares, diamonds, and so on) in a picture; he must pick two of them, put them in two boxes to the right of the picture following a very simple rule: if there is no red square to the left (this part of the rule is called 'antecedent'), then there is a yellow circle to the right (consequent). The strategem is that the subject is asked to *falsify* this rule. Ninety per cent of subjects respond 'red square, yellow circle'. They are wrong, for to falsify the rule, the antecedent must be true—that is, *not* to place the red square but, for example, a green square—and the consequent false, for example, place a blue diamond to the right. In fact, the error stems from a visuospatial solution (one need only change the colour of one of the elements), whereas the correct response is logical (change the colour of the consequent). Learning allows the subject to change his cognitive strategy by teaching him to suppress his perceptual interpretation.

Brain imaging enables visualization of the areas involved in this kind of task and especially demonstrates a marked change in brain activity. Before learning, when subjects make a visuospatial—in other words, primarily perceptual—match, brain activity is predominately parietal and occipital. After learning, when subjects have learned to inhibit the temptation to respond based on their visuospatial perception, activity switches towards the regions of the frontal cortex. In this task, many other areas are activated that respond to its emotional content, and so on. But what is most amazing about this result is the possibility of showing that two cognitive strategies involve distinct areas. This suggests that competition can actually be established between them. As if putting visuospatial and logical strategies in competition recreated a sort of microgenesis similar to what happened over the course of evolution: the appearance of the frontal and prefrontal cortex, whose workings develop over the course of infancy.

In short, decision making is not only implementation of a response, whether it is simple motor behaviour (flee, fight, freeze) or complex (flirting, starting a fight, going to the museum rather than to the movies, and so on) or, finally, a cognitive choice (buying technology stocks rather than blue chip stocks, choosing which political party to vote for). Making a decision also means suppressing action, selectively inhibiting irrelevant solutions depending on the goal, context, past experience. This suppression is part of the basic triad, but it becomes ever more complex the higher one goes in evolution and in the hierarchy and heterarchy of decision making.

To act is to suppress; to think is also to suppress; so to decide is to suppress. As in the case of St Anthony, it entails not only choosing but also resisting temptation. The inclusion of behavioural inhibition in the basic repertoire of all species takes us to the heart of our theme: even the most cognitive decisions are probably also anchored in physical experience; thinking is experienced as the product of an acting body. Choosing among the competing solutions we face when making a decision depends as much on our ability to prefer one solution as to eliminate others. That is what probabilistic theories of decision making cannot really take into account. The problem is not one of assessing a cost and a gain, and of choosing based on some probability of winning more than losing. The opposition pair win/loss is not of the same nature as the opposition pair excitation/inhibition or the competition between behaviours. This competition has a richness that we are nowhere near to understanding, but that goes way beyond the cool calculations of probability, however Bayesian they may be. We need a new neurocognitive theory of decision making.

Experiments that require subjects to carry out tasks combining working memory, aspects linking both emotion and reward, and inhibition in choosing responses relevant to the problem at hand are beginning to shed light on the link between cognition and emotion. For example, a subject is shown a series of letters (one at a time) on a screen [33]. The subject must indicate whether she has already seen a letter. The task is complicated by interposing different letters between repeating letters. The subject must thus simultaneously remember the identical letters and ignore the different ones. The subject is rewarded with money.

This task is very important in all our decision making. On the street, we want to be able to recognize people we know; at the supermarket, we have to decide whether we have already put an item in the shopping cart before taking it off the shelf.

Subjects are asked to remember letters introduced long past to increase the contribution of memory. The stronger the constraint of memory, the more prefrontal the activity. The dorsolateral cortex is activated by simple tasks and the frontopolar cortex

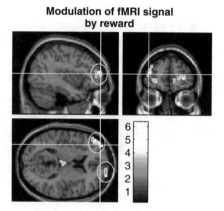
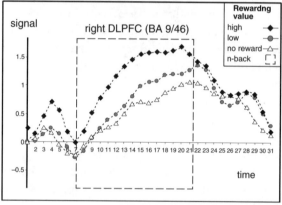

Fig.12.1 Modulation of an fMRI signal by reward. Subjects perform a test called N-back during which they must decide whether a letter that is presented to them in a continuous sequence of letters (at a rate of one letter every 2 s) is identical or different from a letter presented one (1-back), two (2-back), or three (3-back) letters previously. The tests are carried out with three levels of reward: strong, weak, or no reward. Upper panel: Positive effect (increased fMRI signal) of the difficulty of the working memory task (from 1- to 3-back) and of the significance of the reward on activation of the dorsolateral prefrontal cortex and the lateral frontopolar cortex. Lower panel: Evolution of the fMRI signal (y-axis) over time (x-axis) during realization of the N-back task. The signal is modulated by the significance of the reward. This positive effect is associated with a negative effect: lowered activity in the ventrolateral prefrontal cortex (BA 47/11) and right subgeniculate (BA 25) proportional to the significance of the reward (illustrated in the original article). This dissociation may correspond to an emotional filtering phenomenon that enables inhibition of the influence of emotional signals to maximize performance. Thus there may be a balance between the dorsal cognitive regions of the frontal lobe and the ventral limbic regions involved in the control of emotions. After Pochon et al. (2002)

(Brodmann area 10) in tasks requiring a lot of selection (Fig. 12.1). Moreover, this little section of the brain probably plays a very important role in assessing the value of the response-related reward. Sequential and hierarchical organization thus appears to be at the very core of the prefrontal cortex [34], and in humans, Brodmann area 10 must have a strategical role in decision making.

But here also, deactivation is found in areas that are especially involved in processing emotions, such as the subgeniculate part of the prefrontal cortex and the ventral striatum. In Chapter 11 on the physiology of preference I touched on possible competition between the two areas of the anterior cingulum (the 'emotional' part and the 'cognitive' part). This recapitulates the idea of competition between cognition and emotion recently suggested for the ventromedial cortex [35]. What is interesting about these data is that they may explain why depressed patients (who in contrast show strong activation of the areas that process emotion) often focus on the emotional content of difficult (especially negative) situations and struggle to concentrate on the cognitive aspects. Deciding means establishing a delicate balance between the power of emotion and the force of cognition. That does not mean that emotion and decision making are fundamentally antagonistic. Emotion enables us to anticipate the future, to establish a context, to remember, to engage the sentient body. But emotion must be regulated. Keep in mind the metaphor of the fiery steed that the horseman controls with his reins.

References

1 J Gray and N McNaughton, 'The neuropsychology of anxiety', *Oxford Psychology Series*, 35 (2000): 97.
2 See, for example, the chapters on this subject in AM Thierry, J Glowinski, P Goldman-Rakic, and Y Christen, eds, *Motor and cognitive functions of the prefrontal cortex* (Berlin: Springer, 1994).
3 This result is consistent with that of O Houdé, L Zago, E Mellet, S Moutier, A Pineau, B Mazoyer, and N Tzourio-Mazoyer, 'Shifting from the perceptual brain to the logical brain: the neural impact of cognitive inhibition training', *Journal of Cognitive Neuroscience*, 12 (2000): 721–8.
4 EN Sokolov and OS Vinogradova, *Neuronal mechanisms of the orienting reflex* (Hillsdale: Lawrence Erlbaum, 1975).
5 J Baron, *Thinking and deciding* (Cambridge: Cambridge University Press, 2000).
6 R De Jong, MG Coles, and GD Logan, 'Strategies and mechanisms in nonselective and selective inhibitory motor control', *Journal of Experimental Psychology: Human Perception and Performance*, 21 (1995): 498–511; RT Richardson and MR Delong, 'Context-dependent responses of primate nucleus basilis neurons in a go/no go task', *Journal of Neuroscience*, 10 (1990): 2528–40.
7 K Matsumoto, W Suzuki, and K Tanaka, 'Neuronal correlates of goal-based motor selection in the prefrontal cortex', *Science*, 301 (2003): 229–32; BJ Richmond, Z Liu, and M Shidara, 'Neuroscience: predicting future rewards', *Science*, 301 (2003): 179–80; KR Ridderinkhof, WP van den Wildenberg, SJ Segalowitz, and CS Carter, 'Neurocognitive mechanisms of cognitive control: the role of prefrontal cortex in action selection, response inhibition, performance monitoring, and reward-based learning', *Brain and Cognition*, 56 (2004): 129–40.
8 MP Paulus, JS Feinstein, SF Tapert, and TT Liu, 'Trend detection via temporal difference model predicts inferior prefrontal cortex activation during acquisition of advantageous action selection', *Neuroimage*, 21 (2004): 733–43; M Gomez-Beldarrain, C Harries, JC Garcia-Monco, E Ballus, and J Grafman, 'Patients with right frontal lesions are unable to assess and use advice to make

predictive judgments', *Journal of Cognitive Neuroscience*, **16** (2004): 74–89; KR Ridderinkhof, M Ullsperger, EA Crone, and S Nieuwenhuis, 'The role of the medial frontal cortex in cognitive control', *Science*, **306** (2004): 443–47.

9 **K Sasaki and H Gemba**, 'Electrical activity in the prefrontal cortex specific to no go reaction of conditioned hand movement with colour discrimination in the monkey', *Experimental Brain Research*, **64** (1986): 603–6; K Sasaki, H Gemba, A Nambu, and R Matsuzaki, 'Activity of the prefrontal cortex on no go decision and motor suppression', in Thierry, Glowinski, Goldman-Rakic, and Christen, eds, *Motor and cognitive functions*, pp. 139–59.

10 **H Niki and M Watanabe**, 'Prefrontal and singulate unit activity during timing behavior in the monkey', *Brain Research*, **171** (1979): 213–24.

11 **M Sakagami and H Niki**, 'Encoding of behavioral significance of visual stimuli by primate prefrontal neurons: relation to relevant task conditions', *Experimental Brain Research*, **97** (1994): 423–36; M Sakagami and H Niki, 'Spatial selectivity of go/no neurons in monkey prefrontal cortex', *Experimental Brain Research*, **100** (1994): 65–9.

12 **ML Platt and PW Glimcher**, 'Neural correlates of decision variables in parietal cortex', *Nature*, **400** (1999): 233–8.

13 **A Diamond**, 'Close interrelation of motor development and cognitive development and of the cerebellum and prefrontal cortex', *Society for Research in Child Development*, **71** (2000): 44–56; A Diamond, N Kirkham, and D Amso, 'Conditions under which young children can hold two rules in mind and inhibit a prepotent response', *Developmental Psychology*, **38** (2002): 1–20.

14 **A Diamond**, 'Neuropsychological insights into the meaning of objects concepts development', in S Carey and R Gelman, eds, *The epigenesis of mind: essays on biology and cognition* (New York: Lawrence Erlbaum, 1991), p. 67.

15 **K Nakahara, T Hayashi, S Konishi, and Y Miyashita**, 'Functional MRI of macaque monkeys performing a cognitive set shifting task', *Science*, **295** (2002): 1532–6.

16 See also **O. Houdé**, *Rationalité, développement et inhibition: un nouveau cadre d'analyse* (Paris: PUF, 1995); O Houdé, 'Executive performance, competence, and inhibition in cognitive development', *Developmental Science*, **2** (1999): 273–5; O Houdé, 'Inhibition and cognitive development: object number categorization and reasoning', *Cognitive Development*, **15** (2000): 63–73.

17 **A Baddeley**, *Human memory: theory and practice* (London: Lawrence Erlbaum, 1990).

18 **T Shallice**, *From neuropsychology to mental structure* (New York: Cambridge University Press, 1988).

19 **RD Rafal, LS Choate, and J Vaughan**, 'Inhibition of return: neural basis and function', *Cognitive Neuropsychology*, **2** (1985): 211–28; L Riggio, E Scaramuzza, and C Umilta, 'Modulation of inhibition of return by type and number of dynamic changes of the cue', *Psychological Research*, **64** (2000): 56–65.

20 **R Levi and JR Camhi**, 'Wind direction coding in the cockroach escape response: winner does not take all', *Journal of Neuroscience*, **20** (2000): 3814–21.

21 **R Miles**, 'Diversity in inhibition', *Science*, **287** (2000): 244–6.

22 **J Pascual-Leone**, 'Organismic processes of neo-Piagetian theories', in A Demetriou, ed., *The neo-Piagetian theories of cognitive development* (Amsterdam: North Holland Elsevier, 1988), p. 47.

23 **D Dagenbach and TH Carr**, *Inhibitory processes in attention, memory and language* (San Diego, CA: Academic Press, 1994); J Bidault, O Houdé, and J-L Pedinelli, *L'Homme en développement* (Paris: PUF, 1993).

24 **J Piaget and A Seminska**, *La Genèse du nombre chez l'enfant* (Neuchâtel: Delachaux et Niestlé, 1941).

25 K Wynn, 'Addition and subtraction by human infants', *Nature*, **368** (1992): 749–50; O Houdé, 'Numerical development: from the infant to the child. Wynn's (1992) paradigm in 2- and 3-year-olds', *Cognitive Development*, **12** (1997): 373–91.

26 EA Crone and MW van der Molen, 'Developmental changes in real life decision making: performance on a gambling task previously shown to depend on the ventromedial prefrontal cortex', *Developmental Neuropsychology*, **25** (2004): 251–79.

27 SA Huettel and J Misiurek, 'Modulation of prefrontal cortex activity by information toward a decision rule', *Neuroreport*, **15** (2004): 1883–6.

28 A Kerr and PD Zelazo, 'Development of "hot" executive function: the children's gambling task', *Brain and Cognition*, **55** (2004): 148–57.

29 O Houdé and E Guichart, 'Negative priming effects after inhibition of number/length interference in a Piaget-type task', *Developmental Science*, **4** (2001): 119–23.

30 FN Dempster, 'Interference and inhibition in cognition. An historical perspective', in FN Dempster and CJ Brainerd, eds, *Interference and inhibition in cognition* (New York: Academic Press, 1995), p. 15.

31 J Evans, *Biases in human reasoning* (London: Lawrence Erlbaum, 1989).

32 O Houdé and S Moutier, 'Deductive reasoning and experimental inhibition training. The case of the matching bias', *Current Psychology of Cognition*, **18** (1999): 75–85; O Houdé, L Zago, E Mellet, S Moutier, A Pineau, B Mazoyer, and N Tzourio-Mazoyer, 'Shifting from the perceptual brain to the logical brain: the neural impact of cognitive inhibition training', *Journal of Cognitive Neuroscience*, **12** (2000): 721–8; O Houdé, S Moutier, L Zago, and N Tzourio-Mazoyer, 'La correction des erreurs de raisonnement', *Pour la science*, **297** (2002): 48–55.

33 J-B Pochon, R Levy, J-B Poline, S Croziez, S Lehéricy, B Pillon, B Deweer, D Le Bihan, and B Dubois, 'The role of dorsolateral prefrontal cortex in the preparation of forthcoming actions: an fMRI study', *Cerebral Cortex*, **11** (2001): 260.

34 K Christoff and JDE Gabrieli, 'The frontopolar cortex and human cognition: evidence for a rostrocaudal hierarchical organization within the human prefrontal cortex', *Psychobiology*, **28** (2000): 168–86.

35 JR Simpson, AZ Snyder, DA Gusnard, and ME Raichle, 'Emotion-induced changes in human medial prefrontal cortex: I. During cognitive task performance', *Proceedings of the National Academy of Sciences of the USA*, **98** (2001): 683–87.

Chapter 13

The brain as emulator and generator of strategies: the vagabond thought

Before adolescence, the possible is a subset of the real, and after, it is the real that becomes a subset of the possible
(Bärbel Inhelder and Jean Piaget [1])

Action and the sentient body, which construct the perceived world based on our desires, goals, and fears, are at the heart of the mechanisms of decision making, more than language and reason. Or rather, it is the language of the sentient body in its dialogue with the world that remains to be discovered: the gesture of thought. The role of space in the dynamic architecture of thought is also largely unknown. The Italian Renaissance placed at the centre of the world the new man and his freedom, his capacity to choose and to decide, to find several solutions to the same problem with no boundary conditions. This spirit is asserted in the interplay between softness and drama—each triumphing in turn—in the art of Fra Angelico and Donatello. It also found its way into the architecture of the dome of Florence, fruit of the elegant genius of Brunelleschi, an ode to emotional thinking, erected without scaffolding, where thought and gesture meet in a stone 'fountain' launched heavenward. The leaders of Florence were smart enough to bring back and to trust this audacious thinker after initially dismissing him.

The vagabond thought

The tyranny of the disembodied formalist thinking that prevailed during the twentieth century is nowhere better represented than in the abandonment of the roof over houses and public buildings. Only the area around La Défense, illuminated by the dome of the Palais des Expositions, a modern version of the dome of Florence, escapes this tastelessness. At the other end of Paris, the poverty of so-called rational thought is symbolized by the area around Bercy, the big library, and Tolbiac where the roof has disappeared and the implacable logic of order has replaced the finesse that only a supple mind can produce.

The same tragedy plays out in the museum (the former station) of Orsay. The unique gesture, strong design, and combined objectives embodied by this wonderful vault—attractively built with lateral arcades in the form of diadems that welcomed passengers, as in a garden where flower-framed archways line the alleys—has been destroyed by enormous blockhouses. This leaden concept conjures up the fortresses of the Middle Ages, or a mausoleum, factory gangways whose stark crosspieces look like the bridges hastily constructed by the military corps of engineers to withstand the passage of wagons. An absurd central ramp hems in the world in a place where, ironically, everything was intended to follow the direction of the parallel rails, which satisfy our desire to go farther than our eye can see.

Departure is a classic example for a theory of decision making; the train station attests to it. Its space is designed to shelter the delicious and horrible moment when, slowly, precisely without the effort symbolized by the ramp that evokes Sisyphus, the train passes from standstill to motion with an imperceptible caress of wheel along rail, still there and yet gone, gently separating loved ones. A delicate balancing of decision making and hesitation, regret at leaving, but already promising landscapes to be crossed at full speed, vertigo-inducing viaducts, tunnels at whose exit children say to their mothers, 'Mommy! it's already tomorrow!' The train station: locus of choices where anything is possible up to the last minute, where all roads lead to Rome.

All that is destroyed at Orsay, and the objects of an art 'nouveau' whose delicacy of movement, gestural expression, and fanciful colour that once mattered are locked into thick, heavy-walled boxes that look more like a prison than a place where unfettered imagination can, like the water of a fountain, play the vagabond.

In this book, we first briefly covered the major theories about decision making. In the face of mathematical formalism, in particular economics, which would like to reduce the process of making decisions to calculations of probabilities, we moved on to examine the complex ideas of the major forerunners who, ever since Darwin and William James, have shown that in humans, decision making is a complex hierarchical process in which emotion plays a fundamental role. Pathology reveals the multiple forms decision making takes and supports our working hypothesis, namely, that decision making is a process that involves the entire nervous system. After having examined the role of the body in this process and suggested that we decide in concert with our double or sometimes contrary to it, we saw that the process of decision making occurs at all levels, in mechanisms for tracking and capturing prey, in walking and keeping balance. We also saw that perception is not only simulation of action, but also decision making. Finally, we considered the broad outlines of a physiology of preference and the role that inhibition plays in suppressing irrelevant solutions. Now I will attempt, with the aid of a few—admittedly imperfect—examples, to show how neuroscience is finding ways to explain high-level decisions that draw on memory, reasoning, and so on. In particular, I would like to emphasize the role that space plays in decision making, sometimes complementing or even replacing language. For more than a century, too much weight has been given both to calculations of probabilities and to language in explaining decision making; the role of space as a support for the mental basis of deciding needs to be restored, that is, the role of the right side of the brain with respect to the left, 'dominant' side. Don't misunderstand: I am not making a political statement. The left brain processes the data of the visual work on the right side and commands the right-sided limbs; the right side of the brain should pay attention to what is happening on the left side!

Cognitive strategies for finding your way

In the morning, just as you are leaving for work, you remember that you absolutely must go by the post office to mail a letter. The route will be a departure from your usual one. To decide the best way to go, you think of the path you take each day to go from your house to work and construct another to go to the post office. You have at least two possible cognitive strategies for constructing this new path: you can either use the memory of your route (called 'route strategy') or you can use a mental map of your neighbourhood (called 'map strategy' or 'survey'). In fact, there are other possibilities, and we have described several major categories of navigational strategies [2], but we will simplify here.

First let us examine the route strategy. You remember closing the door, going down the elevator (or walking down the stairs), opening then closing another door, walking to the car, the underground, or a bus, and so on. So your memory includes cues linked to the movements of your body (walking right, turning, and so on); but it also includes recollection of visual scenes (the door, the staircase, a tree, a shop) and auditory elements of events, called 'episodes', such as meeting a neighbour, tripping over an obstacle, stopping at a red light.

The brain relies on as many of these cues as structures such as the hippocampus or the parahippocampus, and so on, enable it to retrieve. The problem I would like to pose is the following: How does the brain manage to combine things as different as the length or feel of a step and the sight of a bus? Does the brain mix up kinesthetic and visual information, or does it memorize the information separately and use it as needed [3]? Nearly all mathematical models of multisensory fusion published in the literature over the last 20 years assume that the brain weights visual, vestibular, proprioceptive, and other such information. If that is true, it should be possible to demonstrate experimentally. An alternative view, which I favour, is that the brain decides and selects a sample of landmarks, movements, and episodes for this route memory.

Route memory and conflict processing: weighting or decision making?

To test the idea that in memorizing a route, the brain perhaps does not weight sensory data but separates and chooses what it needs to make what we like to call a 'decision', we have used a device that creates a conflict between vision and the other kinesthetic senses [4]. The experimental setup is very simple (Fig. 13.1): A subject equipped with a virtual reality helmet moves through a virtual corridor that includes several segments. The computer simulates a forward motion at constant linear velocity. At the end of one of the segments, if the subject does nothing, he will bump into a wall because the

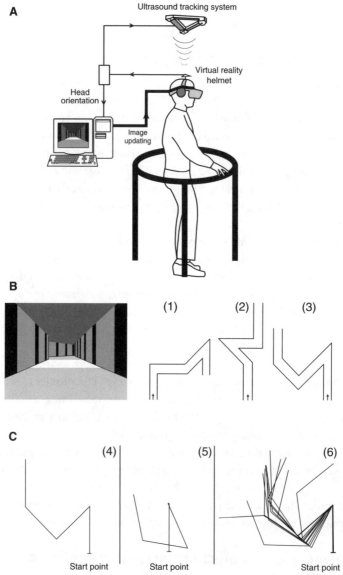

Fig. 13.1 A new paradigm for studying the role of decision making in multisensory conflicts. **(A)** The subject is upright and looking into a virtual reality helmet. Within the helmet, he is asked to navigate in a virtual corridor that comprises several segments. The computer moves the subject through the virtual corridor as in a video game, at a constant speed. When he arrives at the end of the corridor and must turn in the virtual world, if he does nothing, he 'bumps' against the wall. To turn, he has to turn his body in the real world. The rotation is recorded using an ultrasonic sensor. This value is sent to the computer, which updates the image and allows the subject to turn in the virtual world. This device makes it possible to construct a 'trajectory' in the corridor by adjusting the translation imposed by the computer

corridor makes a right angle. To turn in the virtual corridor, the subject has to turn his body in the real world. We measure the rotations of his body with an ultrasonic sensor placed above his head. The advantage of this device is that it allows manipulation of relations between the real rotations of the subject's body and his rotations in the virtual world. For example, if the virtual corridor turns 90°, we can ask the subject to turn 120°, sitting or standing, to carry out this rotation of 90° in the virtual world.

This way we can study which of the two components of the navigation are memorized: vision or movements of the body? For example, after the subject has traversed the virtual corridor, we can ask him to draw the shape of the corridor on a sheet of paper. If what he memorized is what happened in the virtual corridor (visual), he will draw a shape with (in the case of the 90°-turn) a 90°-angle. If what he memorized is the movement of his body, he will draw a 120°-angle. The subject can also be asked to reproduce the movement of rotation. This device was developed to study spatial anxiety (see Chapter 9, note 18).

The results are clear: subjects do not memorize a combination of visual and kinesthetic information. For example, they do not reproduce an angle of 105°. They reproduce either the form they saw or the form that corresponds to the movements of their body. In general, they seem to memorize whichever of the two cues had the greatest amplitude. I think that in this case the brain makes a genuinely perceptual decision and resolves to believe and thus to memorize one of two types of information. Another possibility, which I find plausible, is that both kinds of information are memorized, but that during the task of recollection, the brain chooses one of the two modes of encoding the information. Making a decision is a choice: the brain chooses what it wants to believe is real. Future experiments will probably show that this choice depends on the task and on other circumstances. What is important, for us, is that the

---

and the rotations of the subject's body in the real world. The subject can then be asked to repeat the memory of his walk as we do, for example, when we are imagining our way from the door of our house to the living room. He will not move when he remembers that he went straight, but he will turn his body when he remembers turning. **(B)** Example of the corridor seen from the inside of the virtual reality helmet and an example of three corridors each with four segments seen from above (1, 2, 3). **(C)** A conflict can also be introduced between the two sensory modalities (vision and the information supplied by the bodily perception). For example, if the corridor turns 90°, it is possible to change the gain between the sensor of rotation and the computer so the subject is required to turn 150° in the real world. This is a powerful way of seeing whether the brain, faced with conflict, will combine (weight) visual and somatic information, or whether it will decide to choose (believe?) one or the other. This lower part of the figure shows an example of conflict between vision and kinesthetic information. In this case, the angle of turns in the visual virtual world (4) was very different from the angle of the subject's turn in the real world (5). After 'navigating' in situations of conflict, the subject was asked to draw the path he had followed: he did not draw a combination of the two walks but 'chose' the walk in the visual virtual world. After Lambrey, Viaud-Delmont, and Berthoz (2002)

brain can combine, but it can also choose. Nearly 30 years ago, Laurence Young and his students proposed models of interaction between vision and the vestibular system that contained a module involving choice [5]. More recently, other models have included cognitive control in the priority assigned to one of the senses [6]. Here, too, and doubtless to reduce complexity, faced with such conflict the brain adopts a *cognitive decision-making strategy* in favour of one of the particularly pertinent components of the sensory data. The physiology of this selection of relevant sensors has yet to be worked out.

Space and decision making

In his introduction to *Spatial Schemas and Abstract Thought*, Merideth Gattis writes: 'How has humanity come to develop fundamental abstract abilities such as those seen in science, literature, and art? One answer is familiar to evolutionary theorists: by recruiting old parts for new uses. Stephen Pinker has suggested, for instance, that sensory processes and motor programs may have become adapted to more abstract tasks that share some of the same computational structure as sensorimotor tasks . . . One of the primary candidates for sharing computational structure with abstract cognition is spatial cognition' [7].

Making a decision is literally making a movement of capture towards a goal in space, reaching for a cup, say. This analogy between the spatialized gesture and the cognitive process of decision making refers only to the final phase of deciding, the actual moment when one 'makes' the decision. This analogy, which compares the act of deciding with a gesture, suggests two ideas: first, the laws controlling action probably constitute an important basis for decision making; second, these processes are spatialized—they borrow rules and mechanisms from the mental management of space. Indeed, gesture is essentially a creation of shape (the trajectory of the hand) and grasp is a spatial act involving an object. Extending the analogy further, let us imagine this process of decision making as a destination to which several roads lead. Each road, like any voyage, involves choices, means of transport, different languages. This process is central to cognitive decision making: you have to pick a road. This choice itself means making comparisons, reflecting, weighing the advantages and disadvantages of the various routes.

Before being *made*, a decision must be prepared; it is often said to be 'negotiated'. Here, again, the spatial metaphor is useful: in the car or on a bicycle, one 'negotiates' a turn. Negotiating entails both struggling between opposite constraints and choosing among several solutions. As we saw in Chapter 12, it also means gambling. The mental process of deliberation must come from the processes that control gesture and allow us to pick a road. To deliberate is to envisage several solutions or, rather, several roads that lead to a solution.

The idea that space is used by the brain in reasoning that leads to a decision has been widely debated by psychologists. It is well known that memory uses space to store lists of items such as objects or words. In his splended book titled *The Art of Memory*, Frances Yates [8] reviews mnemonic techniques employed since ancient times. For example, one can use a 'mental palace' and place in each room words one wishes to memorize. But this palace, this space created to hold memory, can be used in many ways. For example, we can 'change our point of view', explore the palace in various ways and manipulate mental objects using means adapted to problems that we hope to solve. Space is not only a place for organizing things; it is also a structure where thoughts can wander.

The two sides of the cerebral cortex do not play the same role in this contribution of space to thought. The difference between the left brain and the right brain has been the subject of a considerable number of studies. For example, spatial relations appear to be treated in two different ways [9]: on the one hand, *categorical* spatial relations, such as the location of objects in the environment, use categories such as 'above', 'to the right', or 'inside', stressing the deictic (demonstrative) aspect; on the other hand, *coordinate-*type relationships specify metric relations, such as the distance of objects, including the body as a specific object that functions as the origin in a space of coordinates. In general, the right brain appears to be especially involved in processing categorical relations (which are easily indicated by one or two words) and the left brain better for coordinate-type relations [10]. This distinction is particularly clear when it is difficult to describe spatial relations [11], which is understandable since a simple relation can be very easily encoded categorically. Here I will present just a few examples of these mental manipulations of space that are, I think, essential in cognitive decision-making strategies.

Decision making and changing point of view from egocentric to allocentric: the penalty

The brain is not just a simulator of action; it makes decisions, chooses with astonishing speed, gambles, and often, to do that, it must switch point of view. The penalty in football is a nice example of the subtle game of quick decision making. The players are two, face to face. The striker must decide to which side he is going to kick. The goalie must decide to which side he is going to dive even before the striker kicks. He gambles on the decision his opponent will make. Deciding also means guessing the decision of another; it means emulating internally not only a double of one's own body for tracking purposes, but also the opponent's decision-making strategy. The penalty is a wonderful model of extremely complex decision making, for the goalie must imagine the possible strategies of the other player. But gambling assumes the ability to put oneself in the shoes of the opponent. It is not enough to apply probabilistic operators; it also requires changing perspective, envisaging the situation by abandoning what we call an 'egocentric frame of reference'. It means being able to manipulate the relations between the elements of the world independent of our perspective.

This transition from egocentric point of view (centred on or with reference to my body) to an 'allocentric' point of view (centred on or with reference to elements or a person in the environment) is fundamental to formulating a decision. It is equivalent to the distinction made earlier on between route, map, and survey strategies.

The penalty is a confrontation. In more complex decision-making situations, the context must be taken into account just as problems must be considered from another point of view. The judicial process with the play of defence and prosecution is a very elaborate and formal form of this requirement. Jurors must be presented with diverse points of view so they can make a decision in cases where most of the time there is a modicum of doubt. Interestingly, human rights advocates think that this contradictory presentation of opposite points of view suffices not to guarantee the impartiality of the verdict of the jurors, but at the very least to ensure that their decision is made 'in their soul and conscience.'

Deliberating to decide thus often entails the same reality considered from different vantages. The arguments underpinning decision making are almost always organized mentally according to a spatial structure. Johnson-Laird was fascinated by the singular effectiveness of diagrams (see Chapter 2). I suggest, moreover, that the ability to compare cognitive points of view is linked to the acquired capacity to recognize places and persons from different perspectives. For example, when I visit an Italian city, as I am leaving the hotel to go to a restaurant, how do I decide which path to take? I often begin by wandering aimlessly at will without necessarily even a map of the city. Drifting through the streets, I encounter a monument that I perceive from a certain vantage. The next day, starting off in another direction and sometimes arriving at the same monument now perceived from another point of view, I always experience a little moment of hesitation: Is it the same monument? What angle did I see it from yesterday? If it happens again, I'm no longer surprised. I have constructed a memory of the monument that allows me to recognize it from various perspectives. It is the same with faces. A face seen in an initial encounter from a certain angle is not always recognizable later from another angle. But we also have the ability to generalize, that is, to recognize the place we have arrived at from any perspective or to recognize a person from any angle.

When space is involved, the ability to choose depends on a shift from an egocentric-type encoding of relations between the body and space to an allocentric-type encoding. For example, if I am in Paris, at the foot of the Eiffel Tower, and I want to know which direction the Opera is in, I can locate it with respect to my body (egocentric) as being at the centre of the axes of Châtelet and the Arc de Triomphe. But I could also reckon it based on the middle-point between Châtelet and the Arc de Triomphe. In that case, I make an abstraction of my body and the Eiffel Tower, and I figure only relations between objects outside my body (allocentric). The advantage of allocentric coding is considerable: it enables one to mentally simulate trips between places, to avoid triangulation based on one's point of view. As I maintained in my 1997 course at the Collège de France [12], this ability may be one of the cognitive bases of geometry. But, apart from acquiring this decentred perception of one's own body, the capacity to extract oneself even more from reality is reinforced by the capacity to construct constants independent of perspective.

Taken together, these properties endow the brain with the capacity to manipulate information about the world in order to act. They also give rise to an even more profound cognitive process, that of rule formation. Indeed, making decisions requires frames of reference just as perception does. These frames of reference, which support both strategic and perceptual decisions, consist of abstract rules that guide behaviour. At the highest level, reasoning enables the working out of new solutions, manipulation of possible actions, definition of strategies. This capacity depends on several properties that appeared over the course of evolution: first, the ability to escape from the *local* point of view and to encode as 'global' perceptions and actions or their function; second, to progress from action to intention, from a component of behaviour to a sequence oriented towards a goal. The mental process that allows us to construct this constant representation of shapes, objects, faces, buildings is fundamental and very complex. We have only the barest idea of its mechanisms.

Night games: the theory of mental pathways

When I was a scout, I adored night games, in particular, walking in the forest without a map and returning to camp at the risk of having to spend the night outside! You might ask, 'What does that have to do with decision making?' A lot. Deliberating is tantamount to mentally navigating a path with no light from the world of facts. It is about going over options, keeping the points of departure and arrival clearly 'in mind'—in the night maze of the thinking brain. This capacity for making one's way by studying mental objects from various points of view is fundamental. I dare propose adding a theory of mental routes to the theory of mental models and schemas [13]. The brain has available many strategies for finding its way along *mental routes* and ensuring that during this process, it keeps a precise 'mental vision' of the objects of reason that it is considering.

Our laboratory carried out a related experiment that shows that several cognitive strategies can be used by the brain to perform such a reconstruction [14]. To help you understand this experiment, imagine that you are facing a statue that, instead of representing a person, depicts a big wooden letter 1 m high, for example, the letter 'F'. You are placed in front of this 'F' and asked to recall the orientation of the room where the experiment is taking place. Then, you are blindfolded and made to walk in the room along a path that follows a straight line, a turn, and another straight line. So you arrive at another spot in the room. Now you are asked to hold a small model of the 'F', 20 cm high, and (still blindfolded) to orient the model parallel to the big 'F' that is in the room, but whose orientation with respect to you has changed. The task thus consists in updating the mental representation of the object during a locomotor movement. This is the task we carry out all the time in moving around a city and trying, for example, to remember the location of a monument or where our hotel is situated.

Subjects use two very different cognitive strategies. In one case, they are asked to *continually think of the object during their walk*, that is to continually update the form they saw. In the second case, they are asked to focus on visualizing the object before starting off, to memorize it, then, during the walk, *to forget the object and concentrate on the movements* they are making—a purely egocentric task of route memorization. At the end of the walk, they are asked to update their perspective and to identify the orientation of the miniature object. The prediction, obviously, is that it will take more time to retrieve the correct orientation of 'F' in the second case since all the cognitive work was done at the end of the walk. The findings clearly show a difference between these two strategies.

This particular exercise illustrates how the brain is able to choose completely different methods of reaching the same end. This is the concept of *vicariance*. For example, if we want to build a house, our brain may choose to build the structure up to the roof (which is what it usually does), then to fill in the interior, lay pipes, and so on. But it can also continually adjust all the components of the house (plumbing, woodwork, masonry), no doubt the best way of embedding the pipes in the walls. I suggest that these cognitive strategies of choice borrow—at least partly—their organization from ways of resolving fundamental problems such as the one I mentioned above for getting oriented in a strange city.

Of course, the cognitive strategies used by humans are much more complex than that. They are founded on methods selected by evolution to solve problems, in particular regarding action and space. On this point, I agree with the proponents of the theory of bounded rationality who propose fast-and-frugal [15] strategies, that is, in the end rather simple.

Route and survey: two strategies for imagining the way to the post office

Let us return to the example of the path to the post office. To imagine this route, we have at least two possible cognitive strategies, defined above. Imagining our route, that is, the movements associated with episodes or landmarks, simultaneously calls into play 'episodic' and 'procedural' memory. This cognitive strategy is egocentric; in the area of thinking required for decision making it corresponds to considering alternatives in succession, traversing a mental route in the first person. But we can also imagine a map, as we do in visiting a museum or planning a vacation. In the latter case, we speak of a 'survey' strategy, which corresponds to what we call 'taking the high ground', 'looking at the situation from above', and so on.

To study the neural basis of these two cognitive strategies, subjects were given the task of memorizing a route in a city. One group of subjects actually went through the streets of the city picking out seven specific landmarks (a petrol station, a door, and so on). Next they were asked to mentally go over certain parts of the route while having their brain activity measured with a positron emission camera. This was a task of egocentric mental navigation using the route strategy.

Another group of subjects simply visualized a map of the same environment projected on a screen. It contained seven dots of colour linked by pathways. At the end of learning, subjects were shown the entire map again and asked to indicate the landmarks on a map without dots. The next day, before the camera, subjects were asked to do mental operations that required them to remember the map they had seen the previous day, as if you were discussing with your family whether to go to Marseille by way of Lyon or Clermont-Ferrand. This is an example of an allocentric survey strategy, which requires manipulating relations between locations in space independent of personal perspective.

The two mental strategies activate a parietofrontal network.[1] This network may be involved in spatial memory for maintaining an image of the environment [16]. The intraparietal sulcus is involved in visuospatial attention, spatial working memory, and mental imagery. Three other series of regions are activated.[2]

More generally, it is now quite clear that in humans, a parietofrontal module, mainly located on the right side, plays a decisive role in egocentric visuospatial tasks. It cooperates with a part of the temporal network in what I have previously called 'topokinesthetic' [17] memory by contrast with 'topographic' (or maplike) memory. The areas that make up this module are also those whose lesions are associated with disturbances observed in patients who suffer from spatial negligence (see Chapter 6).

The areas described above correspond to the parietal and frontal lobes of the brain. Concerning the temporal lobe, in both tasks, the right hippocampus was activated. In

---

[1] Comprising a first component with the bilateral intraparietal sulcus, the bilateral superior frontal sulcus, the right medial frontal gyrus, and the anterior portion of the supplementary motor area (pre-SMA). This group of regions probably constitutes a network specialized in processing visuospatial information when the information is no longer present and must be evoked by spatial memory.

[2] A second bilateral component is located deep in the superior frontal sulcus near the intersection of the precentral sulcus (Brodmann area 6). This region is activated in tasks of spatial working memory and mental imagery. It is located in front of the frontal oculomotor field, which is active during ocular saccades. The third component is composed of frontal activation in the medial frontal gyrus (areas 9 and 46, which is less specifically spatial than area 6). In fact, studies of spatial working memory and object memory show activation in this region. Finally, a fourth component common to both tasks of exploration is pre-SMA. This region is distinct from SMA in the sense that it is activated in complex motor tasks, in particular learning. Recently, we showed specific activation in the area of pre-SMA that is specialized in ocular movements (called pre-SEF) in acquiring and learning sequences of saccadic ocular movements. Generally speaking, pre-SMA participates in preparing selection and learning of motor activities. Its involvement in organizing mental activities simulating visuospatial action give it a fundamental role in processes of visuomotor decision making and even in perception of the temporal aspects of ongoing action.

the task of mentally exploring maps, the mental image included a mental description of the entire map. In the route task, the mental image included only the local views associated with movements. The route perspective thus requires additional work to link all the views into a coherent path.

But the parahippocampus is not always activated in the same way. In the case of route perception, there is bilateral activity of the entorhinal and parahippocampal cortex that is not present in the case of exploring by map. This result is compatible with the role of the parahippocampus in topographical processes [18]. Our results[3] establish a dissociation between memorizing maps, which only involves the right hippocampus, and memorizing the environment in the form of views of buildings, which involves the bilateral parahippocampus.

The posterior cingulate cortex was activated in the mental navigation task and not in the map exploration task. It appears to be involved in transforming mental simulation of a route into a global type of representation. The medial occipitoparietal region (cuneus and precuneus) was activated during the task of mental navigation and not during map exploration. During mental exploration, subjects reproduced visual images that resulted in real navigation and thus very realistic memories of scenes. So this region might be linked to the memory of visual scenes.

The two mental strategies use both common structures and different structures. The door is open to the neurocognitive study of cognitive strategies that the brain can adopt in making decisions.

The neural basis of decisional reasoning: deduction and induction

Of course, deciding also requires reasoning. The left brain, which governs language, is often considered to be dominant in solving problems and verbal reasoning. Yet there are several cognitive strategies for reasoning [19]. For example, we distinguish deductive and inductive reasoning. Are these two forms of reasoning processed in the same areas of the brain? Consider the following statements:

1 All men are mortal. Socrates is a man. [therefore] Socrates is mortal. (Deduction)
2 If Socrates is a cat, he has nine lives. Socrates is a cat. [therefore] Socrates has nine lives. (Deduction)
3 Socrates is a cat. Socrates has 32 teeth. [therefore] All cats have 32 teeth. (Induction)
4 Socrates is a cat. Socrates has a broken tooth. [therefore] All cats have a broken tooth. (Induction [20])

---

[3] Remember that in the card task, the subject is only presented with points of luminous dots and not spatial landmarks in the architectural environment. The parahippocompus is a region specialized in remembering the environment. Aside from the amygdala, memory of objects and faces involves two large regions situated in the superior temporal sulcus and in the fusiform gyrus.

Asking subjects to carry out these different reasoning tasks while being scanned by a positron-emission camera made it possible to observe the differences in brain activity. *Deduction* activated the left inferior frontal gyrus (Brodmann areas 45 and 47) and a part of the left superior occipital gyrus (Brodmann area 19). *Induction* activated a large zone comprising the left medial frontal gyrus, the left cingulate gyrus, and the left superior frontal gyrus (Brodmann areas 9, 24, and 32). To determine the areas specifically involved in these reasoning tasks, the activity of the two strategies recorded by the camera was subtracted out. A robust activation linked to induction remained only in areas situated on the left. The most surprising finding of this study was the total absence of activation on the right, whereas right-sided activation had been described in many tasks of working memory. Moreover, there was no parietal activity, which suggests that in this kind of reasoning, the areas of the brain preferentially concerned with processing spatial information are not activated.

To test this aspect, subjects were asked to carry out three types of reasoning involved in syllogism, spatial relations, and nonspatial relations, respectively, using the following statements:

Syllogism:
    Some officers are generals.
    No privates are generals.
    Some officers are not privates.
Spatial relational:
    Officers are standing next to generals.
    Privates are standing behind generals.
    Privates are standing behind officers.
Nonspatial relational:
    Officers are heavier than generals.
    Generals are heavier than privates.
    Privates are lighter than officers [21].

The three types of reasoning tasks activate different regions.[4] This study reproduces the preceding results as far as dominance of the left hemisphere and involvement of the inferior and medial frontal gyrus. The findings obtained here underscore the absence of activity in the left superior and medial prefrontal cortex (Brodmann areas 8 and 9) during the deductive reasoning task.

Thus, human reasoning involves a distributed network in the left hemisphere, extending from the inferior dorsolateral and medial prefrontal cortex to the temporal

---

[4] *Syllogism* activates the inferior left frontal gyrus (Brodmann areas 45 and 47) and a part of the left medial frontal gyrus (Brodmann areas 45 and 47). *Spatial reasoning* causes activation of the inferior frontal gyrus (Brodmann area 45), a region of the left medial frontal gyrus (Brodmann area 46), and a part of the left cingulate gyrus (Brodmann areas 32 and 24). The only nonfrontal areas activated were a region of the left medial occipital gyrus (Brodmann area 19) and a region of the lateral inferior temporal gyrus (Brodmann area 37). *Relational reasoning* triggers activation of the inferior frontal gyrus (Brodmann area 45) and part of the left cingulate gyrus (Brodmann areas 32 and 24). The only other region of major activity was the left globus pallidus (lateral and medial).

lobe. Yet it is reasonable to associate activity of the dorsolateral gyrus with working memory [22] and the inferior frontal gyrus with language. The medial prefrontal cortex is considered to be critical in so-called executive attention functions, perhaps controlling the areas of working memory.

Deductive memory and probabilistic reasoning

Johnson-Laird's mental-model theory (see Chapter 1) predicts that the brain structures responsible for *deductive* and *probabilistic* reasoning are the same because they involve mental models. They are likely situated mainly in the *right* brain, which seems especially important for the construction of mental models themselves, calling into play—according to the proponents of this theory—visuospatial structuring. In contrast, hypotheses supporting a more logical basis for reasoning lead to the opposite prediction: the activated structures are located in the left brain, as suggested by the studies mentioned above. The difference between the left and right brain has been the subject of much research.

This question was taken up in a positron emission camera study [23] in which brain activity was recorded during three tasks involving 'logical', 'probabilistic', and 'semantic' syllogisms. In the logical, deductive task, subjects had to distinguish between valid and invalid arguments. In the probabilistic task, they had to indicate whether the conclusion of a series of arguments had a greater likelihood of being true than false. Finally, in the semantic task, subjects had to analyse the different arguments and indicate whether they contained anomalies.

The results show a clear variation in activity between the different areas.[5] This result underscores the fact that deductive reasoning about syllogisms is related to visuospatial tasks. Compare the work of Johnson-Laird that we discussed in Chapter 1 on the effectiveness of diagrams in reasoning. The authors conclude that reasoning about syllogisms involves distinct brain areas depending on whether the subject intends to evaluate them deductively or probabilistically. We need to rethink the possible role of the cortex in reasoning that leads to a decision.

The wheel of fortune

I began this book by showing the limits of economic theories about decision making despite the excellent efforts of theoreticians and psychologists such as Daniel Kahneman [24]. In a series of conferences at the Collège de France on decision making

---

[5] The *probability* task activates the *left* dorsolateral frontal regions as well as the insular cortex. The authors link their data to observations of patients who, following ablation due to frontal epilepsy, could no longer estimate the relative frequency of events—a concept that is important in all estimates of probability—whereas patients with temporal lesions had no problem. The *deductive* task activates especially the *right* occipital and parietal regions, as well as the precuneus, the cuneus, and the regions of the superior parietal lobe, known for its involvement in visuospatial processing.

in economics and plans for a 'neuroeconomic' theory of decision making, Massimo Piatelli-Palmarini stressed both the contradictions of normative theories and the fact that recent data—in contrast—do confirm some predictions.

For example, we have all seen at a town fair or at an amusement park those huge spinning wheels with numbers or figures of playing cards on them. As a child, I liked to go spend my pocket money in the (often vain) hope of winning a big Teddy bear, a collection of multicoloured balloons, or of taking home a set of glasses as if I had been off crossing vast mountain ranges and distant continents to bring back treasures that would gain me the respect of the entire family! This wheel, which seemed immense to a kid, set spinning by the stallholder in a gesture that contained all my hopes, caused me to live moments that for gamblers become like a drug, the chance of winning or losing flickering in my mind. I still hear the sound of the little leather or metal tongue tripping along the pegs on the edge of the wheel that seemed to go round forever.

Does our brain really operate like a game of chance similar to the wheel of fortune, as prospect theory predicts? To find out, subjects were placed in an apparatus that measures brain activity (magnetic resonance imager) [25]. They were shown a spinner inspired by wheels at a fair with rewards in the form of money. While waiting for a reward, most of the areas of the brain that we have mentioned (hypothalamus, amygdala, orbitofrontal cortex, reward systems linked to dopaminergic activity, such as the nucleus accumbens, the ventral tegmental area, etc.) were activated. The level of activity in these parts of the brain is linked to the *subjective* and relative value of wins and losses and not to their absolute value.

As a child, if I wanted to win a stuffed bear and only won a sweet, I was more disappointed than if I had hoped to win a sweet and won nothing. In other words, if the subject believes he will win 10 dollars and only wins 5 dollars, he feels he has lost something, and this disappointment is visible in the data supplied by imaging his brain. It is the *difference* between what he hoped for and what he obtained that correlates with the cerebral activity.

The brain is a detector of differences between its hopes and expectations and what it gets; this principle is widespread and probably originates from the fact that the brain is an 'intentional' biological machine, that is, it works by giving itself goals. Moreover, the intensity of responses evolves asymmetrically as a function of winning and losing, as predicted by prospect theory; in the end, losing counts more than winning, as Fig. 13.2 illustrates. The growing field of neuroeconomics systematically examines the neural processes involved in economic choices and reactions to monetary rewards [26]. New cooperation between economists and scientists will probably also lead to development of probabilistic approaches to decision making [27].

Brain-imaging techniques have enabled a further dissociation of the contributions of different parts of the anterior cingulate and other areas of the frontal cortex and subcortical structures in the various phases of a monetary decision-making process. The areas involved during the 'decision phase' and the 'outcome phase' are different [28].

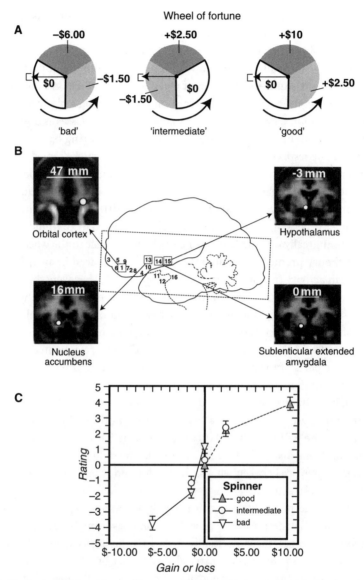

Fig. 13.2 The brain and the wheel of fortune. **(A)** The subject sees in varied order of appearance a wheel of fortune spin and stop at a given position to indicate how much he has won or lost. A moving arrow marks the win or the loss. (This is in contrast to an actual fair, where the wheel turns and the arrow is stationary.) The wheel is divided into three sections that give different wins or losses. Brain activity is measured by fMRI for three different wheels presented in succession. One is 'bad' and gives only zero gain or losses. The other is 'intermediary': the gain is zero, weakly positive, or weakly negative. The third is 'good': it allows zero or positive gains. **(B)** Brain areas active during the wheel of fortune game. Sixteen areas were activated to different extents. Their localization in the brain is indicated by

The brain is a predictor that imagines the future, but it does not necessarily seek to obtain 'objective' truth about the world. It works subjectively, that is, by evaluating differences between its predictions and what it obtains. It assigns itself a goal and assesses reality with respect to the goal. The significance of such findings recalls the work of the Russian psychologist Bluma Zeigarnik, who drew attention to what she called 'levels of aspiration'. She showed that children who are very ambitious in accomplishing a task succeed better than others, and vice versa.

A trip to Egypt

Now let us imagine another situation requiring complex decisions, say, planning a trip to Egypt. How do we get there? In a boat down the Nile or in a car? Making a decision requires simultaneously keeping in memory several past and present facts, for example, the current political situation, and so on. That is the task of working memory. But in reflecting in order to make a decision, we must turn our attention to this or that aspect of the situation and prioritize information. This is attentional supervision [29]. Sometimes people are imprisoned unjustly because the attention of the judge or jurors was focused on the wrong facts.

Context needs to be taken into account, and conflicts and ambiguities clarified. Actions and arguments also have to be planned in a certain order. We cannot think about two things at the same time. In the 1950s, psychologists developed the so-called single-channel theory. The serial ordering of actions is thus basic both for organizing our actions in time, but also for reflecting and making decisions.

Our decisions are a little like trees: not only do we have to foresee the succession of steps, we also have to make binary choices (on arrival, do we take a taxi or a train?). 'Scripts' must be constructed, tasks sequenced and, at precise moments in those tasks, 'branchings' introduced. Mathematicians have invented tools for describing these series (e.g. Markov chains); manufacturers use Pert diagrams in project planning. Where in the brain does this organization take place? What areas are critical to the process? Very simple versions of this type of problem have been constructed to study their neural basis [30].

For example, suppose that I show you a capital letter 'A'. You must say whether it is an A or a B. That's easy. Now, at the same time, I show you the drawing of a shape and you have to say whether it is a square or a circle. Since the two tasks activate different

---

numbers. By way of example, activity is shown in the four brain areas already mentioned in the text. **(C)** This way of presenting probabilities to the subject not only makes it possible to study brain activity while the subject anticipates, predicts, and evaluates his chances but also to show that activity is related to the expectation of a relative gain, not an absolute one, and the nonlinear nature of subjective probability (as predicted by Kahneman and Tversky's theory; see also Fig. 0.1). After Breiter et al. (2001)

pathways (channels) in the brain, there is no conflict. But if at the same time I show you another letter, asking whether it is a consonant, in this case, there is a conflict. The tasks have to be prioritized based on their context. The interesting problem then becomes that of moving from one task to another. We make these kinds of transitions every day. We can initiate them ourselves, or they can be triggered by something in the world outside.

For example, a subject is shown the word 'TABLET'. Then the subject is shown the letters B and E in succession, with a pause, and asked whether they are next to each other in the memorized word. The time of the pause is varied (the subject is asked to take longer to memorize a letter). The subject is also given a double task: when a letter appears, he is asked whether the case (uppercase or lowercase) of the letter has changed. In this event, the subject must choose between the two tasks. The predictive character of the task can also be varied (Fig. 13.3).

The study of brain activity during these tasks suggests a hierarchy of mechanisms involving the prefrontal cortex. According to this scheme, Brodmann area 10 (the frontopolar cortex) enables the choice of goals. This area appears to be critical in carrying out cognitive branchings, in deciding whether we take the taxi or the train. Moreover, the area may itself be divided into two further areas involved either in the processing of information produced in the brain (thoughts) or in information coming from the world outside (perceptions). Comparison with past events is carried out with the contribution of Brodmann area 46, which is important in working memory; the association between the goal, memory, and the present context is carried out in area 44. The result of all this activity is transmitted from the lateral areas to the medial areas. The orbitofrontal cortex and the cingulate cortex (area 32) appraise the value of the reward of the solutions envisaged. Finally, this information is sent to the striatum to be transformed into appropriate motor behaviours if needed (in this case a verbal response, but a motor response if the task had been a decision related to movement).

A role for the frontopolar cortex in predicting the future is also suggested by a study showing that healthy subjects talking about their past or future (far or near) activate more areas in the anteromedial frontal pole when thinking about the future than the past. Most areas of the medial temporal lobes showed a greater or similar level of activation during the future tasks compared with the past tasks [31]. The frontopolar cortex is part of a network that is also involved in inhibiting unwanted motor responses [32].

Patients presenting with lesions of the frontopolar cortex have difficulty organizing their program of action, even if they perform perfectly normally during cognitive tasks, have an intact working memory, and are able to switch tasks. It really is the overall planning of the enterprise that is disrupted. They can be expected to have trouble putting together a trip to Egypt!

But there is more. In deciding, it is not enough to be able to mentally process memories and propositions, make bets, and evaluate wins and losses. The decision *must be rooted in the present reality*. Solomon had before him two different women, each

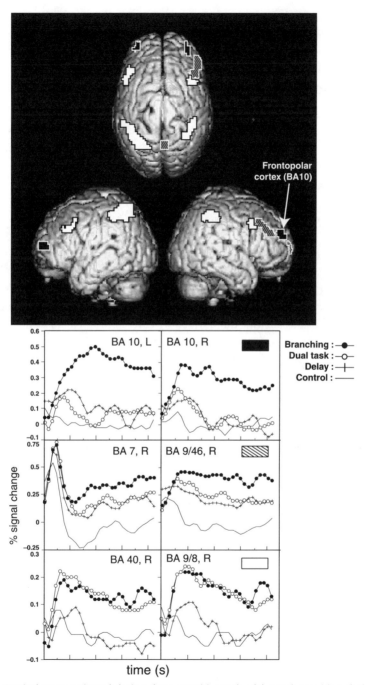

Fig. 13.3 Cortical areas activated during three cognitive tasks: (1) a task requiring choices involving 'branchings' analogous to those made while deliberating and that present to the mind a series of alternatives (see Fig. 1.3); (2) a 'double task'; (3) a delayed task that involves

unique, not reducible to a general notion of 'mother'. As the painting by Poussin (Fig. I.2) shows to perfection, this judge calmly seated on his throne had to consider separately the particular reality of each woman's experience.

Confabulation, a failure to relate to reality?

In making a decision, a judge, a consumer, the leader of an army, a politician, or whoever must take into account the memory of past decisions to adapt them to the present situation, and especially *lived experience*, this amazing synthesis of desire, fact, history, and culture. To decide is to predict the future consequences of action, but it also entails assessing the relevance of those consequences with respect to present reality. I do not like the word 'reality', and its use in pathology in the term 'derealization'. The term 'actuality' is better. I suggest we speak of 'deactualization'. It is just as ugly, but more accurate.

For example, Mr X arrives at a reception, and his host introduces him to his wife, saying: 'Here is Jacqueline! I think you don't know one another!' Mr X feels ill-at-ease because he is very familiar with his host's wife, who is one of his former girlfriends. Yet he decides to acquiesce and to lie out of politeness. Face recognition involves the temporal lobe's task of identification, which we covered in Chapters 2 and 8. But here, Mr X must make an additional decision following recognition. He must adapt his memory to the reality of the moment and either acknowledge his ex-girlfriend or pretend not to know her. We are embarking on a more complex cognitive operation. Social rules must be respected. Past memories must be suppressed. Simple perceptions of similarity or difference must be transcended to enter into the realm of reasoning and of deliberation. How many legal errors are the result of conclusions made too hastily by false witnesses who believed or pretended they recognized someone?

In these complex processes we will encounter the neurons of the prefrontal cortex which, as we have seen, contribute to multisensory integration and participate in suppressing undesired action, recalling memories, and anticipating the consequences of ongoing actions or even only imagined ones. They contribute to the development of emotional states and of motivation. Here, they are also involved in creating and applying social rules, and inhibiting memories that are irrelevant to the present situation.

To show that the neurons of the monkey are activated by abstract rules useful in deciding whether a person is *identical to* or *different from* someone already seen, the monkey is shown an image of an object. After a delay, the monkey is shown an image of the same object or another one and has to compare the content of the two images. This is the classical paradigm of delayed match to sample, but here, in addition, the

working memory; and (4) a control task (see text). The goal of this figure is to show the particular activity of the frontopolar cortex when the subject must perform a branching task. Findings by Gabrieli also suggest involvement of the frontopolar cortex in choices involving internal decisions. After Koechlin et al. (1999)

monkey's response must follow a *rule*. In certain cases, it must respond only if the object is the same and in others, only if it is different. The abstract rule is 'identical or different'. The neurons activated specifically by the requirement to obey a rule are located in the dorsolateral prefrontal cortex [33].

But we don't only have to obey rules. We must evaluate how our memory and the present situation—the reality of the moment—are related. Mr X recognizes Mme Y, but he chooses to pretend not to recognize her if, for example, he had been very much in love with her before she married and wants to suppress that memory. He adapts his response to the reality of the moment. Such a decision involves mechanisms that *confabulation* will illustrate for us [34].

A woman with a brain lesion of the frontal cortex doggedly nurses her 'baby' of 30 years! A clinician just out of hospital returns to his office saying that he had to meet his patients. We are not talking about the obsessional behaviours described in Chapter 3. These patients construct their urgent behaviour based on a memory that has no present reality.

Their concept of 'now' is determined by past memories unrelated to the reality of the moment (Fig. 13.4). This inability to assess the actual real character of a situation is sometimes called 'derealization'. These are spontaneous confabulations. But healthy

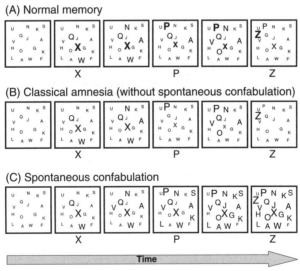

Fig. 13.4 Schnider's model to explain the mechanism of confabulation. Usually, we have many memories, represented here as though they were in a box. Each letter stands for a memory. **(A)** Over time, the subject will choose useful or relevant memories related to the present situation. (Here we show that he is going to choose memories X, P, and Z in succession.) **(B)** Classical amnesia or loss of memory translates into difficulty extracting memories from the brain's storehouse. **(C)** Confabulation is a different process. For example, in the final situation in this figure (last square to the right) the mind is flooded with many memories that are completely unrelated to the context or the present situation

people may also confabulate when they are forced to rely on memories that may be imprecise, for example, when they are serving as witnesses to an accident or a crime [35].

Several explanations have been proposed to account for this strange derangement of rational thought [36]. None of them is satisfactory. For example, memory deficit, the combination of a failure of memory and dysfunction of the so-called executive mechanisms—confusion about the sequence in which events occur—could explain certain confabulations in healthy subjects but not in disturbed ones. In fact, confabulators have trouble defining the present in time intervals of 1 or 2 s. They live past events as if they were happening now. It is possible (Fig. 13.5) that this deficit is abnormality due to a malfunction of the dorsomedial orbitofrontal cortex whose

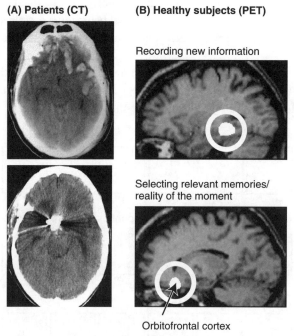

Fig. 13.5 The confabulators. Some patients with lesions of the frontal cortex cannot suppress the traces of old memories, and these memories invade their thoughts, imposing—and justifying—behaviours that have nothing to do with the present reality. They confabulate, that is, they construct a scenario that justifies the appearance of a memory that can be neither repressed nor updated. For example, a patient wanted to go back to the house to give a bottle to her son, now an adult; another, a lawyer, insisted on leaving the hospital for a meeting with a client, only that meeting had taken place long before. **(A)** Examples of frontal lesions in confabulators. **(B)** Healthy subjects were asked to record new information, which activates a region of the temporal cortex. They were then asked to choose memories relevant for the present reality or the situation. This task activated the area of the orbitofrontal cortex that is damaged in confabulators, which demonstrates the critical role of this region in contextualizing memories in the present. After Schnider et al. (1999)

workings we examined in Chapter 11. We have seen that it is concerned with the relation of stimuli or events to associated rewards, and that especially it mediates adaptation in animals and humans to alterations in reward criteria. Animals with lesions lose this flexibility. They continue to react to stimuli that once gained them a reward, even if at present no reward is forthcoming. The lovelorn sometimes manifest this psychorigid behaviour, continuing to frequent the spots where their former paramour made tender love to them. The problem is an inability to extinguish the value of the reward [37].

How might the orbitofrontal cortex affect behaviour? For it is not enough to assume that association remains intact; we also need to understand why it translates into persistent behaviour. Most likely the subcortical frontal loops, which connect the frontal cortex to distinct portions of the basal ganglia (striatum, pallidum, and substantia nigra), and the thalamic nuclei that project in turn towards the neocortex, cause these behaviours. Indeed, transversal links exist between the limbic portion and the motor portions of the loops connecting the basal ganglia, the thalamus, and the cortex [38]. The orbitofrontal cortex, which first projects to the accumbens, can also influence large areas of the cortex by this route, which first descends towards the centre of the brain before going back up towards the cortex. This idea supports the theories of Rodolfo Llinás, who assigns a major role to the centrencephalic structures of the thalamus, in contrast to theories that emphasize intracortical transactions [39]. But perhaps both mechanisms come into play. Time will tell.

In short, the anterior limbic system, in particular the orbitofrontal cortex, is probably the principal mediator of our capacity to tether ourselves to the reality of the moment. It does so by suppressing—inhibiting—memories that are irrelevant for the action happening now. Hence, it is a basic mechanism of decision making since it requires updating the value of the reward or punishment from our memory.

Another interesting dissociation has been revealed by brain imaging. In remembering past events as an aid to making decisions, we may recall the past in two ways. We may simply remember that an event happened without consciously recollecting the content or details ('remember'), or we may fully recollect the content of a scene or episode ('remember and know'). Psychologists have developed paradigms for testing this dissociation. Lateral parietal regions appear to respond preferentially to remember decisions, while anterior and medial regions respond to both remember and know decisions [40].

We may also manipulate the memory of past decisions. For instance, we can develop 'counterfactual' mental representations as alternatives to past events. We often say, 'If I had behaved this way instead of that, I might not be in trouble now.' These alternative thoughts are an important element of the various steps that lead to decision making. In one study, patients with Parkinson's disease, in whom temporal lobe function is impaired, generated fewer of these counterfactual representations [41].

Magical thought

'Don't be superstitious: it brings bad luck,' goes the proverb. Have you ever been tempted by magical thought? For example, 'If you sing out of tune, it will rain'? Some societies perform ritual dances to summon rain. Magical thought creates powerful causal associations between unconnected things or whose connections are very weak. For instance, some people equate cause and correlation. How many mothers use the word 'premonition' in describing the dream they had the night before their child had an accident or was killed in war? How many gurus make a fortune out of exploiting these so-called causes! Everybody knows that before making a decision, many people—even heads of companies and other elites—consult their astrologer.

Some people are more susceptible to magical thought than others. They believe in their horoscope and the magical properties of numbers. The neural basis of this predilection is perhaps close to being elucidated thanks to the work of Peter Brugger's group. Recently, the role of the *right* cerebral cortex in magical thinking was demonstrated [42]. People strongly inclined to magical thinking tend to remember the left side of a complex shape, which is processed by the right cortex, whereas subjects indifferent to such thinking tend to recall the right side. The right cortex may be affected by weak associations. For example, if I say to you, 'castle and king', you find the association a strong one, just as if I say 'bread and butter'. One day, the neurologist Theodor Landis observed a patient with a lesion of the left cortex (which, remember, is involved in processing language). He asked the patient to give him a word associated with 'bread'. The patient answered, 'God'. This association between bread and God is very distant (weak), although everyone raised in a Christian culture would readily understand it. But making this association involves performing a symbolic mental act, a process different than that of simple language association. The right cortex of these patients, who had language disturbances, could make this association very well. Indeed, the right cortex appears to favour weak associations [43], which require processing that is more global, more general, one might even say more abstract.

This idea relates to several observations, first that emotion particularly involves the right cortex, second that psychoses frequently occur in patients with deficits of the right cortex (although psychoses that accompany epilepsy often appear together with left focal seizures), and finally that the right cortex is involved in spatial and global aspects of shapes, objects, and scenes, whereas the left cortex may be more involved in the sequential aspects of the part of the world that surrounds us. So much so that when the left cortex, said to be 'dominant' by neurologists, is weakened, the right cortex is freer to create 'weak' associations. But this may also be the key to creativity [44], which consists precisely in drawing associations between nonobviously related concepts, objects, or ideas. 'The earth is blue like an orange', wrote Paul Eluard, and Rimbaud invented 'correspondences' between vowels and colours.

Liberated from the rational supervision of a left brain constrained by the rules of syntax and semantics, the right brain is freer to make new associations. What matters is that this possibility is acknowledged for so-called normal subjects [45]. Perhaps mathematicians, in any case, geometers, have a very powerful right brain. But it is also clear that dominance of the right brain can lead to serious disturbances by allowing a person to create illusory associations, such as those seen in paranoia, where the least gesture is interpreted by the patient as a conspiracy, and even, as some neurologists think, in schizophrenia [46] as well as in disorders that cause illusions [47]. One might then understand how, in very susceptible subjects, leaders of sects can reinforce associations that cut them off from social norms. It might also serve as a conditioning device for religious fanaticism.

As far as the biological theory of decision making goes, this research is essential, because it suggests that we do not always make decisions as a result of reasoning based on concepts clearly articulable by the left brain. Often, their intuitive character is due to the fact that they result from associations induced by the right brain. If we sometimes make decisions without really knowing why, it is because they are the result of an internal dialogue, maybe a dialogue in sign language, between two brains that do not necessarily agree.

References

1 B Inhelder and J Piaget, *The growth of logical thinking from childhood to adolescence; an essay on the construction of formal operational structures*, trans. A Parsons and S Milgram (New York: Basic, 1958).
2 O Trullier, S Wiener, A Berthoz, and A Meyer, 'Biologically-based artificial navigation systems: reviews and prospects', *Progress in Neurobiology*, 51 (1997): 483–544; A Berthoz, 'Parietal and hippocampal contribution to topokinetic and topographic memory', *Philosophical Transactions of the Royal Society B*, 352 (1997): 1437–48.
3 A Berthoz, I Viaud-Delmon, and S Lambrey, 'Spatial memory during navigation: what is being stored, maps or movements?' in AM Galaburda and S Kosslyn, *The languages of the brain* (Cambridge, MA: Harvard University Press, 2002).
4 S Lambrey, I Viaud-Delmon, and A Berthoz, 'Influence of a sensory conflict on the memorization of a path travelled in virtual reality', *Cognitive Brain Research*, 14 (2002): 177–86.
5 GI Zacharias and LR Young, 'Influence of combined visual and vestibular cues on human perception and control of horizontal rotation', *Experimental Brain Research*, 41 (1981): 159–71.
6 T Mergner and T Rosemeier, 'Interaction of vestibular, somatosensory and visual signals for postural control and motion perception under terrestrial and microgravity conditions—a conceptual model', *Brain Research Reviews*, 28 (1998): 118–35.
7 M Gattis, 'Introduction', M Gattis, ed., *Spatial schemas and abstract thought* (Cambridge: MIT Press, 2001), p. 1.
8 FA Yates, *The art of memory* (Chicago: University of Chicago Press, 1966).
9 SM Kosslyn, 'Seeing and imagining in the cerebral hemispheres: a computational approach', *Psychological Review*, 94 (1987): 148–75.
10 MT Banich and KD Federmeier, 'Categorical and metric spatial processing distinguished by task demands and practice', *Journal of Cognitive Neuroscience*, 11 (1999): 153–66; CF Chabris and SM Kosslyn, 'How do the cerebral hemispheres contribute to encoding spatial relations?', *Current Directions in Psychological Science*, 7 (1998): 8–14; JB Hellege and C Michimata, 'Categorization versus distance: hemispheric differences for processing spatial information', *Memory and Cognition*, 17 (1989): 770–6; SM Kosslyn, O Koenig, A Barrett, CB Cave, J Tang, and JDE Gabrieli, 'Evidence for two types of spatial representations: hemispheric specialization for categorical and

coordinate relations', *Journal of Experimental Psychology*, **15** (1989): 723–35; M Okubo and C Michimata, 'Hemispheric processing of categorical and coordinate relations in the absence of low spatial frequencies', *Journal of Cognitive Neuroscience*, **14** (2002): 291–7.

11 SD Slotnick, LR Moo, MA Tesoro, and J Hart, 'Hemispheric asymmetry in categorical versus coordinate spatial processing revealed by temporary cortical deactivation', *Journal of Cognitive Neuroscience*, **13** (2001): 1088–96.

12 A Berthoz, résumé de cours au Collège de France, *Annuaire du Collège de France* (Paris: Collège de France, 1998), pp. 421–45.

13 J Oakhill and A Garnham, *Mental models in cognitive sciences: essays in honour of Johnson-Laird* (Hove: Psychology Press, 1996).

14 MA Amorim, S Glasauer, K Corpinot, and A Berthoz, 'Updating an object's orientation and location during nonvisual navigation: a comparison between two processing modes', *Perception and Psychophysics*, **59** (1997): 404–18.

15 G Gigerenzer and R Selten, *Bounded rationality: the adaptive toolbox* (Cambridge: MIT Press, 2001).

16 E Mellet, S Briscogne, N Tzourio-Mazoyer, O Ghaëm, L Petit, L Zago, and O Etard, 'Neural correlates of topographic mental exploration: the impact of exocentric egocentric perspective learning', *Neuroimage*, **12** (2000): 588–600.

17 A Berthoz, 'Parietal and hippocampal contribution to topokinetic and topographic memory', *Philosophical Transactions of the Royal Society B*, **352** (1997): 1437–48.

18 M Habib and A Sirigu, 'Pure topographical disoroientation: definition and anatomical basis', *Cortex*, **23** (1987): 73–85.

19 Several recent publications have dealt with this question, of which we will cite only **DJ Koehler and N Harvey**, eds, *Blackwell handbook of judgment and decision making* (Malden, MA: Blackwell, 2004).

20 V Goel, B Gold, S Kapur, and S Houle, 'The seats of reason? An imaging study of deductive and inductive reasoning', *Neuroreport*, **8** (1997): 1305–10.

21 V Goel and B Gold, 'Neuroanatomical correlates of human reasoning', *Journal of Cognitive Neuroscience*, **10** (1998): 293–302.

22 PS Goldman-Rakic, 'Working memory dysfunction in schizophrenia', *Journal of Neuropsychiatry and clinical neurosciences*, **6** (1994): 348–57.

23 D Osherson, D Perani, S Cappa, T Schnur, F Grassi, and F Fazio, 'Distinct brain loci in deductive versus probabilistic reasoning', *Neuropsychologia*, **36** (1998): 369–76.

24 T Gilovich, D Griffin, and D Kahneman, eds, *Heuristics and biases: the psychology of intuitive judgment* (New York: Cambridge University Press, 2002). Recent studies do not always report the asymmetries proposed by this theory between the subjective importance accorded to wins and losses. See, for example, WT Maddox and CJ Bohil, 'Costs and benefits in perceptual categorization', *Memory and Cognition*, **28** (2000): 597–615.

25 HC Breiter, I Aharon, D Kahneman, A Dale, and P Shizgal, 'Imaging of neural responses to expectancy and functional experience of monetary gains and losses', *Neuron*, **30** (2001): 619–39.

26 SM McClure, DI Laibson, G Loewenstein, and JD Cohen, 'Separate neural systems value immediate and delayed monetary reward', *Science*, **306** (2004): 503–7; PW Glimcher and A Rustichini, 'Neuroeconomics: the consilience of brain and decision', Science, **306** (2004): 447–52; AG Sanfey, JK Rilling, JA Aronson, LE Nystrom, and JD Cohen, 'The neural basis of economic decision-making in the Ultimatum Game', *Science*, **300** (2003): 1755–8; W Schultz, 'Neural coding of basic reward terms of animal learning theory, game theory, microeconomics and behavioural ecology', *Current Opinion in Neurobiology*, **14** (2004): 139–47.

27 N Blackwood, D Ffytche, A Simmons, R Bentall, R Murray, and R Howard, 'The cerebellum and decision making under uncertainty', *Brain Research: Cognitive Brain Research*, **20** (2004): 46–53; ND Daw and P Dayan, 'Neuroscience: matchmaking', *Science*, **304** (2004): 1753–4.

28 RD Rogers, N Ramnani, C Mackay, JL Wilson, P Jezzard, CS Carter, and SM Smith, 'Distinct portions of anterior cingulate cortex and medial prefrontal cortex are activated by reward processing in separable phases of decision-making cognition', *Biological Psychiatry*, **55** (2004): 594–602.

29 EK Miller and JD Cohen, 'An integrative theory of prefrontal cortex function', *Annual Review of Neuroscience*, **24** (2001): 167–202.

30 E Koechlin, G Basso, P Pietrini, S Panzer, and J Grafman, 'The role of the anterior prefrontal cortex in human cognition', *Nature*, **399** (1999): 148–51; E Koechlin, G Corrado, P Pietrini, and J Grafman, 'Dissociating the role of the medial and lateral anterior prefrontal cortex in human planning', *Proceedings of the National Academy of Sciences of the USA*, **97** (2000): 7651–6.

31 J Okuda, T Fujii, H Ohtake, T Tsukiura, K Tanji, K Suzuki, R Kawashima, H Fukuda, M Itoh, and A Yamadori, 'Thinking of the future and past: the roles of the frontal pole and the medial temporal lobes', *Neuroimage*, **19** (2003): 1369–980.

32 K Rubia, AB Smith, MJ Brammer, and E Taylor, 'Right inferior prefrontal cortex mediates response inhibition while mesial prefrontal cortex is responsible for error detection', *Neuroimage*, **23** (2003): 351–8.

33 JD Wallis, KC Anderson, and EK Miller, 'Single neurons in prefrontal cortex encode abstract rules', *Nature*, **411** (2001): 95–6.

34 N Berlyne, 'Confabulation', *British Journal of Psychiatry*, **120** (1972): 31–9.

35 PW Burgess and T Shallice, 'Confabulation and the control of recollection', *Memory*, **4** (1996): 359–411.

36 A Schnider, 'Spontaneous confabulation, reality monitoring, and the limbic system: a review', *Brain Research Reviews*, **36** (2001): 150–60.

37 CM Butter, 'Perseveration in extinction and in discrimination reversal tasks following selective frontal ablation in macaca mulatta', *Physiology and Behaviour*, **4** (1969): 163–71.

38 Schnider, 'Spontaneous confabulation', p. 159.

39 S Dehaene, M Kerszberg, and J-P Changeux, 'A neuronal model of a global workspace in effortful cognitive tasks', *Proceedings of the National Academy of Sciences of the USA*, **95** (1998): 14529–34.

40 ME Wheeler and RL Buckner, 'Functional-anatomic correlates of remembering and knowing', *Neuroimage*, **21** (2004): 1337–49.

41 P McNamara, R Durso, A Brown, and A Lunch, 'Counterfactual cognitive deficit in persons with Parkinson's disease', *Journal of Neurology, Neurosurgery, and Psychiatry*, **74** (2003): 1065–70.

42 K Taylor, P Zäch, and P Brugger, 'Why is magical ideation related to leftward deviation on an implicit line bisection task?' *Cortex*, **38** (2002): 247–52.

43 The term is from E Bleuler, *Dementia precox or the group of schizophrenias* (New York: International Press, 1911–50).

44 SA Mednick, 'The associative basis of the creative process', *Psychological Review*, **69** (1962): 220–32.

45 C Mohr, R Graves, L Gianotti, D Pizzagali, and P Brugger, 'Loose but normal: a semantic association study', *Journal of Psycholinguistic Research*, **30** (2001): 475–83.

46 MA Spitzer, U Braun, L Hermle, and S Maier, 'Associative semantic network dysfunction in thought-disordered schizophrenic patients: direct evidence from indirect semantic priming', *Biological Psychiatry*, **34** (1993): 864–77.

47 D Leohard and P Brugger, 'Creative, paranormal and delusional thought: a consequence of right hemisphere semantic activation', *Neuropsychiatry, Neuropsychology, and Behavioural Neurology*, **11** (1998): 177–83.

Epilogue

Let us step back a bit and try to put together the whole puzzle. We have seen that elementary decision-making mechanisms exist at the level of a single neuron. Indeed, one neuron or a small group of neurons are sufficient for some animals to make fairly complex decisions in the face of environmental challenges. The key to these decisions turns out to be the ability to choose among a fairly limited repertoire of behaviours with a certain capacity for combination, exclusion, and so on.

These mechanisms even have predictive power, since their neural organization includes a choice of useful sensory receptors, access to internal models of the physical environment, and—as is the case in locomotion—a certain ability to predict the consequences of actions in response to perturbations and adaptation of the response to environmental conditions. In other words, if the simplest organisms have been able to survive over the course of evolution, it is because they are endowed with already very complex mechanisms of *perceptuomotor* decision making, in which memory of the past plays an important role. We have seen that with the development of the basal ganglia and their double neurophysiological and neurochemical regulation, and the intervention of neuromediators such as dopamine, selection of behaviours allows decision making at a high level of complexity and the involvement of specialized subsystems associated with various modes of action: gaze, gestures, posture and attitude, memory, reward, and so on.

We have also seen that in primates and in humans, visual perception reaches a degree of complexity such as one find in processes that one can call 'decision making' from the earliest sensory relays, and that these processes are guided by the intention to act and by attention. We have also seen the special role played by the most frontal regions of the cerebral cortex, in particular in primates and in humans, that enable deliberation. In these structures, complex processes enable development of strategies and not just choice of behaviours. Incidentally, some people would like to limit application of the term 'decision making' to processes that call into play these structures, a position I do not share. For me, the prefrontal cortex is a useful but secondary annex of the deep structures of the brain. In this matter I am closer to Wilder Penfield and the centrecephalic theory of the brain that he developed. I maintain that the ability to make complex decisions is not only due to hypertrophy of the prefrontal cortex in humans: the mechanisms that developed together, from the ones located in the spinal cord to those found in the cerebellum, the basal ganglia, and the frontal and parietal cortex, are all involved in decision making owing to a principle of harmony and equilibrium that exists among the components of a complex system.

In other words, I think that decision making really is a fundamental property of the nervous system and that the capacity to make complex decisions and to deliberate is due to modification of *all* the stages of organization of the system and not just to the appearance, for example, of modules in the frontal cortex. The brain decides not based on the absolute value of rewards but on the subjective value, the difference between what it expects or desires and what it obtains. This difference is, of course, measured against a yardstick of factors that are social, cultural, and so on. The distinction is important because it opposes two radically different conceptions of the brain: the so-called representational concept of the brain, which holds that the brain constructs an image of the world that guides action, and the idea I propose of a brain that is a part of the world, that has *internalized* its properties and *emulates* some of them but relates them to its own goals, which shape external reality by projecting onto it the brain's preperceptions, desires, and intentions. The brain simplifies the world based on its choices; it only perceives what it wishes to. For example, it randomly perceives patterns in stimuli [1]. This detection of patterns is obligatory and unconscious since it does not require paying attention.

The brain is not only a simulator; it is an emulator that creates the world as in a dream. I am always fascinated by the similarity between the dream world and the perceived world. Aside from a few differences, it is the same world, very coherent, extremely well constructed. Of course, we fly while dreaming, unimpeded by gravity; but almost all the other constraints are respected. That means that the brain has completely internalized the world and our body in the world: it plays out scenarios in this world. For example, allow me to tell you about one of my favourite themes that I live in my dreams almost every week: the frustration of conferences. I dream that I have to give a talk at a conference but that numerous obstacles prevent me from doing so. For the last three years, my brain has produced almost all the conceivable permutations on this theme, from the slide projector that does not work to the wrong schedule, including the building in which I get lost and can find neither the entry nor the exit, and so on. I have collected hundreds of variations during this anxiety-provoking nocturnal life, almost as many as the conferences themselves, since I have never missed one and have never been late for the 300 or 400 talks I have given.

All these situations are perfectly credible: each time my dream universe of failure is perfectly adapted to the external world. Consequently, I have no need of sensory information in making decisions or working out strategies. My double and the internal models I have of my body and the world are all I need to completely emulate a world that in every way resembles the physical world.

In *The Brain's Sense of Movement*, I proposed the idea that action is the very basis of all cerebral activity and that speed is an essential attribute of the ability to survive. I would like to remind those who think that the problem of speed is moot because we no longer live in the jungle or the ocean where we only had 50 ms to elude predators or to pounce, that in the dealing rooms of major banks decisions are made in seconds, and that the pilot of a Rafale fighter plane has scarcely more time to react than did our ancestors in bygone days. In fact, he has much less time because he must process many more data before making a decision.

Rolls's theory states that the cortical pathways for processing sensory data (especially vision) are protected against all interference from emotional conflict up to a very complex stage of perceptual analysis. This stage is where perceptual constants are constructed, which make it possible not to disturb the reconstitution of the unity of an object, a face, or an environmental scene based on diverse points of view. In this approach, emotion—as predicted by Damasio's somatic marker theory—assigns a value, in the full sense of the word, to this reconstituted perception. In contrast, the short pathways through the amygdala discovered by LeDoux enable a rat to process an acoustic stimulus in 12 ms and to trigger a repertoire of behaviours essential to survival, such as immobilization, flight, aggression, submission, and so on. These pathways stamp a provisional emotional value on a specific configuration of stimuli in a specific context.

The brain thus has available at least two, apparently very different, mechanisms. One, which appeared early in evolution and is linked to survival, is rapid. In this mechanism, decision making is the product of close and obligatory cooperation between emotion and perception. The desired object or the enemy must be identified to trigger the appropriate behaviour. Here, preperceptions belong to a precise, species-specific repertoire.

The other mechanism appeared late in evolution. It requires complex processing of sensory information, but especially, it projects purpose, it depends on the past—on the history of every person and each social group belonging to a 'culture'. This mechanism is a discovery of evolution, for it enables prediction, anticipation, selection of sensory data no longer based on a repertoire of innate behaviours but based on intention, on social rules, and on the filtering of sensory information. It has been suggested that this mode, by nature cortical, is protected against emotions because if, for example, anger were to alter the perception that I have of a bouquet of flowers, the perception could be totally perverted, which is not the case. How are such different points of view to be reconciled? By the idea of a heterarchy of decision making—decisions taken at all levels, each level influencing the preceding or the following one—and by changing our ideas of how the brain works. Instead of considering emotion as a reaction, it must be considered as integral to preparing action. Emotion is a tool for making decisions; it is a powerful instrument of prediction for a brain that anticipates and projects its intentions. Accepting these ideas means confronting, for example, Rolls's theory, which holds that perception is completely protected against emotion, with another theory outlined in the following way.

The primary perceptual structures (for example, the visual cortex) may be influenced by emotion but in a diluted and subtle form: that of selective attention. Powerful modulation revealed by brain imaging and neurophysiology of the earliest stages in visual, vestibular, and auditory processing, and especially the role of attention in resolving conflicts between the senses, suggests a possible mechanism of action of emotion on perception. In other words, emotion activates the mechanisms of selective attention and induces not a distortion of the perceived world, but a selection of objects perceived or neglected in the world; it profoundly alters the relation of memory to perception of the present (we say, 'When I am angry, I forget everything'). Emotion, that guide to action, may be a *perceptual filter*. This mechanism is fundamental in decision making since our decisions depend a lot on what we perceive, on what our brain samples in the world and the way in which it connects the objects it perceives with the past.

LeDoux has also touched on this question: 'While many animals get through life mostly on emotional automatic pilot, those animals that can readily switch from automatic pilot to willful control have a tremendous extra advantage. This advantage depends on the wedding of emotional and cognitive functions. So far we've emphasized the role of cognitive processes as a source of signals that can trigger prepackaged emotional reactions. But cognition also contributes to emotion by giving us the ability to make decisions about what kind of action should occur next, given the situation in which we find ourselves now. One of the reasons that cognition is so useful a part of the mental arsenal is that it allows this shift from *reaction* to *action*' [2].

For my part, I will say that this analysis of the difference between a reaction and an action is an excellent illustration of the nature of the latter: an action is indeed an intentional behaviour that predicts its own consequences since it results from a decision whose mechanism involves prediction and even attribution of emotional value. We are also making progress in intuiting the difference between an action and an act. Think about the use of these terms in the theatre: an action is an event, an episode that is produced in a certain context and is dictated by a particular decision. An act, whether in the theatre or in a lawyer's office, is already a history that entails a sequence of actions. As my friend and colleague Jean-Marie Zemb used to say, it is no small matter to translate Goethe's phrase by: 'In the beginning was the Act.'

We still have some way to go in constructing a biological theory of decision making, but I hope I have shown that modern neurophysiology and the combination of efforts of those who record neurons in animals, the psychologists and physiologists who work with modern techniques of brain imaging in humans, and the mathematicians and physicists who help us to devise quantitative theories for approaching these complex processes should enable us to move forward.

Some people like to think that decision making must be the result of structures with specific properties, on which many inputs converge. In all species of animals, and in humans, specialized structures are involved in arbitrating the repertoire of possible actions. But I still maintain that decision making is a fundamental property already present in the Mauthner cell, which decides based on the animal's circumstances, or in the toad, for which the decision to approach or flee is not linked uniquely to a prewired circuit but also to data contributed from the animal's experience and memory.

Understanding how the brain considers context in making decisions, for example, will be a major milestone.

Indeed, having arrived at the end of this exercise, I am convinced that we have to avoid too simplistic a definition of what I have called a 'hierarchical theory of decision making'. The hierarchy I am talking about here is not only a hierarchy of authority but of levels of complexity. Over the course of evolution, increasingly sophisticated processes of prediction and evaluation were developed, and I would like to mention, once more, the thesis of *The Brain's Sense of Movement*: the main function of the brain is to predict the consequences of action based on the results of past action; memory is thus essential not only for remembering the past but for predicting the future. It might be useful to distinguish, as has been done for memory, two major categories of decision making: implicit and explicit.

I would like to end this book with a few more general thoughts. In desiring to discover the sentient body at the heart of the decision-making process, in being sceptical of rational approaches that claim to reduce the complexity of decision making to mathematical formulas, I do not wish to set up an opposition between the obscurity of the body and the light of reason. Restoring the act at the heart of decision making is about revealing the diversity and dynamic range of movement, the subtlety of intentionality, and about critiquing the domination of language (a serial mental tool whose elements are strictly ordered); it is about opening up the richness of space (a parallel mental tool that facilitates examination of simultaneous relations). Bayesian approaches deserve our attention as a modelling tool because Bayes's rule links the present, the past, and the future. The brain is not always Bayesian, but some of its processes probably are.

In this book I have also tried to link emotion and perception. The stakes here are considerable. Indeed, we are emerging from a century that figured out how to explain matter, to dominate energy, and to network information. It invented informatics but forgot the sentient body, separating reason and emotion. And of course, that does not work. The Pentagon's strategists may have had the most powerful computers of the era, but they failed in the face of the Vietnamese, who taught their 4-year-old children how to survive by educating the smallest ones in the midst of strafe bombing. In going to share the life of these children, the great paediatrician Alexandre Minkowski showed that emotional intelligence can triumph over computers.

Today, we have to win another battle. Deprived of emotion and sensitivity, drowning in an avalanche of machines and virtual worlds, people are once again taking refuge in obscurantism and intolerance. Religious fanaticism is carrying the day over moderate democracy. The voting ballot, symbol of a delicate balance between power and will, of decision making that accepts the opinions of others, seems pointless to those who would rather burn cars or slit people's throats. Passion trumps reason, because reason has forgotten to make room for passion.

The twentieth century has seen the victory of the rational over the emotional: a theatre of language without feeling, of formal theory without application, of words without deeds—a victory of reason that desires detachment and that may represent technological success but also serves as a seedbed of fanaticism.

The drift into darkness of people who reject the dictatorship of numbers so dear to profit-oriented societies is deeply worrisome. You can see the urgency of a multidisciplinary scientific program that would reestablish the role of feeling and emotion in combating tyranny and barbarism. We need to reflect seriously on the relation between coldly calculating reason, on the one hand, and on the other the insane retrogression that has led to mass crimes and promoted sectarian leaders who brainwash minds in search of an ideal and the emotions to transform them, as Françoise Sironi puts it, into suicidal henchmen [3]. I have no theory, but perhaps the ideas and observations presented here will contribute towards better understanding and accepting each other.

References

1 SA Huettel, PB Mack, and G McCarthy, 'Perceiving patterns in random series: dynamic processing of sequence in prefrontal cortex', *Nature Neuroscience*, 5 (2002): 485–90.
2 J LeDoux, *The emotional brain: the mysterious underpinnings of emotional life* (New York: Touchstone, 1996), p. 175.
3 F Sironi, Bourreaux et victimes: psychologie de la torture (Paris: Odile Jacob, 1999).

Picture credits

All best attempts to find copyright have been made, but this has not always been possible. Please accept our apologies if we have infringed copyright in any way and please contact the Medicine Editorial department of Oxford University Press.

In1 Reprinted from *Neuropsychology to Mental Structure*, Cambridge University Press, Fig 2, Copyright 1998, with kind permission from Cambridge University Press.

In2 Photo RMN/René-Gabriel Ojeda

1.1 Reprinted from *The Framing of Decision and the Psychology of Choice*. Tversky & Kahneman. Science **211**, pp 453–458. Copyright 1981, with permission from Science.

1.2 Reprinted from *Psychology of Learning and Motivation: Advances in Research & Theory*, no 32, Busemeyer, et al, pp 137–175, Fig 1, Copyright 1995, Academic Press, with permission from Elsevier

1.3 Reprinted from *Judgement and Decision Making: Neo-Brunswickian and process-tracing approaches*, Ch 8 p. 147–173. Fig 8.2, Editions LEA.

1.4 Reprinted from *Cognitive Science*, **20**. Leven & Levine. Multi-attribute decision making in context. p. 271–296. Fig 4. Copyright 1996, with permission from The Cognitive Science Society

2.2 Reprinted from *Le Corps en Jeu*, Odette Aslan, p. 29., CNRS Editions. Copyright 1998. Reproduced with permission.

2.3 Reprinted from *Psychosomatic Medicine*, **11** pp 338–353, Maclean, Psychosomatic disease and the 'visceral brain': Recent developments bearing on the Papez theory of emotion.

2.4 Reprinted from *Nature Medicine* **7**, Manji, Drevets & Charneyin. The cellular neurobiology of depression. p. 541–547, Fig 1, Copyright 2001, with Permission from Nature Publishing Group.

2.6 Reprinted from *The Psychology of Emotion*, 4th edition. K. T. Strongman. 1998. Copyright John Wiley & sons Limited. Reproduced with permission.

2.7 Reprinted from *Traité de perspective*, Paris, Société Nouvelle des Editions du Chêne, p. 102, 1976, Hachette. Every effort was made to locate the original photographer

3.1 Reprinted from Grusser and Landis: *Visual Agnosias and other Disturbances of Visual Perception and Cognition,* Macmillan Press

3.2 Reprinted from *Comptes Rendus*, VIème Congrès Neurologique International. Vol. 3. Penfield. Vestibular sensation. Copyright 1949. With permission from Editions Masson Livres.

3.3 Reprinted from *Annual Review of Neuroscience*, **9**. Alexander, DeLong & Struck. Parallel organization of functionally segregated circuits linking basal ganglia and cortex. pp 357–81. Fig Copyright 1986 Annual Reviews.

3.4 Reprinted with permission from *Science*, **275**, Bechara and Damasio, Deciding advantageously before knowing advantageous strategy, pp 1293–95, Copyright 1997, AAAS.

3.5 Reprinted from *Brain*, **122**, Rahman, et al, Specific cognitive deficits in mild frontal variant frontotemporal dementia, pp 1469–93, Proceedings of The Royal Society, London, Copyright 1999, Oxford University Press, with special permission from Dr B Sahakian, Professor of Clinical Neuropsychology, Addenbrookes Hospital Cambridge, and by permission of Proceedings of The Royal Society, London

4.2 Reprinted from *Handbook of Chemical Neuroanatomy*, Ch 8, Korn, Faber & Triller, Fig 1, copyright 1990, with Permission from Elsevier.

4.3 Reprinted from *Brain Behaviour and Evolution*, **54**, Apomorphine alters prey-catching patterns in the common toad: Behavioural experiments and 14C-2Deoxyglucose brain mapping studies, Glasgow & Ebert, pp 223–242, Copyright 1999, Plenum Publishers, with kind permission from S. Karger AG, Basel

4.4 Reprinted from *Biological Order and Brain Organization*. Chapter 15. Fig 15.12. W. R Hess. Copyright 1981, with kind permission of Springer Science and Business Media.

4.5 Reprinted from *Le Sens du Mouvement*. Alain Berthoz. Copyright Editions Odile Jacob, 1997, with kind permission.

5.1 Reprinted from *Interlimb Coordination: Neural Dynamical and Cognitive Constraints*. Berthoz and Pozzo. Head and body coordination during locomotion and complex movements. pp147–65. Copyright 1994, with permission from Elsevier.

5.2 Reprinted from the *Journal of Physiology*, **93**. P Rudomin. Presynaptic selection of afferent inflow in the spinal cord. pp 329–47. Fig 12. Copyright 1999, with permission from Blackwell Science.

5.3 Reprinted from *Neurobiology*, **51**. Baev. Highest level automatisms in the nervous system. pp 129*–66. Copyright 1997, with permission from Elsevier.

6.1 Reprinted from Grusser and Landis: *Visual Agnosias and other Disturbances of Visual Perception and Cognition*, Macmillan Press

6.3 Reprinted from the *Journal of Neurophysiology*, Vol 8, no 5, Lobel et al, Functional MRI of galvanic vestibular stimulation, pp 2699–2709, Copyright 1998, with permission from The American Physiological Society.

6.4 Reproduced from Iriki, Tanaka & Iwamura, Coding of modified body schema during tool use by macaque postcentral neurones, *Neuroreport*, 7, 14, 2325–30, Copyright 1996.

6.5 Reprinted from 'Anomalous representations and perceptions: implications for human neuroplasticity' by S Aglioti, in *Research and Perspectives in Neurosciences, Neuronal Plasticity: Building a Bridge from the Laboratory to the Clinic*. Grafman & Christen, eds. pp 79–91. Fig 1. Copyright 1991, with kind permission of Springer Science and Business Media.

6.6 Reprinted from *Neuroscience Letters*, Vol 240, Hari et al, Three hands: fragmentation of human bodily awareness, pages 131–134, Copyright 1998, with permission from Elsevier

7.1 Reprinted from *Visions: Aspects perceptifs & cognitifs*. Boucart et al. pp 11–41. Copyright 1998, with permission from Editions Solal.

7.2 Reprinted with permission from Thorpe & Fabre-Thorpe, *Science* **291**:260–263 (2001). Illustration: Carin Cain. Copyright 2001 AAAS.

7.3 Reprinted with permission from *La Recherche*, **248**, no 3, Des muscles pour voir en 3 dimensions, Y Trotter, pp 1320–22, Copyright La Recherche, Societé d'éditions scientifiques

7.5 Reprinted from Kanizsa, V.G. *Organisation in Vision, Essays on Gestalt Perception*, fig 11, 1979. Reproduced with permission of Greenwood Publishing Group, Inc, Westport CT

8.1 Reprinted from JJ Gibson, *The senses considered as perceptual systems* (London: Allen and Unwin, 1968), p. 315, Fig 14.9

8.2 Reprinted from the *Scientific American*, **232**(6) pp 76–87, Johansson, Visual Motion Perception. With permission of Scientific American and the artist, Alan D Iselin

8.3 Reprinted from The Psychological Review, **94, 2:** pp 115–47, Biederman, Recognition by components: a theory of human image understanding

8.5 First published in Duensing & Miller, 'The Cheshire Cat Effect'. *Perception*, **1979, 8**. pp 269–273. Figures 1 and 2. Pion Limited, London.

8.6 Reprinted from Scheinberg & Logothetis, The role of temporal cortical areas in perceptual organisation, *Proceedings of the National Academy of Sciences*, 1997, **94**(7) pp 3408–3413, Fig 3. Copyright 1997 National Academy of Sciences, U.S.A. with permission.

9.1 Reprinted from *Nature Neuroscience* 2(2): 176–85, Kim & Shalden, Neural correlates of a decision in the dorsolateral prefrontal cortex of the macaque

9.2 Reprinted from *Neuroscience Research*, **34**, Sakagami & Tsusui, pp 78–89. Fig 6, Copyright 1999, with permission from Elsevier.

10.1 Albert Philippon, photographer

11.2 Reprinted from *Nature*, 6729 (398) Tremblay & Schultz, Relative reward preference in primate orbitofrontal cortex, pp 704–7, Copyright 1999 Nature Publishing Group.

11.4 Reprinted from *Behavioural Brain Research*, **117**, Albertin, Mulder, Tabuchi, Zugaro & Wiener, Lesions of the medial shell of the nucleus accumbens impair rats in finding larger rewards, but spare reward-seeking behaviour, pp 173–83, Copyright 2000, with permission from Elsevier.

11.4 Reprinted from *Hippocampus* **10**, pp 717–728, Fig 1, Tabuchi, Mulder & Wiener: Position and behavioural modulation of synchronization of hippocampal and accumbens neuronal discharges in freely moving rats. Reprinted with permission from John Wiley & Sons, Inc

11.5 Reprinted from *Trends in Cognitive Science*, **4**, 6, Bush, Luu & Posner, Cognitive and emotional influences in anterior cingulate cortex, pp 215–22, Copyright 2000, with permission from Elsevier.

11.6 Reprinted from *Neuroscience*, **13**, Redgrave, Rescot & Gurney, The basal ganglia: a vertebrate solution to the selection problem?, pp 1009–23, Copyright 1999, with permission from Elsevier.

11.7 Reprinted from *Network: Computation in Neural Systems*, **13**, Humphries & Gurney, The role of intra-thalamic and thalamocortical circuits in action selection, pp 131–156, Fig 2a, Copyright 2002, with permission from The Institute of Physics Publishing.

12.1 Reprinted from *Proceedings of the National Academy of Science*, **99**, Pocho et al, The neural system that bridges reward and cognition in humans: an fMRI study, pp 5669–74, Fig 2, Copyright 2002, with permission from The National Academy of Sciences, USA.

13.1 Reprinted from *Cognitive Brain Research*, **14**, Lambrey, Viaud-Delmon & Berthoz, Influence of a sensorimotor conflict on the memorization of a path traveled in virtual reality, pp 177–86 Copyright 2002, with permission from Elsevier.

13.2 Reprinted from *Neuron*, **30**, Breiter, Aharon, Kahneman, Dale & Shizgal, Functional imaging of neural responses to expectancy and experience of monetary gains and losses, pp 619–39, Figs 1&2, Copyright 2001, with permission from Elsevier.

13.3 Reprinted by permission from Macmillan Publishers Ltd: *Nature*, **399**, The role of the anterior prefrontal cortex in human cognition, Koechlin, et al, pp 148–515, Figs 3 & 5, Copyright 1999.

13.4 & 5 Reprinted from *Nature Neuroscience*, **2**: pp 677–81, Schnider & Ptak, Spontaneous confabulators fail to suppress currently irrelevant memory traces

Index

Note to index: bold page numbers indicate illustrations; *fn* = footnote

abstract rule 271
accumbens nucleus 210–13, 220
   four-bistro experiment **212**
action in decision making
   emotion and cognition 40, **214**, 226–7
   selection by basal ganglia 217–19
      models 223–4, **225**
   simulated xiii–xiv
   subsumption, distribution and central selection 222, **223**
   suppression, no-go potential 235
   suspended actions 233–4
   vs reaction 282
adrenocorticotrophic hormone (ACTH) 33
affective balance 16
agnosias 51–3
   apperceptive 52
   associative visual 52
   geometric apractognosia 126
   tactokinesthetic 124
Aix-en-Provence fountain **193**
alexithymia 216
ambiguity, binocular rivalry 164–7
amphetamines 66–7
amphibians, decision making 77
amygdala 32–3, 200–3, **200**
   gaze control 202
   perseveration 60
   pessimism 202–3
   physiology of fear 200–3, **200**, 235
anterior cingulate cortex *see* cingulate cortex
antithesis, principle of 25
anxiety, sensory conflicts 181–3
appetancy, gourmand syndrome 209–10
appraisal theories 37–9
approach vs flight response 77–9
Aristotle
   'peas illusion' 110
   views on emotion and reason 39
aschematia 125
associative visual agnosias 52
Assyrian character, Dl 74
attention deficit hyperactivity disorder (ADHD) 61
attentional conflict, Stroop tasks 214–15
attentional supervision 267
attentional tasks 216
attributional errors 122
autoscopy 111
   neural basis 113–14
autotopagnosia 123

Baev, KV, motor decision organization **105–6**
balance *see* walking and balance
balance, control 102
Balint—Holmes syndrome 53
basal ganglia
   action selection 217–19
      models 223–4, **225**
   and control of walking 102
   in decision making 217–19
   looped circuits **56**, 240
   prefrontal cortex and dopaminergic system circuit **211**
   walking and balance 102
Bayesian probabilities xii*fn*
behaviour, inappropriate, inhibition 64–5
behavioural circuits 223–4
behavioural theories 41–2
binocular rivalry 164–7
   visual form perception **166**
blind spot, 'filling in' 151
body perception 109–33
   aschematia 125
   autoscopy, neural basis 111, 113–14
   awareness of personal space 126–7
   body extensions and tools 115–17, **117**
   disorders of recognition 123–4
   double images 109–13
      hallucinations 109–13
      mirrors **111**, 113
   'doubles' 127–30
   phantom limbs 118–23, **121**
   types of representations 124
   virtual body hypothesis 114–15, 127–30
brain
   anatomy, shape recognition 160–2
   evolution of decision making 281–2
   gaze control **84–5**
   generation of strategies 251–77
   left/right processes 160, 209, 257, 261, 262, 264*fn*, 274–5
   main function 283
   organisation into looped circuits 55, **56**, 57, 240
   stages of evolution 30–2
   visceral brain theory 30
   see also cerebral cortex; memory; specific regions
brain imaging, modulation of fMRI signal by reward **246**
bulimia, amygdala 202

callosal disconnection syndrome 54, 55, 124
Cambridge Gambling Task 213
Capgras syndrome 53–4
caudate nucleus 106
cenesthesia 113
centrencephalic theory 55
cerebellum
   cognitive function 240
   object control/internal model/control
      law triad **105**
   walking and balance 96–7, 100–1
cerebral cortex
   amygdala 32–3, 60, **200**, 201–3
      speed of response 280–1
   basal ganglia, action selection 217–19, **225**
   cingulate cortex 205, 213–15, 217, 235, 262
   cognition
      activated areas **269–70**
      deduction and induction 262–4
      syllogism and (non)spatial relational 263
   control of balance 102
   coordination 57
   cortical arousal 81
   cortical pathway of visual movement
      174–8
   frontopolar cortex 245–7, 268, **269**
   hippocampal function 236, 261–2
   left/right processes **160**, 209, 257, 261, 262,
      264*fn*, 274–5
   looped circuits **55, 56,** 57, 240
   parietofrontal network 261*fn*
   prefrontal 60–2, **211**
      alternation of percepts 165–6
      intention-to-choose neurons 177–8
      and perceptual decision making 176–8
   right
      and gourmand syndrome 209–10
      and magical thought 274–5
   temporal lobes 261–2
      medial superior temporal (MST) 174–5
   ventromedial frontal lesions 61–2
   walking and balance 102
   see also brain; *specific regions*
Cheshire cat phenomenon 164–7, **165**
cingulate cortex 213–15, 217, 235
   anterior cingulate cortex 213–15, 217, 235
      error detection 217
   in mental navigation tasks 262
   subgenual cingulate 205
cognition
   activated areas **269**
   deduction and induction 262–4
   deductive memory and probabilistic reasoning
      264
   syllogism and (non)spatial relational 263
cognitive algebra 10
cognitive control
   appraisal 37–9
      two-factor theory 38
   locomotive trajectories 102–3

cognitive functions
   competition and inhibition 244
   development 241–3
   and emotion 282
cognitive strategies 8–10
   route finding 253
colliculus 79–81
   fight or flight behaviour 79–81
   function 79–81
colour and movement combinations, perceptual
      choice **181**
compatibility 13–16
competition, decision making 222, 240–1
component-processing model 40
compulsions 58
computational approach 40–1
conditioning, fear 32–3
confabulation (failure to relate to reality)
      270–3
   effects of lesions of frontal cortex **272**
   Schnider's model **271**
conflict processing, and route memory 253–6
corpus callosum, disconnection (syndrome)
      54, 55, 124
Cuban missile crisis 8

Damasio, AR
   modified gambling theory 65–6
   somatic marker theory 29, 62, **63**
danger, responses to 81
Darwin, Charles
   on emotion 23–8
   quoted ix
deactualization 270
decision making
   amphibians 77
   compatibility 13–16
   and competition 222, 240–1
   complex decision making 267–70
   cortical area activation **269–70**
   deciding means choosing 88–90
   decisional reasoning, neural basis 262–4
   deliberation 54–7, 280
   divorce **13, 14**
   dominant structure theory 12–13
   and dopamine 6–7, 78, 219–21
   economists' models 18–19
   and emotion 23–50, **214**
   evolution of neural mechanisms 281–2
   facial expressions 26–9
   formalist theories xi–xiii
   general definition 73–4
   hesitation 224–5
   heterarchy and hierarchy 57–8
   inhibition 90
   and looking 81–2
   mechanisms 88–90
   modified gambling theory 65–6
   non-consequential 8
   Norman and Shallice's model x

as not deciding xiv
as perceiving xi
perception and preference 135–96
postural decision making 103–6
rationality 3–21
as a sequence of operations 9
shape recognition 155–71
as simulated action xiii–xiv
violation 15
voting ballot 283
working memory 63–4
see also action in; emotion and; perceiving and; preference
deduction
and induction 262–4
memory and probabilistic reasoning 264
vs induction 262–3
dementia
frontotemporal dementia syndrome 65–6
subcortical 106
depression
anterior cingulate 216
subgenual cingulate 205
derealization (deactualization) 270
see also confabulation
descriptive theories 4
diagrams 10–11
dipoles 16
disjunction, vs conjunction 11
distance perception 145
divorce, decision making 13, 14
Dl, Assyrian character 74
dominant structure theory 12–13
dopamine
accumbens nucleus 210–13
decision making 66–7
models 220–1
and reward 219–20
see also Parkinson's disease
dopaminergic effects 78, 219–21
double images 109–13
body perception **111**, 113
'doubles' 127–30
dreams 280
drug addiction 211–12

economic man 8–10
economists' models of decision making 18–19, 265
egocentric vs allocentric point of view 257–9, 261
elderly people, falls 95–6
emotion and cognition 282
emotion and decision making 23–50
cause of error 5
and cognition **214**
Darwin on 23–8
James–Lange theory 29
preparation for action 226–7
emotion and movement 34–5

emotion theories 34
phenomenology 36–7
schemata **31**, **32**
typologies **35**
emulation, and simulation 64–5, 280
error 5–7
collective error 5
detection, anterior cingulate cortex 217
and dopamine 219–20
errors in judgement 6
fact vs opinion 6
and fixed belief 6
magical thought 274–5
to predict 220, 221
escape behavoiur, fish 75
evidence, and intelligence 244
excitation–disinhibition, in evolution 238
extinction 173

facial expressions 26–9
facial perception 202
facial recognition 51–3, 162–3
geon theory 157–9
see also shape recognition
fact vs opinion 6
falls, elderly people 95–6
false evidence, and intelligence 244
fear
conditioned 32–3
physiology 200–3
fight or flight 73–94
choice 88–90
colliculus function 79–81
gaze control **80**, **84–5**
inhibition 90
looking and decision making 81–2
Mauthner cells 74–7
network reconfiguration 90–1
neuromodulation 90–1
prey and predator decision making 234–6
scan paths 83–8
fish
brain **74**, 75
flight response 75
flight response 73–94
fountains 185–96
geometry 186–7
and persuasion 194–6
'freezing' (immobilization reactions) 234–6
frontal cortex lesions 61–2
confabulation 272
frontopolar cortex 245–7, 268, **269**
frontotemporal dementia syndrome 65–6

Gabor function 140
gambling 4–5
Iowa Gambling Task 62, 213, 243
gated dipole 16

gaze control **80**, 81–91
  amygdala 202
  brain systems involved **84–5**
  circuit 218*fn*
  gaze avoidance 82
  looking and decision making 81–2
  saccades 82
  scan paths 83–8
  shift decisions 86
  see also visual perception
geometric apractognosia 126
geometrization of thought 12
geon theory 157–60
globus pallidus, gaze control circuit 218*fn*
gourmand syndrome, appetancy 209–10

habits, and expression 25
hallucinations
  body perception 109–13
  phantom limbs 118–23, **121**
Hebb's rule 220
hesitation 224–5
heterarchy 57–8, 225
hierarchical decision making 106–7
hierarchy
  of information 106–7
  and obedience 5
hippocampus, function 236, 261–2
HIV-1 immunocompromise 104–6
horoscope 274
Huntington's disease 237*fn*
hypothalamus 30–3

illusions 155–7
  decisions 156
  height of objects **156**
  moving point trajectories **157**
illusory contours 147–9
immobilization reactions 234–6
immunocompromised patients, postural gamble 104–6
imprinting 54
impulsivity 61
inference, and diagrams 11
inhibition 90, 233–4
  fight or flight 90
  immobilization reactions 234–6
  and logical reasoning 244–7
    modulation of fMRI signal by reward 244–7
  of novelty 236
intelligence, and false evidence 244
internal models 12
Iowa Gambling Task 62, 213, 243

James, W, decision/indecision 42–5
James—Lange theory of emotion 29, 34, 36, 42
Janet, Pierre 45–7
Japan, inhibition and inaction as arts 233–4

Kanizsa, illusory contours 147–9
Kluver—Bucy syndrome 201, 202

Lake Geneva fountain **188**
lateral geniculate body (LGB), thalamus **139**
left/right cortical processes **160**, 209, 257, 261, 262, 264*fn*, 274–5
lexicographic strategy 9
limbic system 30–2
locomotion, hierarchical theory of Baev 105
locomotor trajectories, cognitive control 102–3
long-term potentiation 76
looking, and decision making 81–2

magical thought 274–5
magnocellular pathway 141–2
marker hypothesis *see* somatic marker theory
Mauthner cell 74–7, 115, 282
  fight or flight 74–7
medial superior temporal cortex (MST) 174–5
medial thalamus, function 235
memory
  classical amnesia 271
  confabulation (failure to relate to reality) 270–3
  constructive operators 241
  decision and competition 240–1
  decision and reward 245
  deductive memory and probabilistic reasoning 264
  episodic and procedural 260–2
  route memory and conflict processing 253–6
  topokinesthetic 261
  working memory and decision making 63–4
mental models 10–11
mental pathways, route finding 259–60
Meynert's nucleus 81
mirrors, body perception 110–11, **111**, 113
modularity (heterarchy) 57–8, 225
motivation 206–8
motor decisions, organization **105–6**
Mouffetard, Paris, fountain **190**
movement perception 173–84
  conflict solving 181–3
  cortical pathway of visual movement 174–8
  intention-to-choose neurons 177–8
  mechanism of decision making 175–6
  perceptual choice **178**, 179–81
    colour and movement combinations **181**
  prefrontal cortex 176–7
  spatial anxiety 181–3
movements
  plotting by computer **97**
  see also route finding

negative priming effect 243
neophilia 236
nesting, levels of control 95–7
network reconfiguration 90–1
network simulation 17

neuroeconomics 265
  economic man 8–10
  models of decision making 18–19, 265
neuromediators 91
neuromodulation 90–1
no-go potential, suppressed action 235, 239
normative theories 4
numerical judgement
  in babies 242–3
  in children under seven 242
  and spatial assessment 241–3

object permanence, visual perception 146–7, 239
object recognition *see* shape recognition
obsessional disorders 58–9
obsessive slowness 58–9
ocular systems, gaze control **84**
opponent processing 16
opposite synkinesis 118
orbitofrontal cortex 203–10
  activity, types **207**
  and behaviour 273
  cognitive functions 204–6
  damage in confabulators 272–3
  flexibility and contextualization 203–10
  and inhibition 64
  lesions 204, 205
  localization **200**
  motivation 206–8
  preference 203–10
  rules of association 208
  value of reward 206–8
orientation, vestibular sensors **114**
orientation movements 81
Orsay, museum 251–2

Papez circuit 30, **31**
parahippocampus, function 262
Parkinson's disease
  and postural decision making 103–6
  temporal lobe function impairment 273
parvocellular pathway 141–2
Penfield, W, centrencephalic theory 55
perceiving and decision making xi, 137–54
  fragmentation of visual world 138–40
  magnocellular/parvocellular pathway 141–2
  receptor fields 140–1
  visual perception 142–52
perceptuomotor decision making 279
periaqueductal grey matter 235
perseveration 59–60
personal space 126–7
  awareness 126–7
persuasion, analogies with fountains 194–6
pessimism
  amygdala 202–3
  subgenual cingulate 205
phantom limbs 118–23, **121**
  multiple 121–3, **122**

phenylketonuria 235
Plato, views on emotion and reason 39
position, visual perception 144
posterior cingulate cortex *see* cingulate cortex
postural decision making 103–6
postural gamble, immunocompromised patients 104–6
posture and emotion 226–7
preference 199–231
  accumbens nucleus 210–13
  amygdala **200**, 201–3
  anterior cingulate cortex 213–15, 217
  arbitrating conflict 215–16
  and aversion 208
  dopamine and reward 219–21
  emotion, cognition and action in decision making **214**
  familial decision making 221–4
  gourmand syndrome 209–10
  hesitation 224–5
  heterarchy/hierarchy 225–6
  intention to act 210–13
  orbitofrontal cortex 203–10
  reversal 15
  selection of action
    basal ganglia 217–19, **225**
    subsumption, distribution and central selection 222, **223**
prefrontal cortex 60–2, **211**
  alternation of percepts 165–6
  basal ganglia and dopaminergic system circuit **211**
  in deliberation 235
  hierarchy of mechanisms 267–70
  immobilization reactions 234–6
  and inhibition of novelty 236–7
  intention-to-choose neurons 177–8
  object permanence 146–7, 239
  and perceptual decision making 176–8
  role in thought and suppression 238–40
  see also cerebral cortex
prescriptive theories 4
presynaptic inhibition, spinal cord 99
prey/predator, decision making 234–6
probability testing, in visual perception 142
prospect theory 5, 7–8

reaction, vs action 282
reason, and passion 283–4
receptor fields 140–1
retina
  gaze control **80**
  mapping 79
reversal of preference 15
rewards
  dopamine 219–21
  dopamine and preference 219–21
  evaluation 220–1
  inhibition and logical reasoning 244–7
  modulation of fMRI signal by reward **246**

rewards (*cont.*)
  orbitofrontal cortex 206–8
  value of 206–8, 219
  wheel of fortune 264–7, **266**
Rolls's theory 281–2
route finding
  cognitive strategies 253
  egocentric vs allocentric point of view 257–9, 261
  mental pathways 259–60
route memory, and conflict processing 253–6
route and survey, imagination in route finding 260–2
rules, abstract rule 271, **271**

saccades
  gaze control 82
    frontal eye fields 86–7
    shift decisions 86
Sartre, Jean-Paul, emotion theories 36–7
'satisficing' 8
scan paths, fight or flight 83–8
Schnider's model of confabulation **271**
selection of action: subsumption, distribution and central selection 222, **223**
sensory conflicts
  anxious people 181–3
  movement perception 173–84
sensory data, filtering 98–100, 98–9, **100**
sensory information
  selection **100**
  walking and balance **100**
shape recognition 155–71
  assumed steps in recognition **159**
  binocular rivalry 164–7
  categorization of natural shapes **143**, 167–70
  conscious perception 163
  facial recognition 51–3, 162–3
  geon theory 157–60
  impossible shapes 169
  perceptual decision making 160–3
  trajectories of moving points 157
simulation, and emulation 64–5, 280
simultagnosia 52–3
slowness, obsessive 58–9
somatic marker theory 29, 62–4, **63**
somatognosis 113
somatoparaphrenia 112
somatosensory neurons 116
spatial anxiety, movement perception 181–3
spatial gesture, and decision making 256–7
spatial neglect 115
spatial orientation
  assessment, and numerical judgement 241–3
  categorical vs coordinate-type 257
  egocentric vs allocentric point of view 257–9, 261
  mnemonics 257
  spatial working memory 261*fn*

vestibular sensors **114**
see also route finding
spatial relational, vs non-spatial relational reasoning 263
spinal cord
  presynaptic inhibition 99, **100**
  sensory data filtering 98–100
  walking and balance 96, 98
Stinsteden illusion 46
strategies, generation 251–77
  architecture of thought 251–3
  changing point of view 257–9
  complex decision making 267–70
  confabulation 270–3
  deduction and induction 262–4
  magical thought 274–5
  multisensory conflicts **254–5**
  neural basis of decisional reasoning 262–4
  route, and survey, imagination in route finding 260–2
  route memory
    cognitive strategies 253
    and conflict processing 253–6
    mental pathways 259–60
  space and decision making 256–7
  transition from egocentric to allocentric point of view 257–9, 261
striatum, gaze control circuit 218*fn*
subjective probability 7–8
subjective probability function 4
substantia nigra pars reticulata, gaze control circuit 218*fn*
superior colliculus
  function 79–81
  gaze control 84–5
suppression *see* thought and suppression
sure thing principle 4
suspended action 233–4
syllogism, vs (non)spatial relational 263
synergy, walking and balance 99
synkinetic movement 118

tactokinesthetic agnosia 124
temporal lobes 261–2
  medial superior temporal (MST) 174–5
  new information 272, 273
thalamus
  centrencephalic theory **55**
  function 235
  lateral geniculate body (LGB) **139**
  looped circuits **56**, 57
  Papez circuit 30, **31**
  somatic marker theory 29
  visual processing **139**
thought and suppression 233–49
  decision and competition 240–1
  logical reasoning and inhibition 244–7
    modulation of fMRI signal by reward **246**
  negative priming effects 243–4
  novelty detection 236–7

numerical judgement and spatial assessment 241–3
prefrontal cortex role 238–40
prey/predator decision making 234–6
spatial assessment and numerical judgement 241–3
tics and jitters 237–8
Tourette's syndrome 237

utility function 4
expected-utility theory 18
and theories of normative, descriptive and prescriptive behaviour 3–4

value, and belief 3
value function 7
vestibular sensors, spatial orientation **114**
vicariance 260
virtual body hypothesis 114–15, 127–30
virtual reality, route finding 253
visceral brain theory 30
visual perception 137–52
awareness 167
blindness to change 152–3
conscious perception 163
decision making by comparing patterns 149–50
filling in 150–2
identifying natural shapes (categorization) **143**, 167–70
illusory contours 147–9
image transmission **139**
movement 173–84
object permanence 146–7
position, distance perception 144, **145**
prefrontal cortex 165–6
processing 144–6
receptor fields 140–1
see also shape recognition

walking and balance 95–108
basal ganglia and cortical control 102
cerebellum 100–1
cognitive control 102–3
cortex control 102
hierarchical decision making 106–7
immunocompromised patients 104–6
locomotive trajectories, cognitive control 102–3
motor decision organization **105–6**
movement plotting by computer **97**
nested levels of control 95–7
old people, falls 95–6
postural decision making 103–6
sensory data filtering 98–100
sensory information selection **100**
spinal cord 98
water 186
weighted additive 9